Introduction to Population Ecology

Introduction to Population Ecology

Second Edition

Larry L. Rockwood with Jonathan W. Witt
Professor of Biology, George Mason University
Fairfax, VA, USA

WILEY Blackwell

Registered office: John Wiley & Sons, Ltd, The Atrium, Southern Gate, Chichester, West Sussex, PO19 8SQ, UK

Editorial offices: 9600 Garsington Road, Oxford, OX4 2DQ, UK
The Atrium, Southern Gate, Chichester, West Sussex, PO19 8SQ, UK
111 River Street, Hoboken, NJ 07030-5774, USA

For details of our global editorial offices, for customer services and for information about how to apply for permission to reuse the copyright material in this book please see our website at www.wiley.com/wiley-blackwell.

Library of Congress Cataloging-in-Publication Data

Rockwood, Larry L., 1943-
    Introduction to population ecology / Larry L. Rockwood. –2nd edition.
        pages cm
    Includes bibliographical references and index.
    ISBN 978-1-118-94758-6 (cloth) –ISBN 978-1-118-94757-9 (pbk.) 1. Population biology–
Textbooks. 2. Animal populations–Textbooks. 3. Insect populations–Textbooks.
4. Ecology–Textbooks. I. Title.
    QH352.R63 2015
    577.8′8–dc23
                                                                                                    2014050192

A catalogue record for this book is available from the British Library.

Wiley also publishes its books in a variety of electronic formats. Some content that appears in print may not be available in electronic books.

Cover image: Cover photograph © Yehudi Hernandez

Set in 9.5/12pt Berkeley by Laserwords Private Limited, Chennai, India

1   2015

Contents

Preface

The problem with a textbook, especially in a fast moving science such as population ecology, is that the moment it is published, it is out of date. Given the delays between actually writing a book and its appearance in print, most of the writing for my 2006 first edition was done in 2003–2005. Therefore, the first objective in producing this second edition was to bring it up to date as much as possible. Obviously, this did not mean throwing the first edition into the dustbin and writing everything new from scratch. In many cases it meant finding new examples that illustrate what are still valid ideas or theories. In other cases it meant casting doubt on old favorites such as the theory of r- and K-selection (called by at least one author a "zombie" theory).

In Chapter 1, for example, I have brought the data on human populations as up to date as possible, recognizing that the number of humans in the world has moved from 6.3 billion in my first edition to over 7.1 billion as of 2014. The example of the exponential growth of the gray wolf population in Wisconsin was not available previously. Information on a renewed interest in the Allee effect is new in Chapter 2.

The chapters on population regulation (Chapter 3) and life tables (Chapter 4) use a variety of new and better examples compared to those in the first edition.

Chapter 5, which deals with metapopulations and the MacArthur and Wilson equilibrium theory, benefit from a great deal of new information published in the last few years, much of it reviewed extensively in Losos and Ricklefs (2010).

A great deal has been written recently relevant to life history theory including the metabolic theory of ecology and its spawn (Sibly et al. 2012). At the same time, Dobson and Oli (2007), and Dobson (2007, 2012) have produced a series of papers proposing to replace the theory of r- and K-selection with a theory centered around mass, "life style" and phylogeny. I have attempted to at least introduce the reader to the idea that it may be time to realize the limits of r- and K-selection, if not retire the theory altogether.

In the chapter on competition (Chapter 7), my major contribution is to introduce the reader to the poetic niche theory written by Dr. Seuss (Geisel 1955). A critical evaluation of the Tilman (1982) resource-ratio theory and an example of character displacement among Darwin's finches (Grant and Grant 2006) are other highlights.

The chapter on mutualism (Chapter 8) features an update on the classic Janzen (1966) paper on the Central American ant-acacia mutualism as well as a new long section on the recent brilliant work on ants and acacias in Africa (Palmer et al. 2010). These studies confirm a major point of the chapter, which is, that without constant co-evolution, mutualistic relationships can easily move toward parasitism by one of the partners. Also new in this chapter is a section on geographic variation in coevolution of mutualism (Rickson 1977), which leads into the similar idea of geographic defenses in host-parasite interaction described in Chapter 9 (Foitzik et al. 2003).

The chapter on predator-prey relationships (Chapter 10) brings the reader new information on examples that are engrained in the literature as classics, yet are now being questioned. These include doubts about the hypothesized trophic cascade in Yellowstone after wolf introduction (Middleton 2014) and the apparent failure of the wolf population to either thrive or control the moose population on Isle Royale in Michigan (Mlot 2013). Meanwhile Chapter 11 brings us new information on the potential responses of plants to defoliation.

The last chapter, contributed by my colleague Jon Witt, is entirely new and presents a discussion of multiple trophic level models in much more detail than in the previous edition. Topics examined include trophic cascades, intraguild predation, cannibalism and meso-predator release.

References

Dobson, S.F. and M.K. Oli. 2007. Fast and slow life histories of mammals. *Ecoscience* 14: 292–299.

Dobson, S.F. 2007. A lifestyle view of life-history evolution. *Proceedings of the National Academy of Sciences*. 104: 17565–17566.

Dobson, S.F. 2012. Lifestyles and phylogeny explain bird life histories. *Proceedings of the National Academy of Sciences* 109: 10747–10748.

Foitzik, S., B. Fischer and J. Heinze. 2003. Arms races between social parasites and their hosts: geographic patterns of manipulation and resistance. *Behavioral Ecology* 14: 80–88.

Geisel, T. Seuss ("Dr. Seuss"). 1955. *On Beyond Zebra*. Random House, New York, NY.

Grant, P.R. and B.R. Grant. 2006. Evolution of character displacement in Darwin's Finches. *Science* 313: 224–226.

Janzen, D.H. 1966. Coevolution of mutualism between ants and acacias in Central America. *Evolution* 20: 249–275.

Losos, J.B. and R.E. Ricklefs. 2010. *The Theory of Island Biogeography Revisited*. Princeton University Press, Princeton, NJ.

Middleton, A. 2014. Is the wolf a real American hero? New York Times Monday March 10, 2014: A21.

Mlot, C.. 2013. Are Isle Royale's wolves chasing extinction? *Science* 340: 919–921.

Palmer, T.M., D.F. Doak, M.L. Stanton, J.L. Bronstein, E.T. Kiers, T.P. Young, J.R. Goheen and R.M. Pringle. 2010. Synergy of multiple partners, including freeloaders, increases host fitness in a multispecies mutualism. *Proceedings of the National Academy of Sciences* 107: 17234–17239.

Rickson, F.R. 1977. Progressive loss of ant-related traits of *Cecropia peltata* on selected Caribbean Islands. *American Journal of Botany* 64(5): 585–592.

Sibly, R.M., R.H. Brown, and A. Kodric-Brown. 2012. *Metabolic Ecology*. Wiley-Blackwell, Chichester, UK.

Tilman, D. 1982. *Resource Competition and Community structure*. Princeton University Press, Princeton, NY.

Acknowledgments

Larry L. Rockwood would like to thank his graduate students for keeping him intellectually rejuvenated.

Thanks especially to Christine Bozarth, Jennifer Mickelberg, Jennifer Sevin, and Ryan Valdez. Special thanks to Lorelei Crerar and Naomi Coles who read and gave me feedback on all of the chapters as they were being written. He would also like to thank his office staff, Rebekah Flis, Joanne Zimmerman, and Angel Washington, whose efficiency allowed him time to work on this book.

Jonathan W. Witt would like to thank Kristen Baker, Christopher Clark, and Marlene Cole, who read and provided valuable feedback on Chapter 12. The views expressed in this book are those of the authors and do not necessarily reflect the views or policies of the US Environmental Protection Agency.

Finally, this book is dedicated to Jane Rockwood for her patience, wisdom, support and love.

About the companion website

This book is accompanied by a companion website:

www.wiley.com\go\rockwood\populationecology

The website includes:
- Pdfs of all figures from the book for downloading
- Powerpoints of all tables from the book for downloading

Part 1

Single species populations

Why population ecology? What distinguishes the study of populations from the study of landscapes and ecosystems? The answers lie in scale, focus, and traditions. In population ecology the scale is a group or groups of taxonomically or functionally related organisms. The emphasis is on fundamental properties of these populations: growth, survivorship and reproduction. The tradition is based on the interplay of theory, laboratory testing and, ultimately, field work. The competition and predator–prey equations of Lotka (1925) and Volterra (1926; 1931) stimulated the laboratory work of Gause (1932; 1934); Park (1948; 1954), Huffaker (1958) and others. But Elton (1924), Errington (1946); Lack (1954); Connell (1961a; 1961b); Paine (1966); Krebs et al. (1995) and many others have brought population ecology into the field, where its theoretical underpinnings are constantly tested. In the age of personal laptop computers and the Internet, data can now be analyzed, sent around the world, and experiments redesigned without ever leaving the field site. Increasingly sophisticated experimental design and statistical rigor constantly challenge new generations of scientists. Indeed much of the training of modern ecologists is in methodology.

Yet why do we become ecologists in the first place? Is it because of our love of computer programs and statistics? For most of us, that would be, "No." More likely it is because of a love of the organisms that we find in natural ("wild") places. We love the sounds, the smells, the feel, the being in nature. Perhaps it is also because of our love of the idea of nature and of places not yet under the total domination of man. Nothing quite matches a day (or night) in the field for an ecologist, and we are usually eager to communicate these experiences to other people. Contrast an ecologist to a typical urban dweller like Woody Allen. In one of his movies Woody complains that he hates spending nights in the country because of the "constant noise of the crickets." Undergraduate students at George Mason University often approach the first field trip of the semester with fear and trepidation. Yet, such individuals have little fear of automobile traffic and find traffic noises normal and even soothing. Obviously an ecologist has a different orientation to the world.

Population ecology is, in a primitive sense, an organized way of communicating our ideas about nature to others. Population ecology, with its emphasis on groups of individuals and their survival and reproduction, their relationships with their competitors and their predators, is rooted in both field work and in natural history. As such it appeals to us at a very fundamental level. Instead of (or perhaps in addition to) swapping tales around the campfire at night, we communicate by publishing in journals or books.

Furthermore, without the basic data from population studies, most landscape and ecosystem studies would either be impossible to carry out, or would lack fundamental

Introduction to Population Ecology, Second Edition. Larry L. Rockwood.
© 2015 John Wiley & Sons, Ltd. Published 2015 by John Wiley & Sons, Ltd.
Companion Website: www.wiley.com/go/rockwood/populationecology

meaning. The advantage of ecosystem studies is the comprehensiveness of the approach. However, the disadvantage is the complexity of interactions among species and our lack of understanding of community organization. Everyone can agree, I believe, that we need a better understanding of interspecific interactions, and this is the role of population ecology. To develop laws of ecosystem functioning, we need to comprehend how individual populations behave. From there we can develop an understanding of interactions among populations. Therefore it seems to me that studies at the landscape and ecosystem level must be informed by data first gathered by population ecologists.

But this all sounds rather grand and theoretical. In the real world, knowledge of population ecology is absolutely necessary for conservation biologists, wildlife managers, and resource biologists. They are often faced with problems of preserving biodiversity or a wild living resource without adequate information. How can they best decide whether to limit or even shut down a fishery and for how long? Is it necessary or wise to allow wolf hunting in Alaska in order to increase the caribou herd? How do we control deer populations or is it foolish to attempt to control them in urban environments such as Rock Creek Park in Washington, D.C.? Has the introduction of wolves into Yellowstone actually decreased the elk herds? What are causes of reptile and amphibian declines throughout much of the world? What accounts for the proliferation of Lyme Disease in North America? Although an ecosystem approach may be helpful and necessary to answer many of these questions, basic population data are also necessary. Beyond this we must understand how populations with different life histories grow and/or are limited. We need a fundamental understand of the roles of competitors, parasites and predators and their potential effects on a given population.

When Audubon was in the state of Kentucky in 1813, he witnessed the passing of a great flock of passenger pigeons (*Ectopistes migratoris*). This flock blackened the sky for more than 3 days as they passed overhead. Later Audubon estimated their numbers at between 1.1 and 25 billion birds (Souder 2004). Yet the last passenger pigeon was shot in the wild in 1900, and the last individual in captivity died in 1914. How can a population decline to extinction so swiftly, even if one acknowledges the role of hunting and habitat destruction?

Red grouse go through population cycles every 4–5 years. The numbers oscillate over three orders of magnitude (Hudson et al. 1998), and these oscillations are synchronized over large geographical areas (Cattadori et al. 2005). Yet the population recovers regularly. On the other hand, when tawny owls (*Strix aluco*) were studied in Oxford, the number of mating pairs remained steady, at 17–30 pairs, even though their major rodent prey species oscillated from 10–150 per acre (Southern 1970). What are the differences between red grouse and tawny owls: reproductive parameters, developmental time, or survivorship? Or is it the fact that red grouse are primarily herbivores and owls primarily predators? Are there differences in their competitors, parasites, or predators? These are questions that only knowledge of population ecology allows us to answer.

When the moose population crashed in Isle Royale in Lake Superior, Michigan, in the late 1990s, was the cause wolf predation? Or was it parasites or over-browsing of the vegetation? By 2013 the wolf population was in decline and moose population was increasing, but fitting this interaction to a simple predator–prey model has proven problematic (Mlot 2013). Wildlife scientists have complained for many years that white-tailed deer are over-browsing their habitats and causing changes in the vegetation. If so, why don't these deer populations crash? Is the recent movement of coyotes into the eastern United States and puma into the mid-western United States the result of these large white-tailed deer populations? If not, what explains these dispersals from the "wild west" to the more urbanized areas of the United States, east of the Mississippi River (Bozarth et al. 2011)? One goal of this book is to give you the background and weapons that will allow you to address these questions.

In the twentieth century, the principles of population ecology, as we understood them, were applied to agriculture, forestry, wildlife management, fisheries, and conservation biology. Exploitation of populations in the name of "maximum sustainable yield" was based on the flawed logistic equation and/or inadequate data. Before the days of environmental impact statements, however, politicians and engineers largely ignored advice based on ecological science. While this situation has changed, ecologists, in order to remain credible, must work to develop better theoretical approaches. Applied ecologists must be able to recognize which of several possible theoretical approaches to use for the population or community of concern. The purpose of this book is to help guide undergraduates, graduate students, future wildlife refuge managers, EPA officials, or other applied ecologists through the workings of basic population principles and theory so that they make wise decisions in the future.

In part one of this book we will establish the fundamentals of population growth for single species populations. After determining these basic properties, we will examine how intraspecific competition affects population characteristics. We will also consider the evolution of different types of life histories and discuss whether a biological population is naturally "regulated."

Once we have an understanding of how single populations grow and sustain themselves in particular environments, we can begin to examine how interactions with populations of other species affect their life histories. In part two we will progress to an examination of interspecific interactions such as competition, predation, parasitism, and mutualism. Finally, as we move through these interactions, we can evaluate their relative importance in population growth and regulation.

References

Bozarth, C. A., F. Hailer, L. Rockwood, C. Edwards, and J. Maldonado. 2011. Coyote colonization of northern Virginia and admixture with Great Lakes wolves. *Journal of Mammalogy* 92: ·1070–1080.

Cattadori, I.M., D.T. Haydon, and P.J. Hudson. 2005. Parasites and climate synchronize red grouse populations. *Nature* 433: 737–741.

Connell, J.H. 1961a. Effects of competition, predation by *Thais lapillus* and other factors on natural populations of the barnacle, *Balanus balanoides*. *Ecological Monographs* 31: 61–104.

Connell, J.H. 1961b. The influence of interspecific competition and other factors on the distribution of the barnacle, *Chthamalus stellatus*. *Ecology* 42: 710–723.

Elton, C.S. 1924. Periodic fluctuations in the numbers of animals: their causes and effects. *Journal of Experimental Biology* 2: 119–163.

Errington, P.L. 1946. Predation and vertebrate populations. *The Quarterly Review of Biology* 21(144–177): 221–245.

Gause, G.F. 1932. Experimental studies on the struggle for existence. I. Mixed populations of two species of yeast. The Journal of Experimental Biology 9: 399–402.

Gause, G.F. 1934. *The Struggle for Existence*. Williams and Wilkins, Baltimore. Reprinted in 1964 by Macmillan (Hafner Press), New York, NY.

Hudson, P.J., A.P. Dobson, and D. Newborn. 1998. Prevention of population cycles by parasite removal. *Science* 282: 2256–2258.

Huffaker, C.B. 1958. Experimental studies on predation: dispersion factors and predator–prey oscillations. *Hilgardia* 27: 343–383.

Krebs, C.J., S. Boutin, R. Boonstra, A.R.E. Sinclair, J.N.M. Smith, M.R.T. Dale, K. Martin, and R. Turkington. 1995. Impact of food and predation on the snowshoe hare cycle. *Science* 269: 1112–1115.

Lack, D. 1954. *The Natural Regulation of Animal Numbers*. Oxford University Press, London, UK.

Lotka, A.J. 1925. Elements of Physical Biology. *Reprinted 1956*. Dover, New York, NY.

Mlot, C.. 2013. Are Isle Royale's wolves chasing extinction? *Science* 340: 919–921.

Paine, R.T. 1966. Food web complexity and species diversity. *The American Naturalist* 100: 65–75.

Park, T. 1948. Experimental studies of interspecific competition. I. Competition between populations of the flour beetles, *Tribolium confusum* Duvall and *Tribolium castaneum* Herbst. *Ecological Monographs* 18: 267–307.

Park, T. 1954. Experimental studies of interspecific competition. II. Temperature, humidity, and competition in two species of *Tribolium*. *Physiological Zoology* 27: 177–238.

Souder, W. 2004. Audubon in Kentucky. *Natural History* 113: 46–52.

Southern, H.N. 1970. The natural control of a population of tawny owls (Strix aluco). *Journal of Zoology* 162: 197–285

Volterra, V. 1926. Fluctuations in the abundance of a species considered mathematically. *Nature* 118: 558–60.

Volterra, V. 1931. *Leçons Sur la Théorie Mathématique de la Lutte pour la Vie*. Gauthiers-Vilars, Paris.

1 Density independent growth

1.1 Introduction

1.1.1 What is a population?

The basic definition of ecology, the scientific study of the relationships between organisms and their environment, is rather vague and the word environment requires an explicit definition. An alternative definition of ecology, the scientific study of the distribution and abundance of organisms (Krebs 1994; Andrewartha 1961), is more germane to population ecology. In population ecology we want to know what factors most likely control the growth rates, abundances and distributions of biological populations.

As used here, a population (synonymous with biological population) consists of a group of interbreeding organisms found in the same space or area (i.e., they are sympatric) at the same time. It is presumed that these individuals form a functional unit in that they interact with one another and there is interbreeding among the individuals of the population. A closed population is one in which we expect no immigration or emigration of individuals from outside of the population. In reality, unless we are considering a population on a remote island, a mountaintop, or an isolated cave, populations are not closed to immigration or emigration. Unless we have successfully marked all individuals in a population, we are usually unaware of which individuals might be recent immigrants. Turchin (2003) integrates these ideas in his definition of a population: "a group of individuals of the same species that live together in an area of sufficient size to permit normal dispersal and migration behavior, and in which population changes are largely determined by birth and death processes."

Introduction to Population Ecology, Second Edition. Larry L. Rockwood.
© 2015 John Wiley & Sons, Ltd. Published 2015 by John Wiley & Sons, Ltd.
Companion Website: www.wiley.com/go/rockwood/populationecology

However, as discussed by Waples and Gaggiotti (2006) the term population has a myriad of definitions depending on whether the context is ecological, evolutionary, or statistical. They list six different definitions within the ecological context, but I remain satisfied that they all converge on either the first definition I presented above and/or the Turchin (2003) definition.

A local population differs from a species or a species population, in that we are dealing with a group of individuals interacting in a particular time and space. White-tailed deer from Northern Wisconsin and the Piedmont of Virginia, according to the biological species concept, are the same species as long as they produce viable offspring when they are interbred. But they would belong to different and distinct ecological populations. In actuality, a population is often defined by the investigator(s) and may be somewhat arbitrary.

1.1.2 *Fundamental principles and the use of mathematical models*

What are the fundamental principles that dictate how populations grow? Population ecology is by necessity a quantitative discipline, and in order to answer questions about populations, mathematically-oriented ecologists have derived a variety of predictive models. The first section of this book will examine growth models for populations of single species.

The diversity of life has led to a fantastic array of life histories. Just as the mass of a single bacterium is several orders of magnitude smaller than the mass of an elephant, population characteristics, such as generation time, also differ by several orders of magnitude.

Accordingly, no one model of population growth suits all organisms or all environments. This fact is both frustrating and stimulating. A search for a single set of models that applies to all life forms is pointless. On the other hand, the construction of quantitative models forces us to examine our assumptions about particular populations in an organized and explicit manner. Models, whether quantitative or qualitative, often produce unexpected results that may run counter to our intuitive sense of how things work. The work of Copernicus, Galileo, and others that culminated in the formal quantitative models of Newton showed that the solar system and the universe function in ways that were not at all intuitively obvious. A dissection of the life histories of both the emperor goose (Morris and Doak 2002) and the Amboseli baboon (Alberts and Altmann 2003) populations, using a matrix population model, have shown us that adult survivorship has a greater impact on growth rates than either juvenile survivorship or fertility: a conclusion impossible to reach without the proper population model. As Atkins (1999) commented, "Quantitative reasoning (gives) spine to otherwise flabby concepts, enabling them to stand up to experimental verification." Or as May (2010, p. 8) put it, "Mathematics is ultimately no more … or less than a way of thinking clearly." Thus, models stimulate observations and experiments that allow us to learn more about our natural world.

A general rule of systems is that as one progresses from lower to higher levels of organization, properties are added that were not present at the lower levels. Thus an individual organism is not just a collection of physiological systems. Similarly, a population has properties not evident from the study of individuals. Populations have growth rates, age distributions, and spatial patterns. They also have allelic frequencies and other genetic properties. The first list of properties is within the province of population ecology; the latter is part of the discipline of population genetics. The two areas combined are known as population biology. Although this book deals only with population ecology, much of what I have written is based on the theory of evolution, which relies on principles of population genetics.

The models used here will be largely based on relatively straightforward algebra. However, matrix algebra and differential calculus will be introduced. For more sophisticated mathematical treatments the reader should consult Roughgarden (1998), Case (2000), Vandermeer and Goldberg (2003), or Turchin (2003). I will emphasize the assumptions of the models and discuss them in qualitative terms. Proofs or derivations, where needed, have been minimized, but sample problems and graphs are used to illustrate the workings of the models.

A perfect model would be general, realistic, precise, and simple (Levins 1968). As discussed above, the diversity of life has ruled out the perfect model. In order to attempt generality and simplicity, precision and reality are often sacrificed. If students are able to understand how population models are built, they will then be able to evaluate their reality. It should become evident that most models, while lacking precision, do illuminate basic population trends.

1.1.3 *The general laws of population ecology*

Sutherland (1996) wrote that "population ecology suffers from having no overall *a priori* theory from which explanations and predictions can be devised." He continued that "behavioral ecology has such a theory - evolution by means of natural selection-which yields the prediction that individuals will maximize fitness." I take this to mean that the discipline loosely known as evolutionary ecology has an *a priori* theory. Population ecology, however, should be treated as an extension of evolutionary ecology. Therefore, we should ask ourselves under what circumstances might a characteristic such as the low fecundity of the wandering albatross, or a phenomenon such as the population cycles known for snowshoe hares, have evolved.

By contrast to Sutherland, Turchin (2001; 2003) asserts that population ecology is a vigorous, and predictive science and does have a set of foundational principles that are almost equivalent to the laws of Newton. He has listed these three fundamental concepts: (i) populations tend to grow exponentially, (ii) populations show self-limitation (or bounded fluctuations), and (iii) consumer–resource interactions tend to be oscillatory. In the first case, without density dependent feedback from the environment, all populations show a nonlinear, exponential growth pattern. Turchin (2001) calls this "the exponential law," and sees a direct analogue to the law of inertia proposed by Newton. The exponential law provides a starting point for more complex mathematical descriptions of population dynamics. The second theorem or principle, self-limitation, is based on the idea that per capita population growth decreases with resource depletion. The usual form of this idea, the logistic equation, fails as a law because of its simplistic assumptions (see chapter 2). Nevertheless, it remains useful as a starting point. Finally, the tendency of consumer–resource interactions (such as predator–prey) to produce oscillations is explored at length in later chapters.

It should be noted that Turchin stimulated a number of rebuttals to his proposal of three fundamental "laws" of population ecology (Berryman 2003; O'Hara 2005; Owen-Smith 2005). My particular favorite is O'Hara's "The anarchist's guide to ecological theory. Or, we don't need no stinkin' laws." He states that "law" is much too strong a word, although he is willing to accept the term, "principle." He asserts that, " calling them laws is to give them an epistemological status that they do not deserve ... the laws are not strict enough to judge if a species has broken a law" (O'Hara 2005, p. 393).

Nevertheless, I find these ideas useful as a set of null hypothesis with which to begin our study of population ecology.

1.2 Fundamentals of population growth

If we were trying to understand the growth rate and thus the potential rate of spread of an invasive species, or if we wanted to calculate the potential for long term survival of the Florida panther (Seal and Lacy 1989), what sort of information do we need? How do we gather it? What do we do with the data? What models are appropriate? Here we begin to address these questions.

As a first approximation, population growth is determined by a combination of four processes: reproduction (sexual or asexual), mortality, immigration, and emigration. The addition of new individuals through reproduction, termed fertility or fecundity, may be via sexual reproduction (i.e., live births, hatching of eggs, and seed production) or through asexual reproduction (i.e., binary fission, budding, asexual spores, and clonal spreading of higher plants). The distinction between fecundity and fertility is traditionally as follows:

1. Fecundity is the potential reproductive output under ideal circumstances. This limit is set by the genotype. That is, reproduction is limited by genetic potential, not by the environment.
2. Fertility, by contrast, is the actual reproductive performance under prevailing environmental conditions. The fertility rate, by definition, is less than the fecundity rate and is based on the interaction of the genotype with the biological and physical environment.

The distinction between these two terms is often not rigidly adhered to, but it is useful to keep it in mind.

Both fecundity and fertility are expressed as rates. That is, the mean number of offspring produced per individual (or per thousand individuals in human demography) in the population, per unit time. Often these values are also expressed for a given unit of area. For example, according to the Population Reference Bureau, the fertility rate of the human population of the world declined from 28 per thousand in 1981, to 20 births per thousand in 2012. Meanwhile, the birth rate in North America moved slightly downward from 16 per thousand in 1981 to 13 per thousand in 2013 (Anonymous 2012). In populations such as humans, however, which breed over a period of 30 years without respect to seasons, we need to know the fertility rate for each age category in order to accurately predict population growth.

The second fundamental factor that affects population growth is mortality. Mortality must also be expressed as a rate. That is, the mean number of deaths per individual (or per thousand), per unit time, per unit area. As above, unless the population has a stable age distribution (meaning that the proportion of the population in each age class remains constant over time), in order to predict future population changes, we would need to know the death rate for each age category. Again, using data from the Population Reference Bureau, the human death rate for the world in 2012 was 8 per thousand, a decrease from 11 per thousand in 1981. In North America, the comparable figures are: 9 per thousand in 1981 and 8 per thousand in 2012 (Anonymous 2012).

In populations with age distributions (age structures), growth is also affected by the actual number of individuals in the different age categories. We will explore the effects of age distributions in detail in Chapter 4. At present it is sufficient to note that

Table 1.1 Statistics for human populations of Asia and Europe in 2012.

Region	Population size (millions)	Birth rate (per thousand)	Death rate (per thousand)	Rate of increase per individual (r)	Percent of the population less than 15 years of age
Asia	4,260	18	7	0.011	25
Europe	740	11	11	0.000	16

Birth and death rates are per thousand; r is per individual.
Source: Anonymous 2012. Data from the Population Reference Bureau.

basic data on the overall birth and death rates may not produce an accurate picture of population growth in the short term. For example, examine the population figures for Europe and Asia in 2012 (Table 1.1), again data from the Population Reference Bureau. Not only are the birth and death rates different, but also their age distributions are different. The percentage of the population under 15 years of age in Asia is 25, while that of Europe is a mere 16.

A measure of population growth is the intrinsic rate of increase, r. We will discuss r in more detail later. For now, we define r as the growth rate per individual (or per capita) per time unit (for example, per year) in a population, estimated as $b - d$, where b is the birth rate per individual per year, and d is the death rate per individual per year. The rate of growth per individual is:

$$r = b - d \qquad\qquad\qquad (1.1a)$$

If the birth and death rates are expressed per thousand, as in human demography, the growth rate is;

$$r = \frac{b - d}{1000} \qquad\qquad\qquad (1.1b)$$

From Table 1.1 we see that Asia had a positive growth rate, whereas Europe actually had a projected growth rate of zero in 2012. If the intrinsic rate of increase of these two populations suddenly converged on the same value (a decrease in the Asian birth rate and an increase in Europe's fertility rate, combined with similar changes in the death rates), the population growth of Asia would still be greater than that of Europe for several decades, due to the higher abundance of reproductive individuals. Asia also has a shorter generation time, which would affect population growth for a number of years. The estimated growth rate parameter, r (equation 1.1b), ignores the age distribution and generation time and actually assumes a stable age distribution (defined above). By age distribution we simply mean the proportion of the population in each age category, not the actual number per category.

Two other factors affect population growth: Immigration and Emigration.
- The immigration rate is the number of individuals that join a population per time interval due to immigration. Ideally we should know the ages of individuals as they join the population.
- The emigration rate is the number of individuals that leave the population per time interval, and how old they are when they leave.

Unfortunately, gathering accurate information on immigration and emigration is extremely difficult in biological populations, and these factors are often ignored.

When a population is termed closed, it is thought of as having negligible immigration and emigration. In the last two decades, however, there has been a shift in emphasis from the study of single populations to "metapopulation" ecology. Since the concept of a metapopulation was developed by Levins (1969; 1970), major advances in both theory and field studies have taken place, particularly within the past 20 years (Hanski 1999). Levins (1969) originally defined a metapopulation as a "population of populations." In his view, local populations exist in a fragmented landscape of suitable and unsuitable habitats or "patches." Each local population is prone to extinction, but extinction may be balanced by immigration from other populations in the metapopulation landscape. The long-term survival of the metapopulation depends on the balance and interplay between extinction and immigration. Immigration and extinction are also key elements of the MacArthur and Wilson (1967) theory of Island Biogeography. However, MacArthur and Wilson were primarily concerned with the number of species in the community, while the metapopulation concept focuses on populations of single species. Another difference is that MacArthur and Wilson were concerned with the relationship between islands, where extinction could occur because of small population size or stochastic events, and a source of species (the mainland) in which extinction would not normally occur. By contrast, in a metapopulation, extinction may occur in any patch and colonization can occur from any one patch to another. The applications of metapopulation studies to conservation biology are obvious, and have resulted in an explosion of publications. We will explore metapopulation dynamics in Chapter 5. Suffice it to say that, after decades or being ignored, immigration, emigration, and local extinction are now the subject of many theoretical and field studies (Hanski 1999).

As already noted, a population is rooted in a time and a place. This means that population sizes or population growth rates are scaled for a particular time unit and for a specific spatial unit. When life histories of different organisms are compared (Chapter 6) it becomes obvious that generation times vary across several orders of magnitude. The space needed to sustain one population of elephants may support a metapopulation of butterflies or several separate populations of lichens. Therefore, we are forced to ask, what is the appropriate scale of an ecological investigation (Peterson and Parker 1998)? That is, over what time spans and/or over what spatial scales, should ecological investigations be conducted? As we explore simple models of population growth we should be aware of their limitations and applicability to long periods of time and/or to large landscapes.

In summary, a population is affected by its rates of fertility, mortality, immigration, and emigration, by its recent history (through its age structure), and by its generation time, which is determined by its life history. Growth rate is also determined by the environment, and by how sensitive the population is to changes in the environment. By environment, we mean not only the physical environment, but also interactions of the population with other species in its habitat.

1.3 Types of models

In developing a model of a population we usually begin with the present population; i.e., the population at time $= 0$, and project it t time units into the future. This is expressed as N_0 and N_t, respectively. There are two types of population equations. Each has advantages and disadvantages.

In Difference Equations, populations are modeled using specific, finite, time units. The time units are usually realistic, in that populations are measured in the field once or perhaps, several times per year, but not continuously. Difference equations are most often used to model populations that have "discrete," rather than continuous, growth (see below). A basic equation summarizing the ideas presented in the previous section might look like this:

$$N_{t+1} = N_t + (B - D) + (I - E) \qquad (1.2a)$$

where N_t = the population size at time, t;
$\quad N_{t+1}$ = the population size one time unit later;
$\qquad B$ = the number of births and D = the number of deaths in the population during the time interval $t - t + 1$;
$\qquad I$ = the number of immigrants and E = the number of emigrants during this same time interval.

This equation can be rewritten as:

$$N_{t+1} = N_t + (B + I) - (D + E) \qquad (1.2b)$$

In most population studies it is assumed that immigration and emigration rates are insignificant compared with birth and death rates (Turchin 2003; but see Hanski 1999). Equation 1.2b can be simplified, and the numbers of births and deaths are converted to per capita (per individual) rates, b and d, respectively. The difference between b and d becomes the single growth parameter, R, known as the net growth rate per generation or the net reproductive rate. Alternatively, the difference between b and d also equals λ, the growth rate per time period, usually per year. The λ can be calculated for all types of population models and is known as the finite rate of increase. The usual form for the difference equation (using R) is as shown in equation 1.2c.

$$N_{t+1} = N_t(b - d) = N_t R \qquad (1.2c)$$

In Differential Equations, it is assumed that population growth is "continuous" and populations are being continuously monitored. Models based on differential equations have a long history in the biological literature, including the earliest models of competitive, predator–prey and host–parasite relationships (Lotka 1925). A simple differential equation for population growth is:

$$\frac{dN}{dt} = rN \qquad (1.3)$$

Here dN/dt measures the instantaneous growth of the population, N. On the left side of the equation, the symbol d is used to indicate change in N per change in the time interval, t. The intrinsic rate of increase r (equation 1.1a), measures the per capita birth rate minus the per capita death rate during these same small time intervals. In a sense, r measures the probability of a birth minus the probability of a death occurring in the population during a particular time interval.

1.4 Density independent versus density dependent growth

If a population invades a new environment with "unlimited" resources, no competitors, and no predators, fertility rates will be high (approximating fecundity rates) and death rates will be relatively low. Under these conditions, the population grows "geometrically" or "exponentially" depending upon its life history. This is known as density independent growth. This simply means that the growth rate parameter of the population is not affected by its present population size. In both geometric and exponential models, the growth rate is determined by a fixed parameter (R, λ, or r), which is not modified by competition for resources. Population growth is often curtailed by the environment even if the population is undergoing density independent growth. Major disturbances or catastrophes such as fire, wind storms, landslides, and floods significantly reduce certain populations and may even cause local extinctions. In Chapter 2 we will examine models of density dependent growth. In these models, it is assumed that the population encounters a limiting resource (food, water, nest sites, available nitrogen, space, etc.), which limits its growth. In these models the growth parameter is modified and the net growth rate eventually approaches zero at an environmental limit termed a carrying capacity. The realized growth rate is said to depend on the density of the population, hence the term, density dependent growth.

1.5 Discrete or "geometric" growth in populations with non-overlapping generations

The use of an appropriate model depends first on the life history of the organism. Therefore you first need basic information on the life cycle of the species. In this first model of density independent growth, the population has a life history with discrete, non-overlapping generations. That is, there are no adult survivors from one generation to the next. Examples include: annual plants, annual insects, salmon, periodical cicadas, century plants, and certain species of bamboo. In most of these cases the organism passes through a dormant period as a spore, a seed, an egg, or as a juvenile stage such as a larva or pupa. Once the adults reproduce, they perish, and the future of the population is based on the dormant or juvenile stage of the organism. As noted above, when modeling such populations we usually collapse fertility and mortality into one constant, R, the net replacement rate, or net growth rate, per generation. Alternatively, we use the finite rate of increase, λ, when measuring growth per specific time period. When we are discussing annual plants or insects, λ, the growth rate per year, and R, the growth rate per generation, are identical, since generation time equals one year. However, in some populations, such as the periodical cicada, generation time equals 13 or 17 years, and in these cases it is useful to make a distinction between the growth rate per generation as opposed to a finite rate of increase. That is, $R = \lambda$, when T, the generation time, equals 1 year.

To find R we often count one life stage of the population in successive years. For gypsy moths (Lepidoptera: *Lymantria dispar* L.) we estimate R by counting egg masses in successive years (See example 1, below). R is estimated from the ratio of egg masses at time $t + 1$ versus time t. For the periodical cicada, (Homoptera: *Magicicada septendecim* L.), however, we would have to wait 17 years between generations before we could estimate R. The overall model is based on finding successive estimates of the growth rate based on:

$$R_1 = N_1/N_0$$

$$R_2 = N_2/N_1$$

$$R_3 = N_3/N_2 \text{ etc.}$$

If we find that R remains more or less constant over time (i.e., if these ratios of N_{t+1}/N_t remain constant), then we have:

$$N_1 = N_0 R$$
$$N_2 = N_1 R = (N_0 R)R = N_0 R^2$$
$$N_3 = N_2 R = (N_0 R^2)R = N_0 R^3$$

leading to equation 1.4

$$N_t = N_0 R^t \tag{1.4}$$

or

$$N_t = N_0 \lambda^t \tag{1.5}$$

Note that the population grows whenever $\qquad\qquad$ R or $\lambda > 1$

The population is stationary (there is no growth) whenever $\quad R$ or $\lambda = 1$

The population decreases whenever $\qquad\qquad$ R or $\lambda < 1$

The population grows according to the law of discrete or geometric growth (Figure 1.1), when $R > 1$. Equations 1.4 and 1.5 can be rewritten using logarithms to make the growth curves linear. In equations 1.6 and 1.7 we can use log to the base 10, or we can use natural logs (designated by ln) to the base e. Since other models use natural logs, we will use them in the equations below.

$$\ln N_t = \ln N_0 + (\ln R)t \tag{1.6}$$

or

$$\ln N_t = \ln N_0 + (\ln \lambda)t \tag{1.7}$$

Fig. 1.1 Discrete or "geometric" growth in a population with non-overlapping generations.

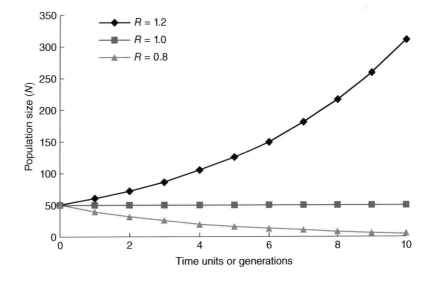

Fig. 1.2 Natural log of growth in a population with discrete generations.

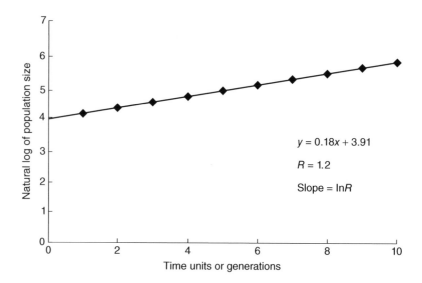

In each case $\ln N_0$ is the y-intercept and $\ln R$ or $\ln \lambda$ is the slope of a linear relationship between $\ln N$ and t (time), with time as the independent variable (x-axis). In Figure 1.2, the value of $R = 1.2$ and the slope is therefore $\ln(1.2)$ or 0.18.

Example 1.1 Gypsy moths (*Lymantria dispar* L.) are annual insects in which breeding takes place in early to mid summer. After the females lay their eggs, all adults die. The eggs hatch the following spring into larvae that feed on the leaves of tree species, especially species of oaks (*Quercus*). After a number of larval stages and a pupal stage, the adults emerge. After mating, females lay their eggs and die, as noted above. Since generation time equals one year, equations 1.4 or 1.5 may be used. In order to determine population growth in this species, we need to determine R. Assume that a local gypsy moth technician makes annual egg mass counts in a local forest. She finds that in 2013 there are, on average, four gypsy moth eggs masses per hectare and each mass contains an average of 40 eggs for a total of 160 eggs per hectare. When she returns to the same forest in 2014, she finds five egg masses with an average of 40 eggs, or a total of 200 eggs per hectare. The local spraying program regulations state that spraying with Bt® begins whenever egg masses reach 1000 per hectare. Assuming egg mass density continues to increase at a constant rate, what is the predicted population for the year 2016? In what year would spraying be required?

Answer:
In order to determine the net growth rate R, we find the ratio of $N_{t+1}/N_t = 200/160 = 1.25$ In the year 2016, 3 years have passed since the original survey in 2013. Using equation 1.4:

$$N_{2016} = N_{2013}R^3 = (160)(1.25)^3 = 312.5$$

We therefore expect around 312 eggs per hectare in 2016.

We can now ask the question, if R continues at 1.25, in what year must spraying commence? Since we wish to solve for t, and time is an exponent in equation 1.3, it is more convenient to use equation 1.6.

$$\ln N_t = \ln(1000) = \ln(160) + \ln(1.25)(t)$$

$$6.91 = 5.08 + 0.223t$$

$$1.83/0.223 = t$$

$$t = 8.2 \, \text{years}$$

Since the population only reproduces once a year, we cannot use a fraction of a year in the answer. Eight years after 2013, i.e., in the year 2021, the number of eggs masses is expected to be 954. By regulation, this does not trigger the spraying regime. One year later, however, the egg mass density would be 1192 and spraying would begin in 2022.

Example 1.2 The periodical cicada (Homoptera: *Magicicada septendecim* L.) has a most unusual life history (Borror et al. 1989). The juvenile stages spend 17 years underground feeding on plant roots. The population in a given area emerges synchronously from the ground as adults. After a great deal of racket, the males and females mate, and females lay their eggs in slits they have made in small branches of trees and shrubs in the forest. The adults then die, leaving the eggs as the next generation. The eggs hatch within a month. The nymphs drop to the forest floor and burrow underground, where they spend the next 17 years feeding and growing. The periodical cicada is obviously affected by disturbances within the forest habitat. Assume that in 1987, a survey found 500 adult female cicadas per hectare. The forest was selectively logged in the 1990s and a survey in 2004 found that the cicada population had dropped to 200 per hectare. More logging is planned during the next 20 years in this forest. Assume the population continues to decline at the same rate. If we define the minimum viable population for cicadas as ten females per hectare, in what year is the population no longer viable (By minimum viable population we mean that the probability of extinction has become unacceptably high (Shaffer 1981; Miller and Lacy 2003)? Random environmental perturbations or inability of males and females to find each other would likely cause this population to become extinct. See the section on Population Viability Analysis, 1.10)?

Answer:
First we must realize that only equations 1.4 and 1.6, using net growth rate per generation are applicable. But we also need to remember that generation time is 17 years. To find R, take the ratio of $200/500 = 0.40$. Since $R < 1$ we note that this population is decreasing. In order to find when the population is not viable, we solve equation 1.6.

$$\ln 10 = \ln 500 + (\ln 0.40)t$$

$$2.3 = 6.2 + (-0.9)t$$

$$-3.9 = -0.9t$$

$$t = 4.3 \, \text{ generations}$$

Again, we cannot use fractions or decimals. After four generations, the population is projected to drop to between 12 and 13. After five generations, it declines to around 5 per hectare and is, by definition, no longer viable. Five generations, times 17 years per generation, equals 85 years. The population is not viable 85 years after the first survey in 1987. That is, in the year 2072. However, action to conserve this forest really cannot wait until 2072.

1.6 Exponential growth in populations with overlapping generations

In the previous section we dealt with a special kind of life history, one in which generations were distinct and non-overlapping. If the adults and juveniles are present simultaneously and they interact with one another, our previous model is inappropriate. Instead we must use a model originally developed for a population capable of continuous growth, such as a *Paramecium* or a human population. That is, a population in which there is no distinct breeding season. Notwithstanding poetry about springtime and theories about phases of the moon, human babies are born throughout the year (although there are major spikes in birth rates 9 months after a major snowstorm). In spite of the fact that this growth model is not strictly applicable for seasonal breeders such as deer, it is general enough that it is used whenever a population has a stable age distribution. (Recall that an age distribution refers to the proportions of the population belonging to different age classes, and that a stable age distribution is one in which these proportions remain constant from year to year.) In order to have a stable age distribution, fertility, and mortality rates must remain constant for an extended period of time. We can approximate human population growth rates using the model, but we should recall that because birth rates around the world increased following World War II and then decreased after 1960, few human populations are in a stable age distribution at this time.

The basic form of this model is the differential equation shown earlier as equation 1.3. $dN/dt = rN$ where r is the intrinsic rate of increase or the instantaneous growth rate.

The r-value is calculated by finding the difference between the instantaneous per capita birth rate and the instantaneous per capita death rate. The parameter, r, can be compared to the interest rate in a bank account which is continuously compounded. Such a rate is the continuous growth rate per dollar in an interest bearing account, while r is the continuous growth rate per individual in a population.

The equation is easily solved by taking the integral from 0 to t of both sides of the equation as follows:

$$\int_{N(0)}^{N(t)} \frac{dN(t)}{N} = r \int_{0}^{t} dt,$$

which becomes:

$$\ln N(t) - \ln N(0) = rt - r0 = rt$$

After exponentiation of both sides of the equation, we have: $\frac{N(t)}{N(0)} = e^{rt}$

Rearranging, we get equation 1.8. This solved form is the one usually used in making population projections to some arbitrary time t in the future.

$$N_t = N_0 e^{rt} \tag{1.8}$$

Fig. 1.3 Exponential growth pattern in a population with overlapping generations and continuous breeding.

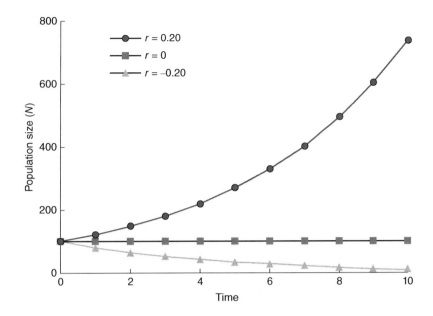

where e is the base of natural logs.

Note that the population grows if $r > 0$
The population is stationary if $r = 0$
Population growth is negative if $r < 0$

When r is positive, the growth is known as exponential; if r is negative the population is in exponential decline (Figure 1.3).

Again, we can make the equation linear by taking the natural logs of both sides of equation 1.8, yielding:

$$\ln N_t = \ln N_0 + rt \tag{1.9}$$

When we graph $\ln N$ versus time, we again have a linear relationship, with $\ln N_0$ as the y-intercept and r as the slope of the line (Figure 1.4).

1.6.1 *Doubling time*

A convenient statistic, often used by population ecologists and human demographers (demography is the study of population statistics), is doubling time. That is, if a population continues to grow at its present rate, how long will it take a population to double from its present population size? Equation 1.8 can be rearranged to: $N_t/N_0 = e^{rt}$. We want to solve for the time at which the ratio, $N_t/N_0 = 2$. So we have: $2 = e^{rt}$. Taking the natural log of both sides of the equation yields, $\ln 2 = rt$, where t is now doubling time. Since $\ln 2 = 0.693$, if we solve for t we end up with:

$$\text{Doubling time} = 0.693/r \tag{1.10}$$

Therefore if we know the intrinsic rate of increase we can easily find the projected doubling time of a population. Remember, however, that we are assuming that the population is not affected by its age distribution, and that r is a constant during this time period. That is, birth and death rates remain unchanged.

Fig. 1.4 Natural log of growth in a population with overlapping generations and continuous breeding.

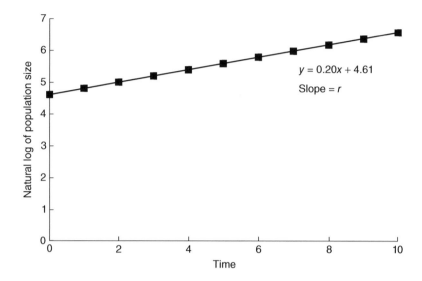

$y = 0.20x + 4.61$

Slope = r

1.7 Examples of exponential growth

During a hurricane in 1962, five captive mute swans (*Cygnus olor*) escaped into the Chesapeake Bay, in Maryland. Since they were pinioned and therefore flightless, their chance of survival during the winter was considered negligible and no attempt was made to capture them. One pair, however, successfully nested. By 1975 the descendents of this original pair numbered approximately 200, and by 1986 totaled 264. By 1999 the estimated population of Mute Swans in the Chesapeake Bay was 3955 (Anonymous 2003; Sladen 2003; Craig 2003). In 2001 the Maryland Department of Natural Resources, in an effort to control the swan population, began shaking (addling) mute swan eggs or covering them with corn oil to terminate embryo development. Mute swans were also removed from Federal National Wildlife Refuges. The result was a decline to 3624 in 2002 (Anonymous 2003). As shown in Figure 1.5, prior to these control efforts, the population was growing exponentially with an intrinsic rate of increase of 0.17 and a doubling time of 4 years! Use equation 1.10 to verify the doubling time.

Fig. 1.5 Mute swan population in the Chesapeake Bay since 1962.

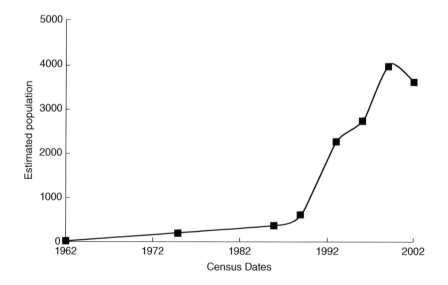

So what's the problem? Swans are considered graceful, even "majestic" and are thought of as harmless by their admirers. However, mute swans, in addition to being a non-native species, have become permanent residents. That is, they do not migrate as do other native swan species. Recent data show that an average adult swan eats eight pounds of *submerged aquatic vegetation* (*SAV*) a day (Craig 2003). This is occurring at a time when biologists are struggling to reestablish SAV in the Bay. Is it necessary to control the Mute Swan population? If so, how?

Gray wolves (*Canis lupus*), after being driven extinct in the lower 48 states in the US, have been making a comeback. In addition to the much publicized population in Yellowstone National Park, wolves have colonized neighboring western states as well as Minnesota and Wisconsin and elsewhere. In Wisconsin the Department of Natural Resources (Wydeven et al. 2012) estimates that the wolf population has grown from a few individuals in the 1980s to well over 800 by 2011 (Figure 1.6a). The intrinsic rate of increase from 1985 to 2011 is 0.16 (Figure 1.6b), similar to that of the mute swan. The response of the Wisconsin DNR has been to open a hunting season on wolves. Is this justifiable? Does the wolf population need control? If so, what are the other options?

1.8 Applications to human populations

Few biological populations grow either geometrically or exponentially for long. As we will explore in the sections on intraspecific competition and logistic growth, as populations grow, resources become scarce. The resultant changes in birth and/or death rates slow growth. The human population of the world, however, has continued to grow since around 1650; it reached 6.0 billion by late 1999, and 7.0 billion by 2012 (Figure 1.7a). Many scientists question how long this growth can be sustained. While most ecologists insist that human population growth must cease in the near future, some economists (Simon, 1996) see no reason for limits to the human population. In the next section we will use data from the Population Reference Bureau (Anonymous 1981–2013) to illustrate how equations 1.8–1.10 may be used in population projections.

Recall from equation 1.9 that if we graph natural log of population growth versus time we can determine the intrinsic rate of increase by finding the slope of the graph. In Figure 1.7b we have plotted the natural log of human population growth against time. The slope of this line, as determined by the statistical technique of linear regression and computed for us in an Excel™ spreadsheet, is 0.007. This is the best fit for the intrinsic rate of increase for the human population from 1650 to 2012.

If we examine Table 1.2, in which human populations in 2012 are broken down by continental regions, the strengths and weaknesses of this simple model become apparent. Most striking are the immense differences among populations. While the human population as a whole was growing about 70% faster in 2012 as compared to the period of 1650 to the present (contemporary $r = 0.012$, historical $r = 0.007$), Europe shows no growth, while Africa's growth is twice the global growth rate. Secondly, over 54% of the human population resides in Asia.

Clearly, although human population growth is of global concern, it is a highly regional problem. From Table 1.2 you should be able to see that r is readily calculated as the difference between the birth and death rates. Secondly, you should try calculating projected doubling times based on equation 1.10. You will find that the data published by the Population Reference Bureau may differ slightly from your calculations. They are using more sophisticated models and are taking age distributions

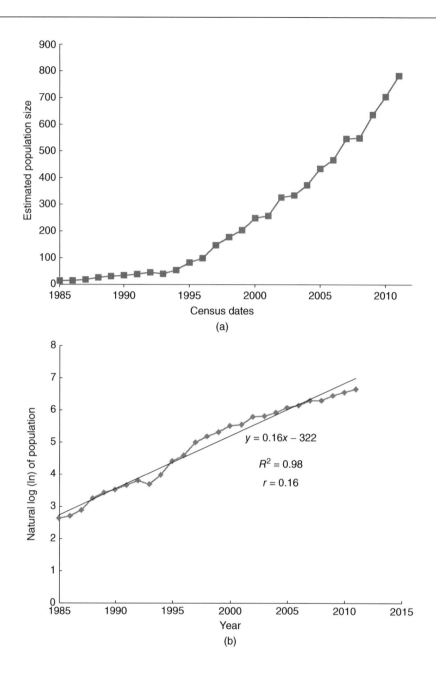

Fig. 1.6 (a) Gray Wolf (*Canis lupus*) population in Wisconsin from 1985 to 2011. (b) Natural log (ln) of Gray Wolf (*Canis lupus*) population in Wisconsin from 1985 to 2011. (Data from Wisconsin Department of Natural Resources (Wydeven et al. 2012)).

into account. Nevertheless, the differences in doubling times are remarkably minor. Finally, if you examine the last column you will also notice another great difference among these populations. The percentage of the population in the pre-reproductive years (15 years or younger) varies from 41% in Africa to a low of 16% in Europe.

In his book, *The Skeptical Environmentalist*, Lomborg (2001) is rather sanguine about human population growth. He accepts the demographic transition model, which states that rapid growth has occurred because of a rapid drop in the death rate (due to

Fig. 1.7 (a) Human population growth since 1650. (b) Natural log of human population growth since 1650.

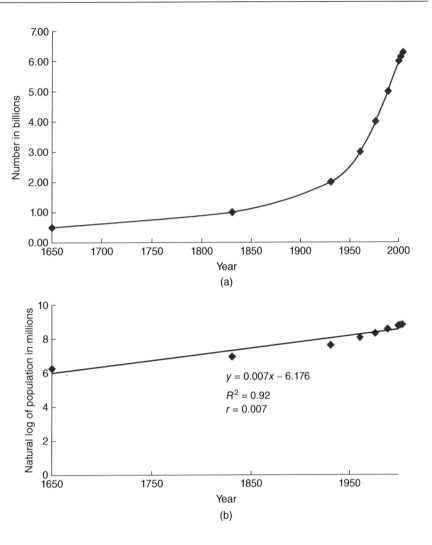

modern methods of sanitation, improved food growth and distribution, better medical care, etc.) and that eventually, with improved standards of living and wealth, birth rates drop to match the low death rates. Indeed, in most European countries, human population growth has slowed, and even gone negative. In 2012, 16 countries out of 45 in Europe had a growth rate of zero or negative, including most of the Eastern European countries. As noted above, the population growth rate (*r*-value) for Europe as a continent is zero.

As for the future, Lomborg (2001) accepts a "medium variant forecast" from the UN. This prediction is zero population growth for the world by the year 2100. However, by then the world population is projected to be 11 billion. Consider that the world population was only one billion in1850, two billion in 1950, and 6.3 billion in 2003. Lomborg is correct when he says that 60% of growth is from just 12 countries. Perhaps the world outside of Africa and Asia will not necessarily suffer a catastrophe from human population density, but what will happen in China, India, Pakistan, Bangladesh, and Nigeria, for example, in the next 100 years? While China's population is predicted to stabilize

Table 1.2 2012 human population data from the population reference bureau.

Region	Population in millions	Birth rate (per thousand)	Death rate (per thousand)	r (per individual)	Doubling time (years)	Percent under 15 years
World	7058	20	8	0.012	58	26
Africa	861	36	11	0.025	28	41
North America[1]	323	13	8	0.005	139	19
Latin America[2]	540	19	6	0.013	53	28
Asia	3830	18	7	0.011	63	25
Europe	727	11	11	0.000	NA	16
Oceania[3]	32	18	7	0.011	63	24

[1]North America includes the United States and Canada
[2]Latin America includes Central and South America and the Caribbean Islands
[3]Oceania includes Australia, New Zealand and the South Pacific Islands
Countries of the former USSR have been distributed between Asia and Europe
Source: Anonymous 2013. Data from the Population Reference Bureau.

Table 1.3 World human demographic trends since 1981.

Year	World population estimate (billions)	Birth rate (per thousand)	Death rate (per thousand)	r (per individual)	Actual average growth per year during specified time period (millions)
1981	4.492	28	11	0.017	
1985	4.845	27	11	0.016	1981–1985: 88.3
1987	5.026	28	10	0.018	1985–1987: 90.5
1989	5.234	28	10	0.018	1987–1989: 104.0
1991	5.384	27	9	0.018	1989–1991: 75.0
1995	5.702	24	9	0.015	1991–1995: 79.5
2000	6.067	22	9	0.014	1995–2000: 73.0
2003	6.314	22	9	0.013	2000–2003: 82.3
2006	6.555	21	9	0.012	2003–2006: 80.3
2009	6.810	20	7	0.013	2003–2009: 85.0
2012	7.058	20	8	0.012	2009–2012: 82.7

Source: All data from the population reference bureau, Anonymous (1981–2013).

at 1.31 billion by 2050, as compared to the present population of 1.35 billion, the PRB predicts a population for India of 1.69 billion (compared to present population of 1.26 billion) (Anonymous 2012). The question on the mind of the concerned biologist: Will there be any room for natural habitats on a planet with eleven billion or, worse yet, fifteen billion people?

Examine Table 1.3, which describes overall human demographic trends since 1981. Lomborg (2001, p.47) states that world population growth, in numbers per year, reached a peak in 1990 at 87 million per year. Population Reference Bureau data agree on the time but not the number (over 100 million added in the period 1987–1989). Absolute growth has averaged about 80 million per year since 2000, according to Population Reference Bureau data. Lomborg used the figure of 76 million, but this applies only to the 1990s. The Population Reference Bureau projects world population as 7.9 billion in 2025 and 9.2 in 2050. Lomborg's comparable numbers are "almost 8 billion" in 2025 and 9.3 billion in 2050.

Table 1.4 Human demographic trends in North America since 1981.

Year	Population estimate (Billions)	Birth rate (per thousand)	Death rate (per thousand)	r (per individual)
1981	0.25	16	9	0.007
1985	0.26	15	8	0.007
1987	0.27	15	9	0.006
1989	0.27	16	9	0.007
1991	0.28	16	9	0.007
1995	0.29	15	9	0.006
2000	0.31	14	8	0.006
2003	0.32	14	8	0.005
2006	0.33	14	8	0.005
2009	0.34	14	8	0.006
2012	0.35	13	8	0.005

Source: Anonymous (1981–2013).

Table 1.5 Human demographic trends in Asia since 1981.

Year	Population estimate (billions)	Birth rate (per thousand)	Death rate (per thousand)	r (per individual)	Actual average growth per year during specified time period (millions)
1981	2.61	29	11	0.018	
1985	2.83	28	10	0.018	1981–1985: 55.0
1987	2.93	28	10	0.018	1985–1987: 50.0
1989	3.06	28	9	0.019	1987-1989: 65.0
1991	3.16	27	9	0.018	1989–1991: 50.0
1995	3.38	24	8	0.016	1991–1995: 55.0
2000	3.68	22	8	0.014	1995–2000: 60.0
2003	3.83	20	7	0.013	2000–2003: 50.0
2006	3.97	20	7	0.012	2003–2006: 46.7
2009	4.12	19	7	0.012	2003–2009: 50.0
2012	4.26	18	7	0.011	2009–2012: 46.7

Population growth in North America (Table 1.4) is rather variable, but reached a relative peak in 1991–1992 when around two million people were added to the population per year. Meanwhile, the US birth rate has fallen to 2.06 births per female (replacement rate is 2.10 births per female) (PRB, Anonymous 1981–2013).

Human population growth is greatest in Asia (Table 1.5). Peak absolute growth was in the period 1987–1989 when around 65 million people were added per year. It has declined unsteadily since 2000 and is now about 50 million people per year. The r-value has declined steadily to 0.011 in 2012.

1.9 The finite rate of increase (λ) and the intrinsic rate of increase (r)

Both the intrinsic rate of increase (r) and the finite rate of increase, λ, are commonly used to track population growth and to compare growth rates among populations. Consequently it is important to understand the relationship between λ and r. As defined in equation 1.5, λ is the growth rate per time period (usually per year) and is based on the ratio: N_{t+1}/N_t. If the population is growing or declining at a constant rate, lacks an age distribution or has a *stable age distribution* (*SAD*), the finite rate of increase, λ, is a constant. The population as a whole and each age class will grow as:

$$\frac{N_{t+1}}{N_t} = \lambda \tag{1.11}$$

Rearranging equation 1.8 and setting $t = 1$, we have: $N_{t+1}/N_t = e^{rt} = e^r$. Thus, when $t = 1$ and when there is a stable age distribution we have

$$\lambda = e^r \tag{1.12}$$

and

$$r = \ln \lambda \tag{1.13}$$

Example 1.3 A *Paramecium caudatum* population is cultured in the laboratory and sampled on a daily basis. Population sizes, based on 0.5 ml samples, are shown below (Table 1.6). The population grows exponentially between days zero and three. Find the intrinsic rate of increase (r) for the population

Answer:

Since we want to know the value of the maximal rate of increase (the density independent rate of increase), we examine growth only during the first 3 days (see Figure. 2.1 in Chapter 2). From equation 1.9, ($\ln N = rt + \ln N_0$) we know that to find r, we need only convert column 2 to natural logs (column 3). Then find the slope between days zero and three. To find the slope we can use the formula:

$$r = (y_2 - y_1)/(x_2 - x_1).$$

Thus,

$$r = (5.26 - 2.64)/(3 - 0),$$

Table 1.6 Growth of a *Paramecium* population.

Time in days	Number (N) per 0.5 ml	Natural log of N ($\ln N$)	Per capita growth $\left(\frac{N_{t+1}-N_t}{N_t}\right)$
0	14	2.64	1.93
1	41	3.71	1.83
2	116	4.75	0.66
3	193	5.26	0.26
4	244	5.50	0.19
5	290	5.67	0.14
6	331	5.80	0.10
7	363	5.89	0.03
8	375	5.93	0.01
9	380	5.94	

Numbers are based on daily 0.5 ml samples.

and

$$r = 0.87$$

Using Excel, a linear regression on the same data yields the value of r as 0.89

Example 1.4 The birth rate for Latin America in 1978 was 33 per thousand, while the death rate was ten per thousand. What was the intrinsic rate of increase, assuming a stable age distribution? If the population size was 344 million, what was the projected population in 1982? Between 1982 and 1990 the population increased from 377 million to 415 million, what was the r during that time? Given this r-value, what was the doubling time? What is λ?

Answer:

(a) Given r = b-d, we have r = 33/1000-10/1000 = 0.023
(b) From 1978 to 1982 is four years. Therefore:

$$N_4 = N_0 * (e^{0.023*4}) = 344 \text{ million} * (e^{0.092})$$

$$= (344 \text{ million}) * (1.096) = 377.15 \text{ million}$$

(c) From 1982 to 1990 is 8 years. Therefore: 415 million = 377 million $* (e^{8r})$
   Simplifying: $415/377 = 1.10 = e^{8r}$
   Taking natural logs: ln 1.10 = 8r
   Or, 0.096/8 = r = 0.012
(d) Doubling time = .693/r = .693/0.012 = 57.7 years
(e) $\lambda = e^r = e^{0.012} = 1.012$

1.10 Stochastic models of population growth and population viability analysis

All of the population models we have examined to this point are deterministic models. The models specify conditions leading to an exact outcome based on the parameters of the models. But natural systems are unlikely to be deterministic; rather they are more likely to be stochastic. In particular, small, isolated populations are subject to stochastic processes because chance events can dominate their long-term dynamics. In stochastic models population parameters vary according to a frequency distribution. This distribution has a "central tendency" (a mean), but also has a range of variability around the mean. For example, in a deterministic model, if we know the present population size and the proper growth parameter, we are able to forecast an exact expected population size for a specific time in the future. In a stochastic model, we would instead predict a range of future population sizes, with assigned probabilities.

Future population size in a small population is strongly influenced by demographic stochasticity, which is driven by variations in the fates of different individuals within a given year. For example, although the average female within a population has 2.5 female offspring, obviously any given female cannot have fractions of offspring. In addition, some individuals may not reproduce at all, while others may have a litter size of 4.0. The same kinds of variations are true when we consider death rates. Demographic stochasticity has effects not only on birth and death processes, but also on sex ratio. In the above example, some females may give birth only to males in a given year.

Another important influence on population growth is environmental stochasticity, which is temporal variation in the population due to unexpected events, often tied to the physical environment, such as droughts, hail storms, fires and volcanic eruptions, but may also include diseases. Environmental stochasticity can affect both large and small populations.

A more realistic growth model, therefore, would make forecasts based on probabilities, rather than predicting a single outcome. For example, weather forecasters no longer simply predict rain, but instead predict a certain probability of rain. Similarly, it would be prudent for population models to predict an expected population size, but allow for other population sizes to occur with particular probabilities. Again, this approach is especially important in small populations, and over short time intervals. If the population and the time frame are very large, the expected population sizes dictated by deterministic models become highly probable.

Stochastic models are the basis for the quantitative approach to conservation biology known as *population viability analysis* (PVA). Although it is beyond the scope of this book to explore stochastic models and PVA in detail, there are excellent discussions of these models in Morris and Doak (2002) and Beissinger and McCullough (2002). For more information on stochastic models, see also Pielou (1977) and Nisbet and Gurney (1982).

PVA is important because many wildlife populations that were once: numerous, widespread, and occupied contiguous habitats; are now: small, restricted in distribution, and isolated from each other. The problem with small, isolated populations is that they are increasing subject to stochastic processes and increasingly likely to go locally, if not globally, extinct. The purpose of PVA is to predict the likely future status of a population or collection of populations (Morris and Doak 2002). PVA is a set of analytical and modeling approaches for assessing the future course and risk of extinction of a population (Beissinger and McCullough 2002). PVA examines how (i) genetic, demographic, and environmental stochasticity, (ii) catastrophes and "bonanzas," and, (iii) spatial variation affect the future of the population. The most commonly used PVA model is "Vortex," which is available free online.

Small populations are also affected by genetic processes such as: (i) genetic drift resulting in the loss of genetic diversity in the population, (ii) inbreeding depression, (iii) monopolization by a small number of males in a polygynous mating system. The biggest concern is the rate of loss of heterozygosity and its effects on the future fertility and mortality rates of the population.

PVA also attempts to anticipate how rare events which result in extremely low survival and/or reproduction (catastrophes) or their opposite (bonanzas) might affect the future course of a population. Catastrophes can be local or regional events of low probability with significant density independent effects. For example, one of two remaining whooping crane populations in the United States was decimated by a hurricane in 1940 and this population went extinct soon thereafter. The only remaining population of the black-footed ferret was being decimated by an outbreak of distemper, while the prairie dog population, its prey species, was suffering from the plague. Fearing extinction, conservation biologists captured the last 18 black-footed ferrets to start a captive breeding program.

Finally, variations in fertility and mortality can also be spatial. That is, if a population is subdivided into different locations, vital statistics can vary depending on the location of the subpopulation. Again, we cannot explore these topics in detail here. However, the following paragraphs explore the consequences of demographic stochasticity to density independent growth.

In a simple stochastic approach we specify probabilities for births and/or deaths rather than using an exact population average. For example, suppose the arithmetic average litter size of a small mammal population is 1.167 females per female per year, but the actual number of females produced per year varies from zero to two (For simplicity, we follow the traditional practice of only counting females). We then must determine the probability that a given female produces zero, one or two female offspring. For a given number of females at time equal zero, we can then make predictions as to the likelihood of various numbers of offspring in the next year.

In the following simple example, assume that adults die after reproduction, but all individuals in a given litter survive. However, litter sizes (B_i) vary from zero to two with the probabilities shown in Table 1.7. The value of λ is based on the arithmetic average of the litter sizes $= \sum p_i B_i$. The expected finite rate of increase for the population as a whole is therefore the sum of the last column ($\lambda = 1.167$).

For N females, there are, therefore, finite probabilities that the next generation will produce anywhere between 0 and $2N$ female offspring in the next generation. The probability that a population of N females goes extinct in the next year, for example, is $(0.167)^N$. For a population of six females the probability that the population will go extinct in the next year is $(0.167)^6 = 2.17 \times 10^{-5}$. For a population of one female, the probability equals 0.167. Similarly, the probability that the population will double in one year is $(0.333)^N$. A radical population shift such as extinction or doubling in one year is likely only in very small populations.

In Figure 1.8 the probabilities from Table 1.7 are applied to a population of three females at time $= 0$. One time unit later ($t = 1$), the population size has a possible range of values from 0 to 6. The probabilities associated with these values are illustrated in Figure 1.8. The most likely outcome is $\lambda N = (1.17) * (3)$, or 3.51. In reality there cannot exist fractions of individuals, so the population, one time unit later, is equally likely to remain at three or grow to four females.

As noted by Pielou (1977) and others, the probability that a population will go extinct can be estimated by equation 1.14.

$$P_{0,t} = \left(\frac{d}{b}\right)^{N_0}$$
(1.14)

where $P_{0,t}$ is the probability of extinction at time, t; d is per capita death rate and b is per capita birth rate.

For any finite population there is a probability of 1.0 that the population will go extinct, given enough time, unless the birth rate is higher than the death rate ($b >$

Table 1.7 Probability that an individual female will have 0, 1 or 2 female offspring and the expected net reproduction.

Probability, p_i, of having a given litter size, b_i	Litter size (B_i) = the number of female offspring per year	Expected net reproduction = $p_i B_i$
0.167	0	0
0.500	1	0.500
0.333	2	0.667
		$\lambda = 1.167$

d, $\lambda > 0$). Even then, there is a finite non-zero probability of extinction in any generation. Again, this chance of extinction is heavily influenced by the size of the population, with the smallest populations the most likely to go extinct.

As pointed out by Morris and Doak (2002), adding variability to population statistics does not simply mean that population growth is more variable; it means that populations do worse than they would without variation. The use of an arithmetic mean, as in the example above, overestimates growth most of the time. As Morris and Doak (2003, p. 25) state, "using simple arithmetic averages to characterize the population growth rate in a variable environment is not just a simplification, it is actually wrong."

When variation is added the most likely result is that the population will grow according to the geometric mean, rather than the arithmetic mean. The geometric mean of a set of numbers is always less than or equal to the arithmetic mean, and the difference between the two increases as the variability in the data increases.

For example, assume that a population with an initial population size of 50 grows for 100 time periods ($t = 100$), with an arithmetic mean value for λ of 1.05. With no variation, using equation 1.8, we get the predicted population size of:

$$N_{100} = N_0\lambda^{100} = 50(1.05^{100}) = 6575.$$

Now assume that we allow λ to vary between 0.90 and 1.20, with equal probabilities ($p_i = 0.50$ for each). We have:

$$N_{100} = 50(0.90^{50})(1.20^{50}) = 50(.005)(9100) = 2345.$$

This is the most likely outcome and is based on the geometric, rather than the arithmetic mean. As shown in Table 1.8, the arithmetic mean $= \sum_{i=1}^{n} p_i\lambda_i$ where $p_i =$

Fig. 1.8 Stochastic growth in a population of three females based on the parameters of Table 1.7.

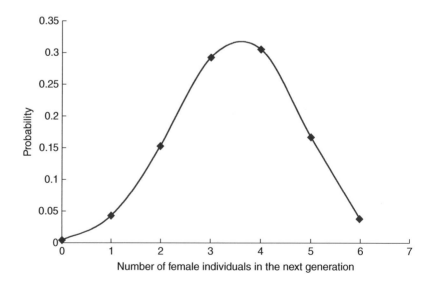

probability of a given λ_i. In the above case, $p_1 = 0.50$ for $\lambda_1 (= 0.90)$, and $p_2 = 0.50$ for $\lambda_2 (= 1.20)$.

Therefore the arithmetic mean $= (0.50 * 0.90) + (0.50 * 1.20) = 1.050$

However, the geometric mean $= \prod_{i=1}^{n} \lambda_i{}^{p_i} = 0.90^{0.5} * 1.20^{0.5} = 0.949 * 1.095 = 1.039$.

As stated above, the geometric mean is always less than or equal to the arithmetic mean, and in this case the geometric mean of 1.039 is less than the arithmetic mean of 1.050. If we use the geometric mean instead of the arithmetic mean in equation 1.8, we have the most likely outcome when λ varies between 0.90 and 1.20 with equal probabilities:

$N_{100} = (50) * (1.039^{100}) = 2345$, which is the same result we found above, but is much less than the projected population of 6575 using the arithmetic mean.

Let us try another example. Assume that $\lambda = 0.60$ 25% of the time, $\lambda = 0.80$ 25% of the time, and $\lambda = 1.40$ 50% of the time (Table 1.8). The arithmetic mean is, again, 1.05. Based on the arithmetic mean, we expect the population to grow since $\lambda > 1.00$. However, the geometric mean is less than one, and the most likely result is that this population will decline.

The geometric mean, however, provides us only with the "most likely" outcome when population parameters vary. In fact, if the population parameters are allowed to vary randomly, many different outcomes are possible. For example, in Figure 1.9 we see the results of one simulation. A comparison of growth using the arithmetic and geometric means yields the expected results. Stochastic growth with low variability (variance around the mean is 0.01), shows growth, but with obvious variation. The end result of growth with high variation (variance of 0.08 around the mean) is a population of only 178 individuals (N_0 was 50) after 100 time units.

Although this result is "typical" there are many other possible outcomes. Table 1.9 presents the results of 20 different simulations of population growth for a deterministic and two stochastic models (low versus high variability). The basic result is that the deterministic model, using the arithmetic mean for λ of 1.05, produced a larger final population size than did the stochastic/low variability model in 15 of the 20 simulations. The final population size for the deterministic model was greater than that of the stochastic/high variability model in 19 of 20 simulations. The low variability result is larger than the high variability result in 18 of 20 simulations.

Table 1.8 Calculating the arithmetic versus the geometric mean for population projections.

Probability, p_i	λ_i	$pi\lambda_i$	$\lambda_i{}^{p_i}$
0.25	0.60	0.15	0.880
0.25	0.80	0.20	0.946
0.50	1.40	0.70	1.183
		Arithmetic mean $= \sum_{i=1}^{n} p_i \lambda$ $= 1.050$	Geometric mean $= \prod_{i=1}^{n} \lambda_i{}^{p_i}$ $= 0.985$

In summary, a stochastic model generates a frequency distribution of probabilities that particular population numbers will appear in the next generation. There will always be a finite probability that the population will go extinct, but the most likely outcome (i.e., the highest probability) will be that $N_{t+1} = N_t \lambda$, using the geometric mean for λ.

1.11   Conclusions

In this chapter we have explored models illustrating the Turchin (2001) first law of population ecology. That is, biological populations tend to grow exponentially. Populations with discrete or continuous generations, as well as populations with age structures all obey the exponential law. As will be detailed in a later chapter, populations with age structures must first achieve a stable age distribution before growing according to the exponential law. The exponential law even applies to populations undergoing demographic stochasticity as described in the previous section (Truchin 2001). In addition, we do not have to assume a constant environment. If the environment varies such that per capita birth and death have a stationary probability distribution, we still obtain exponential growth or decline in the population (Maynard Smith 1974).

Accordingly, if the environment does not affect the population in a systematic manner, all types of biological populations show exponential growth. Traditionally, ecologists have treated populations with discrete generations differently from those with overlapping generations. Difference equations such as 1.4 and 1.5 have been used in the first case. By contrast differential equations (1.3) and their solved forms (1.8) have been employed to describe populations with overlapping generations. In both cases we use the finite rate of increase, λ, or the intrinsic rate of increase, r, as a common currency for comparing population growth potentials.

However, populations do not grow forever. Eventually individuals begin to run out of space, food, water, or other resources and/or become increasingly subject to predation or disease. This is where the second principle, that of self-limitation comes into play. In the next chapter we will examine this principle, and the models, traditionally known as density dependent models, that attempt to implement it.

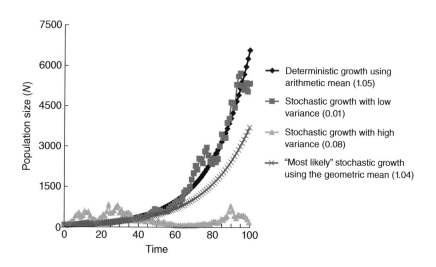

Fig. 1.9 Deterministic versus stochastic growth with high and low variance. Initial population size = 50; $\lambda = 1.05$, except where noted.

Table 1.9 Results of 20 simulations of population growth for a deterministic model versus two stochastic models, one with low and one with high variability.

Simulation number	Deterministic result	Stochastic result low variability	Stochastic result high variability
1	6575	3729	5772
2	6575	4156	28
3	6575	5972	1004
4	6575	3631	13
5	6575	5516	291
6	6575	5700	13
7	6575	2363	201
8	6575	3796	1
9	6575	5821	246
10	6575	2845	44
11	6575	7107	5244
12	6575	2113	3169
13	6575	19561	106
14	6575	3910	640
15	6575	2509	122
16	6575	13731	8
17	6575	3706	1917
18	6575	6304	53
19	6575	15570	4
20	6575	12972	8450
Average for the 20 Simulations	6575.0	6550.6	1366.3

In all cases the initial population size was 50 individuals, the arithmetic mean finite rate of increase (λ) was 1.05, and the simulation was run for 100 time units. In the low variability simulations, λ was allowed to vary between 0.90 and 1.20 ($\bar{x} = 1.05 \pm 0.1$); in the high variability simulation, λ was allowed to vary between 0.55 and 1.55 ($\bar{x} = 1.05 \pm 0.3$). In the stochastic simulations, growth rates were randomly generated using the Excel RAND functions. Note that the deterministic result is greater than the stochastic result/low variability in 15 of 20 simulations and greater than the stochastic result/high variability in 19 of 20 simulations.

References

Alberts, S.C. and J. Altmann. 2003. Matrix models for primate life history analysis. Pages 66–102 in P.M. Kappeler and M.E. Pereira, Editors. *Primate Life Histories and Socioecology*. University of Chicago Press. Chicago.

Andrewartha, H.G. 1961. *Introduction to the Study of Animal Populations*. University of Chicago Press. Chicago, IL

Anonymous. 1981–2013. *Population Data Sheets*. Population Reference Bureau. Washington, D.C.

Anonymous. 2003. *Atlantic Flyway Mute Swan Management Plan*. 2003–2013. Atlantic Flyway Council. Maryland Department of Natural Resources. Posted July 28, 2003. http://www.dnr.state.md.us/wildlife/afcmuteplan.html.

Anonymous 2012. *Audubon, Christmas Bird Counts*. www.audubon.org and www.christmasbirdcount.org.

Atkins, P. 1999. *London Sunday Times Supplement* (October 24).

Beissinger, S.R. and D.R. McCullough. 2002. *Population Viability Analysis*. University of Chicago Press, Chicago, IL.

Berryman, A.A. 2003. On principles, laws and theory in population ecology. *Oikos* 103 (3): 695–701.

Borror, D.J., C.A. Triplehorn, and N.F. Johnson. 1989. *Introduction to the Study of Insects*. 6th Edition. Saunders College Publishing, Philadelphia, PA.

Case, T.J. 2000. *An Illustrated Guide to Theoretical Ecology*. Oxford University Press, Oxford, UK.

Craig, T.. 2003. *Wildlife Agency Offers to Halt Swan Killings*. Washington Post. September 18, 2003; B01.

Hanski, Ilkka. 1999. *Metapopulation Ecology*. Oxford University Press, Oxford, UK.

Krebs, C.J. 1994. *Ecology*. 4th Edition. Harper Collins College Publishers, New York, NY.

Levins, R. 1968. *Evolution in Changing Environments*. Princeton University Press, Princeton, NJ.

Levins, R. 1969. Some demographic and genetic consequences of environmental heterogeneity for biological control. *Bulletin of the Entomological Society of America* 15: 237–240.

Levins, R. 1970. Extinction. Pages 77–104 *in* M. Gerstenhaver, Editor. *Some Mathematical Problems in Biology*. American Mathematical Society, Providence, RI.

Lomborg, B. 2001. *The Skeptical Environmentalist*. Cambridge University Press, Cambridge, UK.

Lotka, A.J. 1925. *Elements of Physical Biology*. Reprinted 1956. Dover, New York, NY.

MacArthur, R.H. and E.O. Wilson. 1967. *The Theory of Island Biogeography*. Princeton University Press, Princeton, N.J.

May, R. 2010. Page viii. *in* J.B. Losos and R.E. Ricklefs, Editors. *The Theory of Island Biogeography Revisited*. Princeton University Press. Princeton,NJ.

Maynard Smith, J. 1974. *Models in Ecology*. Cambridge University Press, Cambridge, UK.

Miller, P. S., and R. C. Lacy. 2003. Appendix I: An overview of population viability analysis using VORTEX. Pages 101–121 *in* VORTEX: a Stochastic Simulation of the Extinction Process. Conservation Breeding Specialist Group (SSC/IUCN), Apple Valley, MN.

Morris, W.F. and D.F. Doak. 2002. *Quantitative Conservation Biology, Theory and Practice of Population Viability Analysis*. Sinauer Associates, Sunderland, MA.

Nisbet, R.M. and W.S.C. Gurney. 1982. *Modeling Fluctuating Populations*. John Wiley and Sons, New York, NY.

O'Hara, R.B. 2005. The anarchist's guide to ecological theory. Or, we don't need no stinkin' laws. *Oikos* 110 (2): 390–393.

Owen-Smith, N. 2005. Incorporating fundamental laws of biology and physics into population ecology: the metaphysiological approach. *Oikos* 111 (3): 611–614.

Peterson, D.L. and V.T. Parker. 1998. *Ecological Scale*. Columbia University Press, New York, NY.

Pielou, E.C. 1977. *Mathematical Ecology*. 2nd Edition. Wiley-Interscience, New York,NY.

Roughgarden, J. 1998. *Primer of Ecological Theory*. Prentice Hall, Upper Saddle River, NJ.

Seal, U.S. and R.C. Lacy. 1989. *Florida Panther Viability Analysis and Species Survival Plan*. Cooperative agreement #14-16-0094-90-902. Captive Breeding Specialist Group, U.S. Fish and Wildlife Service, Gainesville, FL

Shaffer, M.L. 1981. Minimum population sizes for species conservation. *Bioscience* 1: 131–134.

Simon, J. 1996. *The Ultimate Resource 2*. Princeton University Press, Princeton, NJ

Sladen, W. 2003. The feral Mute Swan problem in Virginia and Maryland. *Cygnus* 16: 1–3.

Sutherland, W.J. 1996. *From Individual Behavior to Population Ecology*. Oxford University Press, Oxford, UK

Turchin, P. 2001. Does population ecology have general laws? *Oikos* 94: 17–26.

Turchin, P. 2003. *Complex Population Dynamics: a Theoretical/Empirical Synthesis*. Princeton University Press, Princeton, NJ.

Vandermeer, J.H. and D.E. Goldberg. 2003. *Population Ecology: First Principles*. Princeton University Press, Princeton, NJ

Waples, R.S. and Gaggiotti, O. 2006. What is a population? An empirical evaluation of some genetic methods for identifying the number of gene pools and their degree of connectivity. *Molecular Ecology* 15: 1419–1439.

Wydeven, A., J. Wiedenhoeft, R. Schultz, J. Bruner, and S. Boles. 2012. *Wisconsin Endangered Resources Report #143*. Status of the timber wolf in Wisconsin. Performance report 1 July2011 through 30 June 2012 (also progress reports for 15 April 2011–14, April 2012, and 2011 summaries. 37 pp.

2 Density dependent growth and intraspecific competition

2.1 Introduction

One of the great philosophical divides between ecologists and many economists is the application of the ecological principle of self-limitation to human populations. The late University of Maryland economist, Julian Simon, long the bête noir of the environmental movement, was no believer in the ecological notion of a *carrying capacity* for humans. In his book, *The Ultimate Resource*, Simon (1966) proposed that human ingenuity and technology would always triumph over any limiting resource. He had public disputes with ecologists such as Norman Myers and Paul Ehrlich. Simon famously won a series of ongoing bets with Ehrlich on whether certain raw materials would run out by specific dates. Now, as human population growth has ceased or gone negative in many European countries (Anonymous 2012), publications decrying the coming "population crash" and its ramifications have materialized in the popular media. Basically, ecologists see self-limitation of all biological populations as inevitable, while most economists, especially those in the United States, see economic growth as both certain and beneficial.

The concept of a carrying capacity for biological populations is connected with the logistic equation, found in all ecology text books. See equation 2.8. The logistic was originally formulated by the French mathematician Verhulst in pre-Darwinian times (1838), but was not applied routinely to biological populations until Pearl and Reed (1920) rediscovered it. Pearl (1927) then promoted the application of the logistic to a variety of biological populations. See Kingsland (1995) for an interesting review of this history.

Introduction to Population Ecology, Second Edition. Larry L. Rockwood.
© 2015 John Wiley & Sons, Ltd. Published 2015 by John Wiley & Sons, Ltd.
Companion Website: www.wiley.com/go/rockwood/populationecology

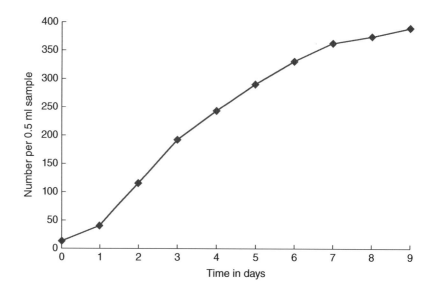

Fig. 2.1 Population growth in a *Paramecium* population.

Yet Turchin (2003) and many other population biologists now assert that, although the logistic is useful as a general framework, this equation is fundamentally flawed when applied to biological populations. The logistic model is not a general law of population growth, but is rather a special case. If an ecologist wants to win the argument with an economist about human (or any) population, limitation he/she needs to understand the assumptions and flaws of logistic or logistic-like models. The goals of this chapter are to first describe density dependent growth models for both discrete and continuously breeding populations; second, to examine the assumptions of these models; third, to investigate how violations of these assumptions shape the behavior of populations.

In Chapter 1, we assumed *density independent growth*. That is, population growth unlimited by competition for resources. Most biological populations, however, do not long sustain such growth. Even in an isolated laboratory population, growing without competing species or predators, realized growth slows and ceases. Examine the *Paramecium* population history presented as Example 1.3 and plotted in Figure 2.1. Population growth slows after day 2 and almost ceases by day 7. Our experience in the laboratory is that a *Paramecium caudatum* population will stop growing at about 400 per 0.5 ml sample. Based on data such as this, one of the basic assumptions of most ecological models is the populations do not have unlimited resources and eventually the population encounters a limiting resource (or perhaps, a parasite or predator) which restricts population growth. This is by no means a new idea. In 1855 Liebig, in his Law of the Minimum, asserted that under steady state conditions, the population size of a species is constrained by whatever resource is in shortest supply. According to the *logistic model*, population growth ceases when the population reaches the *carrying capacity* of the environment for that population. This is a *density dependent growth* model in which the carrying capacity is identified by the symbol *K*. For a given species, in a specific environment, carrying capacity is defined as the number of individuals that can be maintained indefinitely.

Fig. 2.2 Per capita growth in a *Paramecium* population.

One way to visualize density dependent growth is to graph the per capita growth rate versus population size. Figure 2.2 plots the growth rate per individual in the *Paramecium* population versus population size. From Example 1.3, per capita growth is found by dividing the growth between time intervals t and $t + 1$ by the population size at time t. Notice that even though the population is growing through day 6, the trend per individual is steadily downward. Although a linear model with a negative slope can be fitted to the data, a negative exponential curve is actually a better fit. As we should already know, although linear models are more convenient, many population functions are non-linear. Where these lines intersect the x-axis, per capita growth has fallen to zero. The value of this point $(N, 0)$ is an estimate of the carrying capacity, K. The two models reach $N, 0$ at between 325 and 400. The R^2 value means that the models have explained 83–93% of the variance in the data. Again the R^2 is greater for the negative exponential model.

Since the logistic model is really based on competitive interactions, we should define *competition* before proceeding. A formal definition of competition is: *a biological interaction between two or more individuals for a resource in short supply*. When the interaction is between individuals of the same species it is termed *intraspecific* competition; when between individuals of different species it is known as *interspecific* competition. A *resource* is any substance or factor in the environment that determines growth, survivorship, or reproduction of individuals in the population. Therefore, depletion of this resource decreases growth, survivorship, or reproduction. For competition to be meaningful, the resource must be in short supply now, or in the immediate future. Plants may compete for space, light, water, or nutrients, while animals often compete for food, nesting sites, hiding places, or mates. Certain aspects of the environment, such as temperature, are not resources, per se, and cannot be competed for. On the other hand, if a lizard needs to raise its body temperature it will seek out a rock on which to bask in the sun. If there is a limited number of basking sites, they become resources in short supply, and may be competed for.

Paradoxically, perhaps, the ultimate effect of competition is a decrease in fitness. Thus competition is said to be a *reciprocally negative interaction*. All individuals that engage in competition may lose energy and/or time that they could have invested in their own growth, survivorship or reproduction. When sports teams or animals engage in competition, we identify a winner and a loser (throwing out the occasional tie). Male elk, deer, or big horn sheep engage in some amazing combats. The winner mates with the female(s) and his fitness is increased, *relative to the losing male*. But, if that male had been able to mate with the females without combat, his long-term fitness would be greater still, since he would have conserved the energetic costs (and risks of injury) associated with combat. This theoretical point, however, ignores situations where competitive interactions between males are necessary for stimulation of reproductive activities.

Competition also differs in its manifestations. We recognize here two basic forms of competition: interference and depletion. The term interference competition seems to have originated with Park (1962). A similar concept is *encounter interference* (Schoener 1983). In interference competition access to the resource is blocked by behavioral or chemical means. Interference competition applies to territoriality, guarding behaviors, and, by this definition, allelopathy. In allelopathy, plants secrete chemicals that accumulate in the environment and prevent other plants from germinating or growing within this area. A similar phenomenon is the secretion of antibiotics by fungi that prevent growth of bacteria within a certain radius of the colony. In ants, when a high quality bait such as a chunk of tuna is placed on the forest floor, one species often recruits soldiers to form a ring around the tuna. If they deny access to all workers and soldiers except those for those from their own colony, they are engaging in interference competition.

Depletion competition involves the simple removal of the resource without active interference. This is the same idea as *exploitation* competition (Park 1962) and *consumption* competition (Schoener 1983). All of these terms refer to situations in which plants or animals consume resources to the detriment of competitors, but without directly interfering with access to the resources. This is a sort of "first come, first served" type of competition. Sutherland (1996) compares depletion competition to "drinking the pub dry." We will avoid here the terms "scramble" and "contest" competition (Nicholson 1954), neither of which is biologically realistic.

Intraspecific competition manifests itself through density dependent modifications in: (i) birth and death rates; (ii) growth rates; and (iii) adult size, especially in organisms with determinant life cycles. That is, the eventual size of an adult beetle, for example, is largely determined by the feeding rates and sizes of the larval stages (within genetic constraints). Finally, intraspecific competition is reflected in complex behavior patterns such as male–female interactions. These latter topics will be explored near the end of this chapter.

In the following sections (2.2 and 2.3) we will do a simple derivation of equations that describe density dependent growth for populations with discrete and with continuous growth. We will analyze how these equations work and what they might tell us about how populations behave in nature.

Fig. 2.3 Reciprocal of growth per generation versus population density.

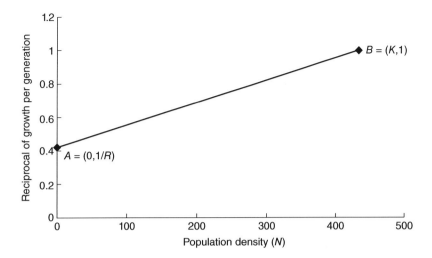

2.2 Density dependence in populations with discrete generations

As we saw in Figures 2.1 and 2.2, in a density dependent population we expect growth to slow and eventually stop as a population increases, and reaches the carrying capacity. For populations with discrete generations, we can begin with equation 1.4: $N_t = N_0 R^t$.

To incorporate intraspecific competition into a model, we simply modify the growth rate factor, R. One approach is to graph the reciprocal of increase per generation, N_t/N_{t+1} versus N_t (Figure 2.3). If a population is very small (virtually zero), the population is assumed to grow at the maximal rate, R. Rearranging equation 1.3, we have $N_t/N_{t+1} = 1/R$. Point A is therefore $(0, 1/R)$. The carrying capacity, K, occurs when $N_t/N_{t+1} = 1.0$. That is, when there is no population change from one generation to the next. Point B, then, is $(K, 1)$. If we assume that population growth per generation follows a straight line between points A and B, we have that the y-intercept is $1/R$ and the slope is therefore

$$\frac{1 - (1/R)}{K - 0} = \frac{1 - (1/R)}{K}$$

The general linear equation, $y = a + mx$ becomes

$$N_t/N_{t+1} = 1/R + \left\{ \frac{1 - (1/R)}{K} \right\} N_t$$

Rearranging and providing the common denominator RK,

$$N_t/N_{t+1} = \frac{(K) + \left(1 - \dfrac{1}{R}\right)(N_t R)}{RK} = \frac{(R - 1)(N_t) + K}{RK}$$

Therefore:

$$N_t = N_{t+1} \frac{(R - 1)(N_t) + K}{RK}$$

And,

$$N_{t+1} = (N_t) \frac{RK}{(N_t)(R - 1) + K}$$

If we divide the numerator and the denominator of the right side of the equation by K, we get:

$$N_{t+1} = (N_t) \frac{(RK/K)}{((N_t)(R-1)/K) + (K/K)}$$

Finally:

$$N_{t+1} = \frac{N_t R}{1 + \frac{(N_t)(R-1)}{K}} \tag{2.1}$$

Equation 2.1 is known as the Beverton–Holt (1957) Model, well known among fishery scientists, and is very similar in behavior to the traditional logistic equation (Gurney and Nisbet 1998).

By convention and to simplify equation 2.1, we let $a' = (R-1)/K$. Equation 2.1 becomes:

$$N_{t+1} = \frac{N_t R}{1 + a' N_t} \tag{2.2a}$$

A good way to see how see this equation encompasses density dependence is to distinguish R_1, the density independent growth parameter, from R_A, the density dependent or "actual" growth parameter. In this case, equation 2.2a becomes:

$$N_{t+1} = N_t R_A \tag{2.2b}$$

$$R_A = R_1 (1 + a' N_t)^{-1} = \frac{R_1}{1 + a' N_t} = \frac{R_1}{1 + N_t \left(\frac{R_1 - 1}{K} \right)} = \frac{R_1}{1 + \left(\frac{N_t R_1 - N_t}{K} \right)} \tag{2.3a}$$

$$R_A = R_1 \left[1 + \left(\frac{N_t R_1 - N_t}{K} \right) \right]^{-1} \tag{2.3b}$$

Equation 2.3b tells us that the maximal or density independent growth rate, R_1, is modified by the population size at time t relative to the carrying capacity, K. For example, if N is very small (try replacing N_t by 0), the actual growth rate, R_A, is essentially equal to R_1.

If $N = K$, however (replace N_t by K in equation 2.3b), the expression inside the bracket collapses to R_1. Therefore, $R_A = R_1 {}^* R_1{}^{-1} = 1.0$. This means that: $N_{t+1} = N_t(1.0) = N_t$. So if $N = K$, there is no growth in the population and $N_{t+1} = N_t$.

In Table 2.1 notice how equation 2.3b modifies the R_A and the actual population size with time. The population size after 15 generations for the density dependent population is about 40% of the density independent population, and the actual R steadily drops toward the no growth value of 1.00 by 30 generations.

We must remember, however, that equation 2.3b is based on the two points, $(0, 1/R)$ and $(K, 1)$, from Figure 2.3. Furthermore, we assumed a straight line would connect these two points. This, in turn, is based on the assumption of *exact density dependence*

Table 2.1 Density independent growth compared to density dependent growth using equation 2.3b. $N_0 = 100$, $R_1 = 1.2$, $K = 1000$ in all cases.

Time, t	Density independent net reproductive rate, R_1	Density dependent net reproductive rate, R_A	N_t in the case of density independent growth	N_t in the case of density dependent growth
0	1.20	1.20	100	100
1	1.20	1.18	120	118
2	1.20	1.17	144	138
3	1.20	1.17	173	161
4	1.20	1.16	207	187
5	1.20	1.16	249	217
6	1.20	1.15	299	249
7	1.20	1.14	358	285
8	1.20	1.14	430	323
9	1.20	1.13	516	364
10	1.20	1.12	619	408
11	1.20	1.11	743	452
12	1.20	1.10	892	498
13	1.20	1.09	1070	543
14	1.20	1.08	1284	588
15	1.20	1.07	1541	631
30	1.20	1.01	23738	963

or "exactly compensating" density dependence (Silvertown and Doust 1993). This assumption is unrealistic. As we saw in Figure 2.2 a non-linear model was a better approximation of per capita growth versus population density in a laboratory *Paramecium* population. Hassell (1975) therefore proposed that we could relax this assumption of exact or linear density dependence by simply modifying equation 2.3b and replacing -1 with the exponent: $-b^*$ (equation 2.4). Exact compensation (linear density dependence) occurs when $b^* = 1$, producing a slope of -1, but *overcompensation* ($b^* > 1$, means a slope < -1) is the result when plant yield, for example, drops more rapidly than expected with increases in density.

$$R_A = \left[1 + \left(\frac{N_t R_1 - N_t}{K} \right) \right]^{-b^*} \qquad (2.4)$$

Undercompensation occurs when $b^* < 1$, and means that population size drops more slowly than expected (as compared to exact compensation) as density rises. From Figure 2.4 you can see that while the actual value of R declines in the density dependent model, it declines fastest when $b^* > 1$ (that is, *overcompensation*), and declines more slowly when $b^* < 1$ (*undercompensation*).

Figure 2.5 illustrates the time path of population growth under the conditions specified in Figure 2.4. Population growth is obviously most rapid with density independent growth and is slowest with density dependent growth and overcompensation ($b^* = 1.6$).

The preceding model can be applied directly to the *Law of the Constant Final Yield*, well known from botanical and agricultural research. The law essentially states that agricultural yield per area will increase with plant density up to the maximum or "final" yield. Thereafter, increasing the number of plants per area simply reduces the average size per plant (or animal) without increasing total yield (Figures 2.6a and 2.6b).

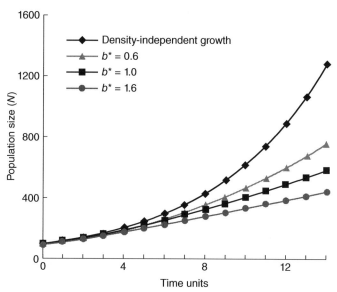

In simple terms we can write:

$$C = N\overline{w} \tag{2.5}$$

where C = the final constant yield in kilograms per area,
N = the density of plants. That is, the number per unit area, and
$\overline{w}$ = mean weight per plant in kilograms.

This equation, however, only provides information about the end point of a dynamic
process. Just as $R_A = R_1(1 + a'N_t)^{-1}$ in equation 2.3a, the actual mean mass per plant,

Fig. 2.6a The law of the final constant yield for plants and sessile animals. Yield (mass) per unit area increases with density until it reaches a threshold, after which there is no increase in total yield.

Fig. 2.6b As density increases, mass per individual decreases linearly on a log–log scale.

$\overline{w}$, can be expressed as a function of its maximum potential mass when grown under density independent conditions.

$$\overline{w} = w_m(1 + a'N)^{-b^*} \tag{2.6}$$

where w_m = maximum potential mass per plant and
        a' = a carrying capacity parameter.

In this case, a' is often interpreted as the amount of area needed for each individual plant to achieve its maximum growth potential, and b^* provides a mechanism for different reaction rates to density. As before, when $b^* = 1$, there is exact density dependent compensation and the law of the final constant yield is obeyed in a linear fashion. Combining equations 2.5 and 2.6, we have an equation that describes the effects of density dependent growth on yield under a variety of conditions. Plant populations may vary between plots, and different plant parts (grain yield or above ground biomass, for example) may respond differently to changes in density. Equation 2.7 produces different shaped curves as the value of b^* is varied.

$$C = Nw_m(1 + a'N)^{-b^*} \qquad (2.7)$$

The shapes of the curves would be similar to those of Figure 2.5, except that the y-axis would be output in mass rather than population density. Part of the dynamic process leading to the law of the constant yield involves an increase in mortality (self-thinning) over time as populations increase in density. As a new population is established in a suitable habitat, self-thinning follows several steps. (i) As individuals grow, they increase in size (mass). (ii) When a critical density is reached, known as the thinning limit, density dependent mortality begins. This step occurs earlier in populations with higher initial density. (iii) Eventually the population reaches a stage where any increase in the mass of some individuals is offset by mortality of other members of the population. Total mass no longer increases and the final constant yield in mass per unit area has been reached (Figure 2.6a). The point of final yield is reached more quickly in populations with higher initial densities.

Although the increase in mortality with population density is usually assumed to be linear in models such as the logistic (see below), we can introduce non-linear responses as shown in equations 2.4, 2.6, 2.7, 2.11, and 2.13.

2.3 Density dependence in populations with overlapping generations

2.3.1 The logistic equation

The more familiar treatment of density dependence is to examine growth curves such as Figure 2.1, and apply a modification of the differential equation: $\frac{dN}{dt} = rN$. The resultant equation, known as the logistic, can be derived as follows. Examine Figure 2.2 once more. The y-axis is per capita growth rate $(dN/dt)(1/N)$. Since the per capita growth rate $= r$ when N is very small, we can identify a point A as the y-intercept $(0, r)$. When $N = K$, according to theory, per capita population growth stops. Therefore point B, along the x-axis, would be the point $(K, 0)$. As in the previous example, if we assume that populations respond in a linear manner to population density, we use these two points to describe a straight line. The slope of this line is:

$$(0 - r)/(K - 0) = -r/K$$

The y-intercept is r and we have:

$$(dN/dt)(1/N) = (-r/K)N + r$$

Rearranging, we now have:

$$dN/dt \, (1/N) = r\left[1 - \frac{N}{K}\right] = r\left[\frac{K-N}{K}\right]$$

Multiplying both sides of the equation by N reveals the usual form of the logistic.

$$dN/dt = rN\left[\frac{K-N}{K}\right] \qquad (2.8)$$

Although the logistic is in the form of a differential equation, it is fairly easy to understand how it affects population growth. Again, it is useful to examine what the equation does to the growth rate, r. As above, we will distinguish between r_a, the actual growth rate as modified by carrying capacity, and r_m, the density independent growth rate. The variable, r_m, has also been called r-max or the Malthusian parameter. The r-max represents the maximal growth rate of a genotype as it interacts with the environment without competition.

$$r_a = r_m \left[\frac{K - N}{K} \right]$$ (2.9)

When the population is very small, $N \cong 0$, and $\left[\frac{K-N}{K} \right] \cong 1$. Therefore, $r_a \cong r_m$.

When $N = 0.5K$, then the expression, $\left[\frac{K-N}{K} \right] = 0.5$, and $r_a = (0.5)r_m$.

When $N = K$, the expression $\left[\frac{K-N}{K} \right] = 0$ and $r_a = 0$.

Finally, when $N > K$, the expression $\left[\frac{K-N}{K} \right] < 0$. Therefore r_a is negative and the population drops back toward the carrying capacity, K.

The differential form of the logistic equation can be integrated and solved, resulting in the following.

$$N_t = \frac{K}{1 + e^{a-rt}}$$ (2.10a)

where $a = $ constant of integration.

Dividing both sides of the equation by K, yields:

$$\frac{N_t}{K} = \frac{1}{1 + e^{a-rt}}$$

Taking the inverse:

$$\frac{K}{N_t} = 1 + e^{a-rt}$$

Manipulating, we get

$$\frac{K - N_t}{N_t} = e^{a-rt}$$

Finally, taking the natural log of both sides gives us:

$$\ln \left[\frac{K - N_t}{N_t} \right] = a - rt$$ (2.10b)

This expression is useful because it becomes the equation for a straight line with $a = $ y-intercept and the slope equal to $-r$ (see Figure 2.7). When $t = 0$ the y-intercept, a, becomes,

$$a = \ln \left[\frac{K - N_0}{N_0} \right]$$

Fig. 2.7 Estimate of actual r for a Paramecium population using the logistic equation.

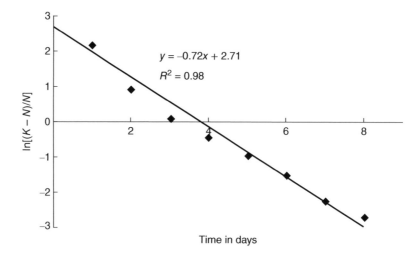

Time in days

Table 2.2 Estimating r_a, (r actual) from the solved form of the logistic equation.

Time in days	Number per 0.5 ml sample, N	$\ln\left[\frac{K-N_t}{N_t}\right]$
0	14	3.32
1	41	2.17
2	116	0.90
3	193	0.07
4	244	−0.45
5	290	−0.97
6	331	−1.57
7	363	−2.28
8	375	−2.71
9	380	−2.94

Source: Data are for the *Paramecium* population in Example 1.3. K is approximated at 400. See Figure 2.7.

This gives us another form of equation 2.10a, in which a, the constant of integration, is replaced by the y intercept.

$$N_t = \frac{K}{1 + \left[\left(\dfrac{K - N_0}{N_0}\right)(e^{-rt})\right]}$$

(2.10c)

Furthermore, if we graph $\ln\left[\frac{K-N_t}{N_t}\right]$ versus t (Figure 2.7), the absolute value of the slope of the line approximates r. This allows us to estimate the actual value of r over a specified time period. For example, if we use the example, once again, of the *Paramecium* population in Example 1.3 and Figure 2.1, we can determine its actual growth rate, r_a, over the eight days of the experiment. We simply add a column for the expression $\ln\left[\frac{K-N_t}{N_t}\right]$ (Table 2.2). To do this, however, we must have an approximation for the carrying capacity, K. In table 2.2, K is estimated as 400.

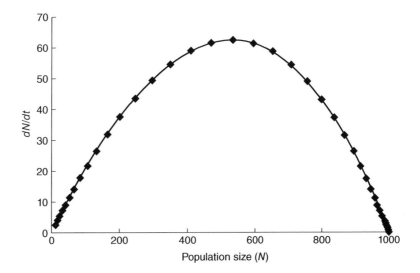

Fig. 2.8 Population growth as a function of N based on the logistic equation.

In Figure 2.7 a linear regression identifies the slope as -0.72 (R^2 is the proportion of the variance explained by the linear model). The actual r is therefore 0.72, as compared to the r_{max} of 0.89 calculated in chapter one.

When trying to understand how the logistic model affects population growth, it is instructive to examine population growth, dN/dt, as a function of population size. Since $dN/dt = rN\left(\frac{K-N}{K}\right)$, it also equals $rN\left(1 - \frac{N}{K}\right)$, which equals $rN - (rN^2)\left(\frac{1}{K}\right)$. If we set dN/dt equal to zero, there are three solutions to this equation: $r = 0$; $N = 0$; or $N = K$. If we assume that $r > 0$ we are left with two solutions ($N = 0$ and $N = K$). That is, $dN/dt = 0$ when $N = 0$ and when $N = K$. The result of plotting dN/dt versus N results in a parabola (Figure 2.8). Maximum growth (dN/dt) occurs where $N = K/2$, which is 500 in this case, since we have set $K = 1000$. The problem with this solution is that, since maximum growth theoretically occurs at half carrying capacity, harvesting of wild living resources was managed with that number as a goal. This has lead to the decimation of many populations since stochastic and density-independent mortality are not accounted for. Furthermore, neither the carrying capacity nor the actual population sizes are known in many wild populations. The notion that natural populations are resilient enough to withstand heavy harvesting persists even in general biology textbooks. For example, Sadava et al. (2011, p. 1180) state: "Some fish populations can be harvested heavily on a sustained basis because a relatively small number of females can produce sufficient numbers of eggs to maintain the population." Do we know if that "small number of females" is sufficient to withstand stochastic perturbations? Declining fish stocks around the world suggest that we do not.

Assumptions of the logistic equation

How much trust can we put in either the traditional logistic equation or the Beverton-Holt equations? Is the typical logistic growth curve actually found in biological populations? Laboratory studies on growth of populations such as *Paramecium caudatrum*, yeast, *Drosophila*, grain beetles and diatoms (Gause 1932, 1934; Vandermeer 1969; Pearl 1927; Park et al. 1964; Crombie 1945; Tilman 1977) do consistently show a logistic growth curve. A number of field populations have also followed logistic growth fairly closely. Examples include Tasmanian sheep (Davidson 1938), wildebeest (Deshmukh 1986), willows (Alliende and Harper 1989) and barnacles (Connell 1961). However, there are many more cases where populations grow cyclically or unpredictably and generally do not display logistic growth. An examination of the assumptions of the logistic equation explains why many populations display non-logistic growth patterns.

1. The carrying capacity is a constant.
2. Population growth is not affected by the age distribution.
3. Birth and death rates change linearly with population size (it is assumed that birth rates and survivorship rates both decrease with density, and that these changes follow a linear trajectory).
4. The interaction between the population and the carrying capacity of the environment is instantaneous. That is, the population is "sensitive" to the carrying capacity with no time lags.
5. Abiotic, density independent factors, do not affect birth and death rates (no environmental stochasticity).
6. Crowding affects all members of the population equally.

Considering all of the above, it is not surprising that many populations in the field do not often stay at a given density for long periods of time. In the laboratory, when we grow a *Paramecium* population, its growth curve often fits the logistic since: (i) it is maintained in a constant environment and hence should have a constant carrying capacity; (ii) it reproduces via binary fission and has no age structure; (iii–vi) are seemingly irrelevant or satisfied. Once we step into the field and work with insects, vertebrates, or plants, several of these assumptions are violated.

Since these assumptions cannot be met, a natural population is unlikely to long remain "at equilibrium" with the environment. In the next two sections we will examine assumptions 3 and 4 in some detail, since their effects on population growth are less obvious and more interesting than might be expected.

2.4 Nonlinear density dependence of birth and death rates and the Allee effect

As mentioned above, density dependent birth and death rates are assumed to vary linearly with density (Figure 2.9). Although it is known from both laboratory (Smith and Cooper 1982) and field studies (Arcese and Smith 1988) that birth and death rates are often non-linear (Figure 2.10), such differences seem to have a minimal impact on natural populations. One major exception, however, is known as the Allee effect (Allee 1931). The Allee (1931) effect can be defined as: "positive density dependence" (Courchamp et al. 2008, p. 9.). That is, a very low densities the growth or fitness of a population is positively correlated with density. Allee proposed that many species have a minimum viable population (MVP) size. Below the MVP level the population shows positive density dependence, but above this level the population is negatively density dependent as expected. See Figure 2.11. To conservation biologists the concern is that the probability of extinction becomes unacceptably high in small populations (Shaffer 1981; Miller and Lacy 2003), but there is no one specific number described as an MVP.

As pointed out by Courchamp et al. (2008), although the concept proposed by Allee has been part of the literature since 1931, it did not inspire a great number of publications until the late 1990s when reviews by Courchamp et al. (1999) and Stephens and Sutherland (1999) helped spark an exponential increase in publications over the

Fig. 2.9 Linear response of birth and death rates to population density.

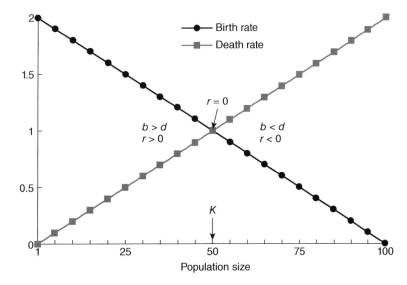

Fig. 2.10 Nonlinear response of birth and death rates to population density.

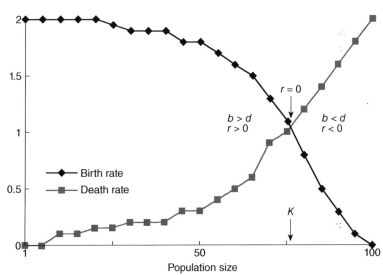

next 5 years (Courchamp et al. 2008). The new interest was motivated in part by the large number of declining natural populations throughout the world and the desire by scientists to understand the dynamics of small populations.

The Allee effect can be demographic or genetic. At the demographic level, the Allee effect can influence reproductive success or survivorship. At the genetic level, inbreeding and genetic drift can cause loss of fitness in small populations.

With regard to reproductive fitness various reviews (Berec et al. 2007; Stephens and Sutherland 2000) have identified situations in which the Allee effect would operate such as pollination failure and "broadcast spawning." In these cases, low reproductive success in small populations could result from high pollination failure in plants, low probability of sperm and egg meeting in a water column, male and female animals unable to locate each other, or the chance of a very unequal sex ratio

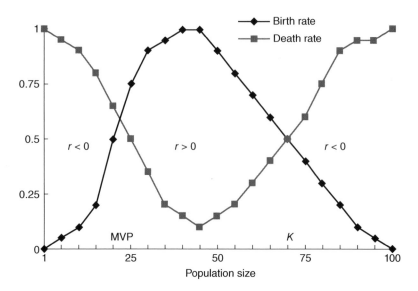

Fig. 2.11 Birth and death rate versus population density showing the Allee effect. MVP = Minimum viable population size; K = carrying capacity.

(large number of males, few females). In other instances such as in African wild dogs, breeding success is lower in small populations (Courchamp et al. 2000).

With regard to survivorship, why should there be higher death rates in very small populations, or conversely, lower death rates in larger populations? Proposals include: (i) group cooperation reduces losses from predators; (ii) group foraging for food is more efficient (foraging facilitation); (iii) better thermoregulation is possible in larger numbers (examples include alpine marmots (*Marmota marmota*) and bobwhite quail (*Colinus virginianus*)); and (iv) small populations are more subject to density independent or stochastic extinctions. For a more complete and detailed list see Courchamp et al. (2008).

For example, the common guillemat (*Uria aalge*) in South Wales nests in colonies. Breeding success is only 25% in the least dense populations as compared to an average of 75% in the densest populations (Birkhead 1977). The reason for this appears to be that predation on eggs and chicks by gulls is reduced in dense guillemat populations. A similar result was obtained in a study on lapwings (*Vanellus vanellus*) in which egg clutches lost to avian predators declined with an increase in the number of close neighbors (Berg et al. 1992). Nest parasitism also appears to increase in small populations. For example, small dickcissel (*Spiza americana*) populations are particularly hard hit by brown-headed cowbirds (*Molothrus ater*) (Fretwell 1986).

Other studies have found that cooperative hunters such as lions, hyenas, wolves and various fish species have much higher success rates when hunting in large groups and do poorly when population sizes fall (Major 1978; Caraco and Wolfe 1975). This translates into a higher mortality rate in the smaller groups. From the perspective of the prey, a dense population is harder to surprise, and mortality from predation is lower in larger prey populations (Kenward 1978; Jarman and Wright 1993). Colonial nesting sunfishes have even been found to suffer lower rates of fungal infections on their eggs as compared to solitary sunfish (Cote and Gross 1993). In all of these cases higher density leads to higher, not lower, survivorship.

Although this is an oversimplification, we can illustrate the Allee effects graphically by identifying a minimum viable population size, MVP (Figure 2.11). When the population is below the MVP, the population declines to extinction. When the population is greater than the MVP, the population increases rapidly before slowing down as it approaches K. The value of r is positive above MVP and below K, but is otherwise negative (Figure 2.11).

The extinction of the heath hen is a likely example of the Allee effect. By 1870, hunting and habitat loss had restricted it to Martha's Vineyard off the coast of Massachusetts. In 1908 a 650 hectare (ha) refuge was established and the population grew to about 2000 birds. In 1916, however, a fire swept across the island, destroying nests, eggs, and females on the nests. The following winter was severe and an unusually heavy concentration of goshawks arrived. The population was reduced to less than 150, of which most were probably males. By 1928 the population had declined to a single male that died in 1932. In this case, abiotic effects contributed significantly to the extinction of the heath hen.

Although we can never be sure, it appears likely that something of a similar nature happened to the passenger pigeon described in chapter one. They were mercilessly hunted and no one imagined the possibility of extinction. But once their flocks were drastically reduced, they appeared unable to recover (Souder 2004). Although there is speculation that passenger pigeons could be brought back from extinction through modern molecular methods (Novak 2013), what we know about their nesting behavior indicates that they survived through the mechanism of "predator satiation." That is, they had few defenses against predators and, in an excellent illustration of the Allee effect, small populations are likely to be decimated by predators (Clark 2013). If this is the case, bringing the passenger pigeon back from extinction would be an interesting curiosity, but there would be small probability that they could successfully be reintroduced into North American ecosystems.

2.4.1 Non-linear Modifications to the Logistic

In order to evaluate the potential for a non-linear feedback on the logistic population response, we can modify equation 2.10c by adding the term b^* as we did in equation 2.4 and Figures 2.4 and 2.5. Recall that a value for b^* of 1.0 described "exact compensation" and depends upon the linear response by the population to a carrying capacity. A $b^* > 1$ illustrates overcompensation and a $b^* < 1$ describes undercompensation. We can modify equation 2.10c by adding a b^* value as shown below.

$$N_t = \frac{K}{1 + \left[\left(\dfrac{K - N_0}{N_0} \right) (e^{-rt/b^*}) \right]}$$

(2.11)

Figure 2.12 illustrates the effect of a nonlinear feedback on logistic growth. The population with the b^* value of 0.6 grows the most rapidly whereas the population with the b^* value of 1.6 grows the most slowly. As in Figure 2.4, when $b^* > 1$, small increases in density result in a rapid drop in the actual value of the growth rate, r, as a result of *overcompensation*. Similarly a b^* value < 1 results in *undercompensation* and a higher actual r-value.

2.4.2 The Theta Logistic Model

As we will see in Section 2.8, the life history of a population frequently does not respond to increases in density in a linear fashion, as assumed by the logistic equation. A well-known variation of the logistic model, known as the theta logistic, more elegantly introduces non-linear density dependence than in the previous section. First, we

Fig. 2.12 Effect of
non-linear feedback on
logistic growth.

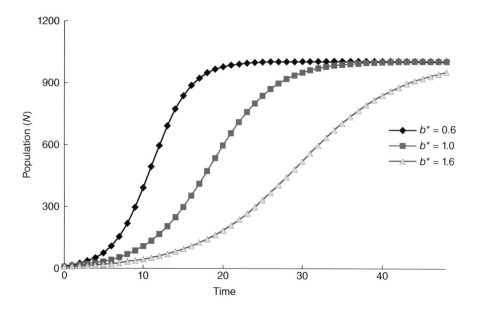

must introduce another model from fishery science, the Ricker (1952) Model, which
is a useful discrete form of the logistic.

To find the Ricker, we begin by making a distinction between the actual rate
of increase, r_a and the exponential rate of increase, r_m or r_{max} as we did above in
equation 2.9.

$$r_a = r_m \left[\frac{K - N}{K} \right] = r_m \left(1 - \frac{N}{K} \right)$$

Now let us substitute r_a for r in the equation for exponential growth (equation 1.8),
which gives us:

$$N_t = N_0 e^{r_a t}$$

Next we transform this to a simple difference equation for adjacent time intervals N_{t+1}
and N_t. Since this is one time step and $t = 1$, we can remove t from the exponent in
the above equation. Simultaneously we will substitute $r_m \left(1 - \frac{N}{K} \right)$ for r_a. The result is
(equation 2.12), the Ricker equation.

This equation can be modified with the parameter θ (theta) as a superscript of the
ratio N/K (equation 2.13). The theta logistic was originally proposed by Gilpin and
Ayala (1973). When $\theta = 1.0$, we have the traditional logistic growth response to density, with the assumption of linearity. When $\theta < 1.0$ density dependence is strong even
when the population is far below the carrying capacity. By contrast, when $\theta > 1.0$ density dependence is weak until the population is close to the carrying capacity. When
$\theta \neq 1.0$, then, we have non-linear responses of the population to increases in population size.

$$N_{t+1} = N_t e^{r\left[\frac{K - N_t}{K} \right]} = N_t e^{r\left(1 - \frac{N_t}{K} \right)} \tag{2.12}$$

$$N_{t+1} = N_t e^{r\left(1 - \left(\frac{N_t}{K} \right)^{\theta} \right)} \tag{2.13}$$

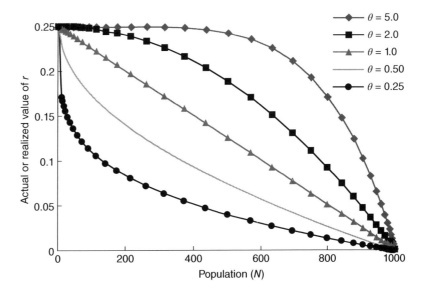

Fig. 2.13 Behavior of the Theta logistic. Effect of population size on the realized value of r. $r_{max} = 0.25$ and $K = 1000$ in all cases.

For example, in Figure 2.13, we can examine the effect of θ on an actual or realized r-value. In each case the r_{max} is 0.25, the carrying capacity is 1000 and the initial population size equals 10. As predicted from the logistic, when $\theta = 1$ the decline in the actual r-value is linear as the population increases. When θ is less than 1.0 the actual r-value decreases rapidly with population size. By comparison, if θ is greater than 1.0 we can see that r remains close to r_m until the population gets much closer to the carrying capacity.

If we examine population growth versus time for the same theta values, the expectation is that growth will be suppressed at low population levels when theta is less than 1.0, but that the population will approach the carrying capacity quickly when theta is greater than 1.0. These predictions are borne out by Figure 2.14.

As shown by Saether et al. (2002) the theta logistic is a powerful model for analyzing variation in density dependence among bird populations, and is the basis for other population models (e.g., predator–prey interactions), in which we do not want to assume a linear relationship between population density and survivorship, fertility, or r.

2.5 Time lags and limit cycles

Assumption four of the logistic equation, in which populations are assumed to respond immediately to carrying capacity, is highly unlikely for populations with great reproductive potential. In order to explore this possibility, we can introduce a "lag time" effect into the logistic equation. Using the discrete time form of the logistic (equation 2.1), substituting λ for R, and remembering that $\lambda = e^r$, equation 2.12 is an equivalent to equation 2.1 and 2.10c. To introduce time lags, equation 2.14 is modified as shown in equation 2.15 (Pielou 1977).

$$N_{t+1} = \frac{\lambda N_t}{1 + \dfrac{N_t(\lambda - 1)}{K}} \tag{2.14}$$

$$N_{t+1} = \frac{\lambda N_t}{1 + \dfrac{N_{t-T}(\lambda - 1)}{K}} \tag{2.15}$$

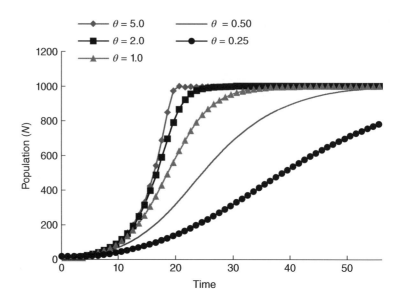

Fig. 2.14 Behavior of the Theta logistic. $r_{max} = 0.25$ and $K = 1000$ in all cases.

A more familiar form of this same equation is simply:

$$dN/dt = rN_t \left(\frac{K - N_{t-T}}{K} \right)$$ (2.16)

in its continuous form, and

$$N_{t+1} = N_t e^{r\left[\frac{K-N_{t-T}}{K} \right]}$$ (2.17)

in its discrete form. This last equation (2.17) is a simple modification of equation 2.12, the Ricker Model.

Where N_t appears, it is modified by the Greek letter tau (τ). The population responds to the carrying capacity based on what the population size was Tau-time units in the past. Robert May and others (May and Oster 1976; May 1981) have shown that lag time, combined with the intrinsic rate of increase (r), produces a predictable and interesting series of modifications to logistic growth. The product of r and τ determines the behavior of the population. As summarized below, long time delays before the population reacts to carrying capacity, combined with a high growth potential, led to population behaviors that wander further and further from the stable point at K predicted by the logistic equation.

If $0.37 > r\tau > 0$, the population follows the logistic equation, and the population achieves a stable number (or *stable point*) at the carrying capacity with no oscillations.

If $1.57 > r\tau > 0.37$, the population is temporarily oscillatory, but the oscillations dampen to a stable point at the carrying capacity.

If $2.0 > r\tau > 1.57$, the population undergoes permanent oscillations around the carrying capacity. This is called a *limit cycle*.

If $r\tau > 2.0$, the oscillations are so violent that the population goes extinct.

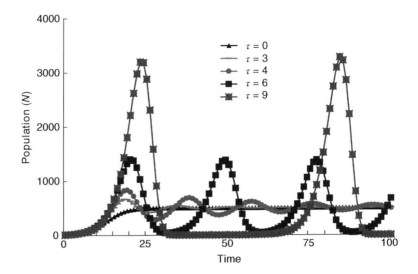

Fig. 2.15 Logistic growth with time lags. $r = 0.30$ and $K = 500$ in all cases.

Figure 2.15 displays five simulations based on equation 2.17. In all cases, $r = 0.30$. In series 1, $\tau = 0$ and we have the usual logistic growth curve. In series 2, with $\tau = 3$, the product of $r\tau = 0.90$, and in series 3, with $\tau = 4$, the product is 1.20. We expect temporary oscillations converging on a stable point at the carrying capacity in both of these cases. In series 4, $\tau = 6$ and the product of $r\tau = 1.80$. Here we expect a stable limit cycle. Finally, in series 5, with $\tau = 9$, the product of $r\tau = 2.70$. Now we expect extinction, and although the simulation shows two population cycles, in reality the population is extinct after 30 time units (Figure 2.15).

2.6 Chaos and behavior of the discrete logistic model

Time lags are implicit in the discrete logistic model. We can actually remove Tau from equation 2.17 and return to equation 2.12: $N_{t+1} = N_t e^{r\left[\frac{K-N_t}{K}\right]} = N_t e^{r\left(1-\frac{N_t}{K}\right)}$. As May (1975b, 1981), and May and Oster (1976) have shown, if the growth rate (r or λ) is very large, populations behave in unusual and unexpected ways.

Recall that if $b^* > 1$, a population shows overcompensation. That is, there is a larger than expected reduction in growth rate or biomass due to density dependence. As shown in Table 2.3, if the combined values of r and b^* produce a net rate of increase with a large reproductive potential, the population moves from a stable equilibrium at the carrying capacity to fluctuations which ultimately reach chaos when the net r is large enough. For example, when r is less than 2.0 (R or $\lambda < 7.39$), the population moves to a stable point (Figure 2.16), although note that when $r = 1.5$ there is a small oscillation before the population settles in at the carrying capacity. At r-values between 2.0 and 2.53 (Figure 2.17, Table 2.3) the population regularly cycles between two points. For r-values between 2.53 and 2.66, the population cycles among four points (Figure 2.18). An eight point cycle is produced by r-values between 2.66 and 2.69. Finally, at r-values > 2.69 (R or $\lambda > 14.761$) the behavior of the population is known as *chaos* (Figure 2.19). That is, the population never enters into a predictable pattern. Over short periods, these chaotic fluctuations would be indistinguishable from seemingly random responses to the environment. A deterministic model, then, can produce results that appear to be stochastic and, if one were looking for biological causation for the behavior of such populations, one would be confused indeed.

Table 2.3 Behavior of the discrete logistic model based on the net growth parameter, r with equivalent values of R or λ. Net growth is also influenced by the non-linear feedback parameter b^*, when $b^* \neq 1$.

Net growth rate, r	Equivalent value of R or λ	Behavior of the discrete logistic model
$2.000 > r > 0$	$7.389 > \lambda > 1.000$	Stable equilibrium point
$2.526 > r > 2.000$	$12.503 > \lambda > 7.389$	Two point cycle
$2.656 > r > 2.526$	$14.239 > \lambda > 12.239$	Four point cycle
$2.685 > r > 2.656$	$14.658 > 14.239$	Eight point cycle
$r > 2.692$	14.761	Chaos

Source: Adapted from May, 1975, May and Oster (1976), and Alstad (2001).

Fig. 2.16 Behavior of the discrete logistic model: stable equilibrium point when $r < 2.0$.

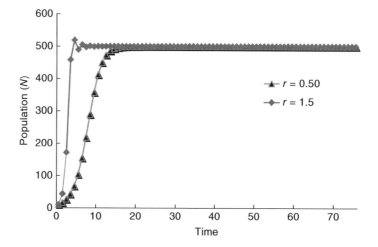

2.7 Adding stochasticity to density dependent models

Just as we did in chapter one, we can perform stochastic simulations with density dependent models. Using the Beverton–Holt Model (equation 2.1), Figure 2.20 shows a deterministic growth curve for an initial population size of 100, a carrying capacity of 1000, and a deterministic value for lambda of 1.1. In order to simulate the effects of demographic stochasticity, we can add a random function in Excel™ that allows lambda to vary with a mean of 1.1 but with a variance of 0.03. One such result is shown in Figure 2.20. In this particular case, when the population was small it did not grow very quickly, but it eventually reached the carrying capacity. Notice that it took over 100 time units to reach carrying capacity even though the deterministic population reaches K at around 50 time units. Running 25 stochastic simulations in this manner produced a range of population sizes after 100 time units of 128–1000 with a mean of 904 individuals. The lessons are basically the same as in the previous chapter, adding variability leads, in most cases, to a smaller population than that expected from a deterministic model.

Using the Ricker Model (equation 2.12), we subsequently simulated both demographic stochasticity (adding variability to r) and environmental stochasticity (by allowing K to vary with time) in Figure 2.21. In all cases the initial population size is 50, the deterministic r is 0.1, and the carrying capacity is 1000. We have also allowed both r and K to vary simultaneously in one series of simulations. In Figure 2.21 we see that the population in which both r and K were allowed to vary went extinct. The population with the stochastic r eventually reached carrying capacity and the population with the stochastic K has gone through several crashes. The results

Fig. 2.17 Behavior of the discrete logistic model: two point cycle when $r = 2.20$.

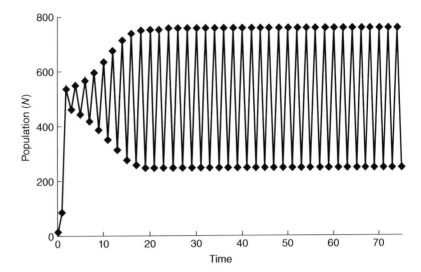

Fig. 2.18 Behavior of the discrete logistic: four point cycle when $r = 2.60$.

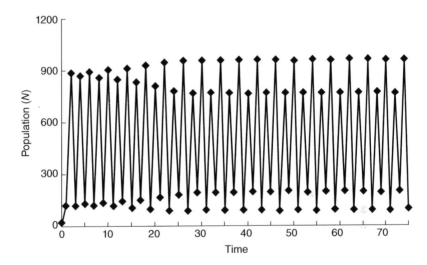

of these simulations depend on how much variability we allow for demographic versus environmental stochasticity. Suffice it to say to the combined effects of demographic plus environmental stochasticity raised the probability of extinction and, as above, variability normally produces smaller populations. Again, you can run these simulations yourself by logging onto the web page associated with this book.

2.8 Laboratory and field data

According to the assumptions discussed above, an increase in population density should lead to one or more of the following:

(a) A linear increase in mortality;
(b) A linear decrease in fertility;
(c) A reduction in average growth rate; and
(d) A reduction in the average size of adults.

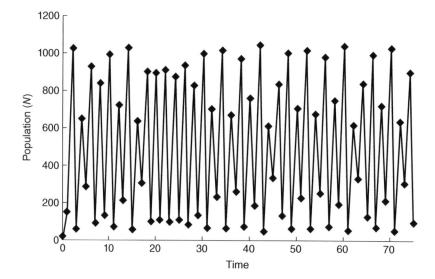

Fig. 2.19 Behavior of the discrete logistic model: chaos when $r = 2.75$.

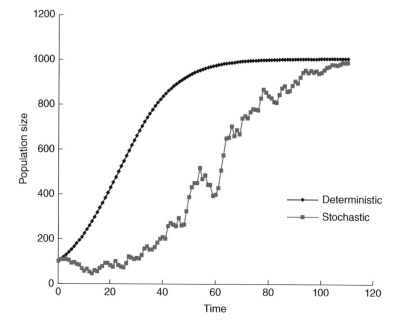

Fig. 2.20 Deterministic versus stochastic growth in a population with an initial population size of 100, a carrying capacity of 1000 and a deterministic lambda of 1.1. In the stochastic model, the deterministic average lambda is 1.1 with a variance of 0.03.

However, recent research has shown that an increase in population density can cause an increase in growth of the offspring in red squirrels (*Tamiasciurus hudsonicus*; Dantzer et al. 2013). But first we will review studies confirming points a–d above.

Crombie (1942, 1944) showed that flour beetles (*Rhizoperta dominica*) raised in the laboratory were negatively affected by density. Both survivorship (Figure 2.22) and fertility (Figure 2.23) decreased with density, although in both cases the effects were non-linear. Moreover, many large mammal populations have fertility and mortality patterns, which show density dependence but are also non-linear (Figures 2.24 and 2.25, Fowler 1981; Gross et al. 1973). Non-linearity extends to bobwhite quail (Roseberry

Fig. 2.21 Effect of demographic stochasticity (r) and environmental stochasticity (K) on behavior of the Ricker model. Initial population size = 50, deterministic $r = 0.1$, carrying capacity, $K = 1000$.

Fig. 2.22 Survivorship of the grain beetle, *Rhizopertha dominica*, versus initial density of first-instar larvae.

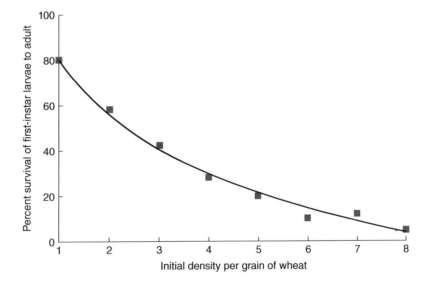

and Klimstra 1984) and cladocerans (Smith and Cooper 1982). On the other hand, both elk (Figure 2.26, Houston 1982) and grizzly bears (McCullough 1981) show a linear decrease in fertility with population density.

Two examples of reduction in growth rate with density will suffice here. The first is that of the growth of tadpoles of the frog, *Rana tigrina*. Whereas it takes only two to three weeks for tadpoles to develop into mature frogs at densities of 5–10 (in a two liter aquarium), it takes almost ten weeks when there were 160 frogs in the same space (Dash and Hota 1980). In harp seals (*Phoca groenlandica*) sexual maturity is achieved when an individual reaches 87% of mean adult body weight. In low populations this occurs at between 4 and 5 years, whereas in dense populations sexual maturity is reached at between 6 and 7 years (Lett et al. 1981).

Fig. 2.23 Fertility versus density in *Rhizopertha dominica.*

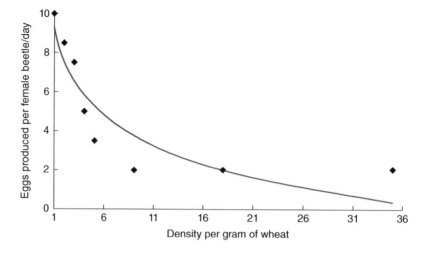

Fig. 2.24 Survivorship of Northern Fur Seal (*Callorhinus ursinus*) pups as a function of pups born. (Source: Data from Fowler 1981.)

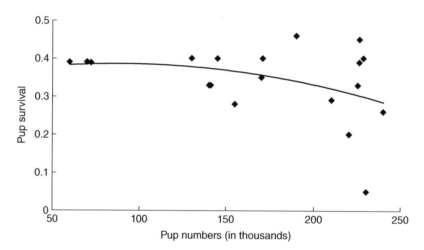

The body size of adults is also affected by density in many populations. For example, in highly dense reindeer populations, mean jaw size of adults is 23 cm, whereas at low density the mean size is between 24 and 25 cm (Skogland 1983). As we saw in Section 2.2, the response to an increase in density among plant populations is a reduction in mean weight per individual. The same principle applies to sessile animal populations. When Branch (1975) examined populations of limpets (*Patella cochlear*), the most common diameter was 60 mm when there were 125 individuals per square meter. When the density was increased to 1225 per square meter, the most common size class was 20 mm. The total biomass obeyed the law on the constant final yield (Figure 2.6a). Biomass increased with density up to 400 individuals per square meter, but then leveled off at 125 g per square meter for all densities from 400 to 1225.

An interesting counterexample and an illustration as to why the logistic equation cannot be used uncritically is the study by Dantzer et al. (2013) on red squirrels (*Tamiasciurus hudsonicus*). Based on a 23 year study in the Yukon, Dantzer et al. (2013) found that population density varies greatly, based on fluctuations in white spruce

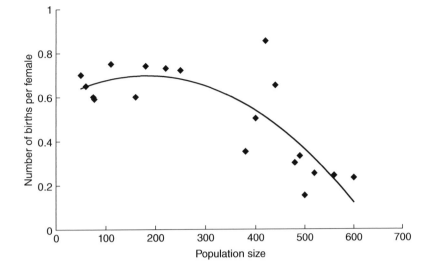

Fig. 2.25 Birth rate of the American bison (*Bison bison*) as a function of population size. (Source: Data from Fowler 1981.)

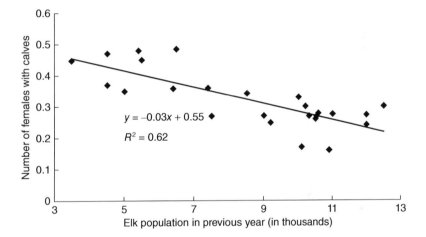

Fig. 2.26 Recruitment of elk calves in Yellowstone National Park as a function of adult population size in the previous year. (Source: Data from Houston 1982.)

(*Picea glauca*) cone production. An increase in cone production leads to a larger squirrel population the next year. Under the resultant high density conditions, faster growing offspring are selected for in the next year, since they are more likely to find a territory and to survive the winter. When population density is low there are no advantages to faster growing offspring. In the natural environment red squirrels use territorial vocalizations to defend their territories. The frequency of these calls indicates population density. In their experiments, Dantzer et al. (2013) exposed pregnant mothers to audio playbacks that simulated high density conditions. The response of the female squirrels was to increase glucocorticoid production, which then resulted in faster growth rates among the young. As indicated above, fitness is enhanced since the faster growing juveniles have a higher chance of surviving the winter. The ultimate chain of events is that high density related social cures causes an increase in offspring growth rates mediated through hormonal changes in the females. Whether female adult survivorship is negatively affected is not discussed.

2.9 Behavioral aspects of intraspecific competition

2.9.1 Castes in social insects

Passera et al. (1996) have shown that intraspecific competition may affect the ratio of castes among social insects. Colonies of the ant, *Pheidole pallidula*, increase their relative investment in soldiers when exposed to other colonies of the same species. The non reproductive individuals are divided into two basic castes: small-headed minors (workers) who do most of the labor, and large-headed majors (soldiers) specialized for defense. In a laboratory experiment, when colonies were exposed to odors from a con-specific colony, they increased the number of soldier pupae. After seven weeks, those colonies exposed to con-specific odors averaged 40.1 soldiers as compared with an average of 22.6 soldiers in the control colonies.

2.9.2 Male–male competition in horned beetles

The complexities of intraspecific competition are illustrated through elegant work on horned beetles (Emlen 2000). Beetle horns are rigid extensions of the exoskeleton and have evolved repeatedly within this order of insects. These horns are generally only expressed in males and are used in combat with other males for access to resources and/or females. These resources are in discrete, readily defensible patches, and the horns allow males to defend these sites and to mate with the females found there. The jousting contests between males can be dramatic and the winner is rewarded with mating privileges.

These horns, however, do not come without cost. They constitute a large investment in energy, and as much as 10% of body mass may be devoted to them. Horn growth prolongs development time and risk of larval mortality, and there is a trade-off between horn development and the ability to fly. Finally, the development of horns leads to lowered visual acuity and to smaller eyes. Nevertheless, it appears that since larger horns are useful in gaining access to females, the reproductive benefits gained from large horns offset the costs of production and maintenance.

The ability of males to grow horns is based on larval nutrition. Both final adult body size and the size of horns in males depend on the amount of food they consume as larvae. Males exposed to poor conditions as larvae are small and do not produce horns, whereas well-fed males become large adults able to produce horns. Horn production shows an insignificant level of heritable genetic variation (Moczek and Emlen 1999).

These complexities and trade-offs are illustrated through studies of dung beetles of the genus *Onthophagus*. The basic life history of these beetles is as follows. After finding a patch of dung, a female digs a tunnel in the soil beneath the dung. She then buries dung below ground to provide nutrition for the larvae. Females spend days inside a single tunnel, pulling down pieces of dung to various blind ends of tunnel branches, each with its "brood ball" of dung. A single egg is laid at the top of a brood ball, and a single larva develops in isolation within the brood ball.

Females mate repeatedly with males during the time of egg laying. Male reproductive behavior consists of securing their unique access to females in the tunnels. The large, horned males guard the tunnels and fight other males trying to approach the females. Larger males with larger horns win these fights. For two males of the same size, the one with the larger horns usually wins. Thus large horns provide males with significant advantages.

However, all is not lost for a small male. Although they are not adept at guarding entrances to tunnels or winning jousts, small males have other tactics. They attempt to slip undetected past the large males, or they dig side tunnels that intercept a guarded tunnel well below ground. Thus hornless males may manage to sneak undetected into guarded tunnels and mate with the female. If such a "sneaker" male is caught, he is chased out by the resident male, who then returns to the female and mates with her. This dilutes or displaces sperm from the sneaker male. Sneaker males actually do best when they have no horns at all, since horns get in the way of sneaking! Success for a small male depends on rapid and undetected entry into tunnels, and horns cause vibrations as they scrape against tunnel walls, which would alert the resident male.

Finally, as mentioned earlier, beetles with very large horns tend to have undeveloped eyes and/or a diminished flying ability due to smaller wing size and greater body weight. Since dung is a temporary resource, another component of fitness is the ability to disperse and locate new dung heaps. As the dung resource diminishes, some of the males, as larvae, will receive a less than optimum amount of food. They will be small and lack horns. Thus, although they would be poorly equipped to defend a female in a burrow, they are well adapted to disperse and find a new dung pile.

Thus, both intraspecific competition and sexual selection are played out on the fields of manure.

2.9.3 Male–female competition in dunnocks

Another aspect of intraspecific competition is competition between the sexes as expressed through mating systems. Obviously ecological conditions can influence the mating behavior of individuals. As conditions change, a cooperative pair or an animal society may quickly dissolve into a set of competing individuals. The following study (Davies 1992, 1995) illustrates some of the complexities in what appears to be a simple pair bonded mating system.

The dunnock (*Pruella modularis*) is not a true sparrow but an accentor. In Old English "dun" means brown and "ock" means little. Thus *P. modularis* is the archetypal little brown bird. It seems unremarkable as it shuffles about under the bushes collecting tiny insects for its young. The impression gained is of harmonious cooperation. The Reverend Morris (1856, in *A History of British Birds*) admired this species so much that he urged his parishioners to emulate its behavior. The Reverend, as quoted by Davies (1992), found this bird to be "Unobtrusive, quiet and retiring, without being shy, humble and homely. … sober and unpretending in its dress … while still neat and graceful … " Due to its extraordinary mating system, however, the Reverend Morris would hardly have been pleased had his congregation followed the example of these birds. Although a dunnock territory can contain one seemingly harmonious pair, a female in the territory next door may be mating with two males, or a male may be mating with two females.

The conventional view of pair formation in songbirds is that males first set up territories and advertise for mates by singing. Females then choose among male territories. However, in dunnocks females defend their own territories against other females, occupying exclusive areas with little overlap among neighbors. Females settle independently of the males, based on the quality of the territory. Males then compete to defend the females from other males.

The result is that dunnock mating behavior includes several combinations. In some cases, a single male defends one female territory (producing monogamy); at other times a male defends two adjacent females (producing polygyny). In other cases two males share the defense of one female (polyandry), or several adjacent females (polygynandry).

If two males share a territory with a single female, the dominant, or alpha male (usually the older one), tries to evict the beta male. If he is unsuccessful, the two share defense of the joint territory. Some males wander for weeks before finding a permanent home. Sometimes a beta will overlap with two alphas. The larger the territory of the female, the harder it is for a male to monopolize her. The larger the male territory, the more likely he is to have two females.

Once a female has built her nest, she solicits copulations from males. Mating begins 3–7 days before the first egg is laid, and lasts up to completion of the clutch of 3–5 eggs. One egg is laid per day during this period. Monogamous males chase off neighboring males who are interested in mating. When two males share a territory, the alpha male follows the female everywhere to prevent her from copulating with the beta male. A female often maneuvers to throw off the alpha male. She then solicits the beta for mating. Females have many tricks, and seem intent on preventing exclusive mating by the alpha.

The act of copulation was described by Selous (1933) as "bizarre." "The hen elevated her rump and stood still, when the male, hopping up, made little excited and very wanton-looking pecks in this region, that is to say the actual orifice. There was actually no mistaking the nature and significance of the actions, rather lecherous, as it seemed to me. This is a very remarkable thing … but I do not understand it" (Selous 1933, p. 107–109). During the pecking described by Selous, the female cloaca makes pumping movements, resulting in the ejection of a small droplet of fluid. As soon as this occurs the male copulates with her.

Davies (1992) analyzed the droplets and discovered that they contain masses of sperm from the previous matings. Like other birds, female dunnocks store sperm. Females with two males may copulate up to six times per hour. As a result, there will be a pool of sperm in the female's cloaca and vagina for much of the mating period. The male's pecking stimulates the female to eject this pool of sperm to make way for his insemination and give his own sperm a better chance of being stored.

After the excitement of the mating period, life on a dunnock territory becomes peaceful during incubation. Females incubate the clutch of eggs alone for 11–12 days. Males help with chick feeding for 11–12 more days in the nest and for two weeks after fledging until the young become independent.

The results of DNA tests on the blood of the chicks and the parents showed that the female is the mother of all chicks in the nest. In monogamous and polygynous territories, the male was the father of all chicks. When two males guarded a territory with one female, however, if the female mated with both males, paternity was mixed. The alpha male fathered 55% of the brood and the beta male 45%.

From the female perspective, when both males mated with her they both helped feed the brood. If she mated exclusively with the alpha male, only he helped feed the brood. Males, however, cannot recognize their own young. If they have mated with the female, they will help raise the young, even in cases where they did not happen to father any of the chicks. If a beta male gained a large share of the copulations he more readily helped with feeding of the chicks. Similarly, if an alpha male was removed experimentally, allowing the beta a larger share of the matings, the beta male worked harder to feed the chicks than did the alpha. Thus, males varied their parental effort in relation to their chance of paternity, not simply according to dominance rank.

For a female, polygyny is the least desirable situation. She must share the help of one male with another female, and some of her chicks often starved to death. A polyandrous female was the most successful because she had the help of two males in raising her young. This explains why the females sneaked around and tried to get the beta males to mate with them. They hoped for future help in rearing of the young.

For a male, the situation is reversed. In polygyny, although each female is less productive, the combined output of two females often exceeds that of one female in monogamy.

Thus the mating system reflects an intraspecific competitive battle between the sexes in dunnocks. Behavioral and genetic studies of other bird species show that the dunnock mating system is not unusual. However, extra-pair copulations and fertilizations vary across species for reasons only poorly understood (Petrie and Kempenaers 1998; Blomquist et al. 2002). Extrapair paternity is much less frequent in non-passerine than in passerine birds (Birkhead et al. 2001). For example, Blomquist et al. (2002) found that in Western sandpipers (*Calidris mauri*), only 5% of all chicks were the product of extra-pair matings. Nevertheless the application of molecular techniques to behavioral studies has allowed us to ask new questions about the mechanisms of intraspecific competition

2.9.4 Competition versus cooperative behavior within a group

Research on lions by Heinsohn and Packer (1995) illustrates the behavioral complexities displayed by animals that are simultaneously territorial and cooperative group-foragers. African lions (*Panthera leo*) engage in a wide variety of group-level activities from hunting to communal cub rearing. At the same time, the group defends a territory from other groups of lions. When prey is difficult to capture they hunt cooperatively, but cooperation breaks down when prey is easy to catch. Female lions nurse each other's young, but more importantly, they jointly protect their young from males that are intent upon infanticide. The threat of attack by con-specifics is a driving force in lion sociality. Large prides dominate smaller ones, and solitary lions are often killed or injured during attacks by lions of the same sex.

Using playbacks of recorded roars, Heinsohn and Packer (1995) found that lions are able to distinguish pride members from strangers. They also found that certain females show a consistent behavior of lagging behind their companions during group activities, including hunting.

Female lions live in social groups (prides), which contain 3–6 related adults, their dependent offspring, and a group of immigrant males. The males defend the pride against incursions by other males; females defend their young against infanticidal males, and the territory from other females. At least two females are needed for a territory, and they advertise ownership by roaring. Using broadcasted roars, the investigators showed that some females become "laggards" early in life, and this behavior persists into adulthood. Laggards were those individuals that hung back, and approached the audio speaker only after the leaders had already responded. The order in which the individuals approached the speaker was the same throughout the playbacks. Because territorial fights often led to injury or death, laggards were ensuring their safety, at least from initial attacks. They typically followed the leaders by 30–120 s.

In the theoretical game or model known as "prisoner's dilemma," in any single task two individuals benefit when they work together (mutual cooperation) but both lose when neither contributes (mutual defection). However, in this game, the greatest payoff for one individual comes from providing no help (cheating) to a partner who cooperates, while the lowest payoff results from helping out (cooperating) while the partner cheats. In a repeated series of encounters, however, cheaters are eventually punished by withdrawal of further cooperation by other individuals. In large groups the game gets more complicated, but cheaters can eventually be detected and punished.

In the case of the lion pride, female leaders approached the speaker slowly and stopped to look behind at the laggards. The leader females mistrust the laggards, but they do not directly punish them. Although lions might be tempted to reduce their risk of injury from territorial defense by withdrawing or not cooperating, they still need companions for territorial defense, to share the protection of the young from males, and for the capture of large prey. In addition, females are often closely related, usually sisters. Given their shared genes, laggards appear to be tolerated, and leaders continue to arrive alone at the speaker 30 s to 2 min before the laggards arrive.

Heinsohn and Packer (1995) classified the behavior of individual female lions as: (i) unconditional cooperators, who always led the response; (ii) unconditional laggards, who always lagged behind; (iii) conditional cooperators, who lagged least when needed most; (iv) conditional laggards, who lagged most when needed most.

The parallels with human societies are evident. How these behaviors developed and why they are tolerated in animal (as well as human!) societies will remain a fascinating topic for future investigations.

2.10 Summary

Models can be derived for density dependent populations using both difference and differential equations for populations with discrete and overlapping generations. Although the equations differ in detail, all presume that the growth rate parameter is dampened as the population approaches a carrying capacity. Modifications of the logistic include the introduction of the Allee effect and the inclusion of time lags. In populations with discrete generations there is an inherent time lag which produces an overshoot of the carrying capacity when the net growth rate is large enough. In continuously breeding populations, if we introduce a time lag variable (Tau), a large value of r combined with even a modest time lag, can cause populations to exhibit a variety of behaviors commonly found in nature. These include limit cycles and dramatic growth phases followed by spectacular population crashes ("boom and bust" cycles). Therefore, when we study populations in nature we should never be surprised when many of them, particularly those populations with high growth potentials, do not remain constant from one year to the next. Finally, the logistic equation is merely a starting point for encompassing the idea of limitation to population growth. It is, as suggested by Turchin (2001), a special case assuming a linear relationship between population density and vital rates. Most importantly it does not allow for time lags, which we have seen have powerful effects on the behavior of populations.

Intraspecific competition has a major influence on the life history of a population. Fertility, mortality, growth and development rates, as well as behavioral characteristics are all affected by intraspecific competition, often in subtle and surprising ways.

References

Allee, W.C. 1931. *Animal Aggregations. A Study in General Sociology*. University of Chicago Press, Chicago, IL.

Alliende, M.C. and J.L. Harper. 1989. Demographic studies of a dioecious tree: colonization, sex, and age-structure of a population of *Salix cinera*. *Journal of Ecology* 77: 1029–1048.

Alstad, D. 2001. *Basic Populus Models of Ecology*. Prentice Hall, Upper Saddle River, NJ.

Anonymous. 2012. Audubon, Christmas Bird Counts. www.audubon.org and www.christmasbirdcount.org.

Arcese, P. and J.N.M. Smith. 1988. Effects of population density and supplemental food on reproduction in song sparrows. *Journal of Animal Ecology* 57: 119–136.

Berec, L., E. Angulo, and F. Courchamp. 2007. Multiple Allee effects and population management. *Trends in Ecology and Evolution* 22: 185–191.

Berg, Å, T. Lindbergh, and K. Gunner. 1992. Hatching success of lapwings on farmland: differences between habitats and colonies of different sizes. *Journal of Animal Ecology* 61: 469–476.

Beverton, R.J.H. and S.J. Holt. 1957. On the dynamics of exploited fish populations. *Fisheries Investigations* 19: 1–533.

Birkhead, T.R. 1977. The effect of habitat and density on breeding success in the common guillemot (*Uria aalge*). *Journal of Animal Ecology* 46: 751–764.

Birkhead, T.R., B.J. Hatchwell, R. Lindner, D. Blomquist, E.J. Pellatt, R. Giffiths, and J.T. Lifjeld. 2001. Extra-pair paternity in the Common Murre. *Condor* 103: 158–162.

Blomquist, D., B. Kempenaers, R.B. Lanctot, and B.K. Sandercock. 2002. Genetic parentage and mating guarding in the Arctic-breeding Western Sandpiper. *The Auk* 119: 228–233.

Branch, G.M. 1975. Intraspecific competition in *Patella coclear* Born. *Journal of Animal Ecology* 44: 263–281.

Caraco, T. and L.L. Wolf. 1975. Ecological determinants of group sizes for foraging lions. *American Naturalist* 109: 343–352.

Clark, P. 2013. Resurrection for the passenger pigeon? *Washington Post* (July 9, 2013). http://www.washingtonpost.com/wp-srv/special/health/passengerpigeon/index.html.

Connell, J.H. 1961. The influence of interspecific competition and other factors on the distribution of the barnacle, *Chthamalus stellatus*. *Ecology* 42: 710–723.

Cote, I.M. and M.R. Gross. 1993. Reduced disease in offspring: a benefit of coloniality in sunfish. *Behavioral Ecology and Socioibiology* 33: 269–274.

Courchamp, F., T.H. Clutton-Brock, and B.T. Grenfell. 1999. Inverse density dependence and the Allee effect. *Trends in Ecology and Evolution* 14: 405–10.

Courchamp, F., T.H. Clutton-Brock, and B.T. Grenfell. 2000. Multipack dynamics and the Allee effect in the African wild dog, *Lycaon pictus*. *Animal Conservation* 3: 277–85.

Courchamp, F., L. Berec, and J. Gascoigne. 2008. *Allee Effects in Ecology and Conservation*. Oxford University Press. Oxford, UK.

Crombie, A.C. 1942. The effect of crowding upon the oviposition of grain-infesting insects. *Journal of Experimental Biology* 19: 311–340.

Crombie, A.C. 1944. On intraspecific and interspecific competition in larvae of graminivorous insects. *Journal of Experimental Biology* 20: 135–151.

Crombie, A.C. 1945. On competition between different species of graminivorous insects. *Proceedings of the Royal Society of London, Series B* 132: 362–395.

Dantzer, B., A.E.M. Newman, R. Boonstra, R. Palme, S. Boutin, M.M. Humphries, and A.G. McAdam. 2013. Density triggers maternal hormones that increase adaptive offspring growth in a wild mammal. *Science* 340: 1215–7.

Dash, M.C. and A.K. Hota. 1980. Density effects on the survival, growth rate, and metamorphosis of *Rana tigrina* tadpoles. *Ecology* 61: 1025–8.

Davidson, J. 1938. On the growth of the sheep population in Tasmania. *Transactions of the Royal Academy of South Australia* 62: 342–6.

Davies, N.B. 1992. *Dunnock Behavior and Social Evolution*. Oxford University Press, Oxford, UK.

Davies, N.B. 1995. Backyard battle of the sexes. *Natural History* 104: 68–73.

Deshmukh, Ian. 1986. *Ecology and Tropical Biology*. Blackwell Science, Oxford, UK.

Emlen, D.J. 2000. Integrating development with evolution: a case study with beetle horns. *Bioscience* 50: 403–418.

Fowler, C.S. 1981. Density dependence as related to life history strategy. *Ecology* 62: 602–10.

Fretwell, S.D. 1986. Distribution and abundance of the dickcissel. *Current Ornithology* 4: 211–242.

Gause, G.F. 1932. Experimental studies on the struggle for existence. I. Mixed populations of two species of yeast. *Journal of Experimental Biology* 9: 399–402.

Gause, G.F. 1934. *The Struggle for Existence*. Williams and Wilkins, Baltimore. Reprinted in 1964 by Macmillan (Hafner Press), New York, NY.

Gilpin, M.E. and F.J. Ayala. 1973. Global models of growth and competition. *Proceedings of the National Academy of Sciences* 70: 3590–3.

Gross, J.E., J.E. Roelle, and G.L. Williams. 1973. *Program ONEPOP and information processor: a systems modeling and communications project*. Progress Report to the United States Department of Interior, Colorado Cooperative Wildlife Research Unit, Colorado State University, Fort Collins, Colorado, USA.

Gurney, W.S.C. and R.M. Nisbet. 1998. *Ecological Dynamics*. Oxford University Press, New York.

Hassell, M.P. 1975. Density-dependence in single species populations. *Journal of Animal Ecology* 44: 382–95.

Heinsohn, R. and C. Packer. 1995. Complex cooperative strategies in group-territorial African lions. *Science* 269: 1260–2.

Houston, D.B. 1982. *The Northern Yellowstone Elk: Ecology and Management*. MacMillan, New York, NY.

Jarman, P.J. and S.M. Wright. 1993. Macropod studies at Wallaby Creek. IX. Exposure and responses of eastern grey kangaroos to dingos. *Wildlife Research* 20: 833–43.

Kenward, R.E. 1978. Hawks and doves: factors affecting success and selection in goshawk attacks on wood pigeons. *Journal of Animal Ecology* 47: 449–60.

Kingsland, S. 1995. *Modeling Nature: Episodes in the History of Population Ecology*. 2nd Edition. University of Chicago Press, Chicago, IL.

Lett, P.F., R.K. Mohn, and D.F. Gray. 1981. Density-dependent processes and management strategy for the northwest Atlantic harp seal populations. Pages 135–158, *in* C.W. Fowler and T.D. Smith, Editors. *Dynamics of Large Mammal Populations*. Wiley, New York, NY.

Major, P. 1978. Predator–prey interactions in two schooling fishes, *Caranax ignobilis* and *Stalephorus purpureus*. *Animal Behaviour* 26: 760–77.

May, R.M. 1975. Biological populations obeying difference equations. Stable points, stable cycles and chaos. *Journal of Theoretical Biology* 49: 511–24.

May, R.M. 1981. Models for single populations. Pages 5–29, *in* R.M. May, Editor. *Theoretical Ecology. Principles and Applications*. 2nd Edition. Blackwell Scientific, Oxford, UK.

May, R.M. and G.F. Oster. 1976. Bifurcations and dynamic complexity in simple ecological models. *American Naturalist* 110: 573–89.

McCullough, D.R. 1981. Population dynamics of the Yellowstone grizzly. Pages 173–196, *in* W. Fowler and T.D. Smith, Editors. *Dynamics of Large Mammal Populations*. Wiley, New York, NY.

Miller, P.S., and R.C. Lacy. 2003. Appendix I: An overview of population viability analysis using VORTEX. Pages 101–121, *in VORTEX: A Stochastic Simulation of the Extinction Process*. Conservation Breeding Specialist Group (SSC/IUCN), Apple Valley, MN.

Moczek, A.P. and D.J. Emlen. 1999. Proximate determination of male horn dimorphism in the beetle *Onthophagus taurus* (Coleoptera: Scarabaeidae). *Journal of Evolutionary Biology* 12: 27–37.

Nicholson, A.J. 1954. An outline of the dynamics of animal populations. *Australian Journal of Zoology* 2: 9–65.

Novak, B. 2013. How to bring passenger pigeons all the way back: Ben Novak at TEDxDeExtinction. http://longnow.org/revive/tedxdeextinction/how-to-bring-passenger-pigeons-all-the-way-back/.

Park, T. 1962. Beetles, competition, and populations. *Science* 138: 1369–1375.

Park, T., P.H. Leslie, and D.B. Mertz. 1964. Genetic strains and competition in populations of *Tribolium*. *Physiological Zoology* 37: 97–162.

Passera, L., E. Roncin, B. Kaufman, and L. Keller. 1996. Increased soldier production in ant colonies exposed to intraspecific competition. *Nature* 379: 630–1.

Pearl, P. 1927. The growth of populations. *Quarterly Review of Biology* 2: 532–548.

Pearl, P. and J.L. Reed. 1920. On the growth of the population of the United States since 1790 and its mathematical representation. *Proceeding of the National Academy of Sciences* 6: 275–288.

Petrie, M. and B. Kempenaers. 1998. Extra-pair paternity in birds: Explaining variation between species and populations. *Trends in Ecology and Evolution* 13: 52–58.

Pielou, E.C. 1977. *Mathematical Ecology*. 2nd Edition. Wiley Inter science, New York, NY.

Ricker, W.E. 1952. Stock and recruitment. *Journal of the Fisheries Research Board of Canada* 11: 559–623.

Roseberry, J.L. and W.D. Klimstra. 1984. *Population Ecology of the Bobwhite*. Southern Illinois University Press, Carbondale, IL.

Sadava, D., D.M. Hillis, H.C. Heller, and M.R. Berenbaum. 2011. *Life, The Science of Biology*. 9th Edition. Sinauer Associates, Inc. Sutherland, MA.

Saether, B, S. Engen, and E. Matthysen. 2002. Demographic characteristics and population dynamical patterns of solitary birds. *Science* 295: 2070–3.

Schoener, T. 1983. Field experiments on interspecific competition. *American Naturalist* 122: 240–285.

Selous, E. 1933. *Evolution of Habit in Birds*. Constable, London, UK.

Shaffer, M.L. 1981. Minimum population sizes for species conservation. *Bioscience* 1: 131–134.

Silvertown, J.W. and J.L. Doust. 1993. *An Introduction to Plant Population Biology*. Blackwell Scientific, Oxford, UK.

Simon, J. 1966. *The Ultimate Resource 2*. Princeton University Press, Princeton, NJ.

Skogland, T. 1983. The effects of density dependent resource limitation on size of wild reindeer. *Oecologia* 60: 156–168.

Smith, D.W. and S.D. Cooper. 1982. Competition among Cladocera. *Ecology* 63: 1004–1015.

Souder, William. 2004. Audubon in Kentucky. *Natural History* 113: 46–52.

Stephens, P.A. and W.J. Sutherland. 1999. Consequences of the Allee effect for behavior, ecology and conservation. *Trends in Ecology and Evolution* 14: 401–5. Stephens and Sutherland 2000.

Stephens, P.A. and W.J. Sutherland. 2000. Vertebrate mating systems, rarity and conservation. Pages 186–213, *in* M. Apollonio, M. Festa-Bianchet, and D. Mainardi, Editors. *Vertebrate Mating Systems*, World Scientific Publishing, London.

Sutherland, W.J. 1996. *From Individual Behaviour to Population Ecology*. Oxford University Press, Oxford.

Tilman, D. 1977. Resource competition between planktonic algae: an experimental and theoretical approach. *Ecology* 58: 338–348.

Turchin, P. 2001. Does population ecology have general laws? *Oikos* 94: 17–26.

Turchin, P. 2003. *Complex Population Dynamics: A Theoretical/Empirical Synthesis*. Princeton University Press, Princeton, NJ.

Vandermeer, J.H. 1969. The competitive structure of communities: an experimental approach with protozoa. *Ecology* 50: 362–371.

Verhulst, P.F. 1838. Notice sur la loi que la population suit dans son accroissement. *Correspondance Mathématique et Physique* 10: 113–121.

3 Population regulation

3.1 Introduction

Are populations regulated? If so, how? What does population regulation really mean? These questions have been debated for many years and were at the core of a hotly contested difference of opinion between ecologists who emphasized the importance of density-dependent factors versus density-independent factors in population regulation (Turchin 1995). The logistic equation is, of course, based on density dependence. But how do we differentiate density independent from density-dependent causation in populations with time lags, chaotic behavior or in those dominated by stochastic processes?

The ecologists who emphasized the primary role of density-independent factors in population regulation were often those who worked with small animals, especially insects and rodents and/or in habitats characterized by drought or short growing seasons. For example, early influential ecologists such as Andrewartha and Birch (1954), worked on Australian grasshoppers, the distribution of which was determined by the length and intensity of the wet season. The northern boundary was determined by conditions that were too dry for their food plants, and the southern boundary was said to have too much moisture for the grasshoppers.

In the first major ecology textbook, Andrewartha and Birch (1954) concluded that abundance of a population was limited by the same conditions that limited its distribution. A major example used by Davidson and Andrewartha (1948) was the distribution and abundance of thrips, tiny insects that feed on pollen and soft tissues of flowers in southern Australia. Populations were said to increase unchecked in the spring, with a sharp decline during the summer drought when flowers were scarce. Andrewartha and Birch (1954) asserted that thrip populations were checked by rainfall, not by their food supply, a potentially density-dependent factor. Advocates of density dependence such as Nicholson (1957), Solomon (1957), and Lack (1954, 1966), often cited the importance of competition in population regulation, especially in vertebrate populations.

Turchin (1995, 1999, 2003) and others (Ricklefs 1990) have asserted that this debate is largely artificial and that density dependence has been repeatedly and convincingly demonstrated (see examples from Chapter 2). At the same time we know that the

Introduction to Population Ecology, Second Edition. Larry L. Rockwood.
© 2015 John Wiley & Sons, Ltd. Published 2015 by John Wiley & Sons, Ltd.
Companion Website: www.wiley.com/go/rockwood/populationecology

physical environment and other stochastic factors often reduce population size before density-dependent factors become operational.

When population regulation takes place, what is the dominant mechanism? Competition? Predation? Parasites? We will explore this question in the second half of the book. Note, however, that the answers may differ by trophic level. For example, Hairston et al. (1960) proposed that while herbivore populations were mainly limited by predation, producers (green plants), decomposers and predators were usually limited by competition. This is such a broad generalization that it is difficult to imagine how to test it. Nevertheless, this theory resurfaces regularly in modified forms.

3.2 What is population regulation?

One of the problems with this debate is a lack of agreement as to what a "regulated population" is. Given what we learned in the previous two chapters about the behavior of populations with time lags, high reproductive potentials, or under the influence of demographic and environmental stochasticity, it is not realistic to expect a population to show a simple attraction to a specific number called the carrying capacity. In chapter two we defined a *stable point* as a stable number at the carrying capacity with no oscillations over time. But we also recognized that the population could be temporarily oscillatory with oscillations dampening and moving toward a stable number or point. We also discussed populations that show regular oscillations around the carrying capacity (*stable cycles*), oscillations between two, four and eight points, and chaotic behavior. All of these population behaviors are based on variations of the density dependent, deterministic, logistic model.

Population regulation does not depend on a specific stable equilibrium point, but rather on a "long-term stationary probability distribution of population densities" (Turchin 1995), or a "stochastic equilibrium probability distribution" (May 1973). The key concept is that there is some mean population level around which a regulated population fluctuates. In addition, over time the population does not wander increasingly far away from this level. The variance of the population density is bounded (Royama 1977). All of these definitions relieve us of the expectation of a fixed carrying capacity or of a fixed stable point. The population is not expected to seek a stable point from which it does not wander. Instead, we allow for stochastic variation around the carrying capacity, *which can itself vary over time*. Hence, an ecological equilibrium is not a fixed or stable point, but a cloud of points. However, since it is often difficult to distinguish density dependence from stochastic noise (see below), many populations have been described as "density vague" (Strong 1986).

Fluctuations in the Dow Jones Industrial Average have been used as an example of an unregulated system, with rainfall patterns on Barro Colorado Island in Panama as an example of a regulated system (Turchin 1995). Economists and meteorologists might argue with both of these examples, but the general idea is that the Dow Jones Average, in spite of short term ups and down, shows a trend upward over long time spans. A better example might be the carbon dioxide content of the atmosphere from 1850 to present. Although there are annual fluctuations, CO_2 concentration has steadily increased from around 280 ppm to over 400 ppm. By contrast, ice cores have shown that carbon dioxide concentration in the atmosphere from the year 1000 to 1750 was a good example of a regulated system. With regard to rainfall, records from 1900 to 2000 showed that rainfall fluctuated from year to year in Panama, but showed no overall pattern of increase or decrease.

An example of an apparently regulated population was described by Rodenhouse et al. (2003). The number of adult black-throated blue warblers remained relatively stable for three decades (1969–2001) at about 10 per 10 ha census plot at the Hubbard Brook Experimental Forest, N.H. The density-dependent mechanisms responsible for this relative stability are discussed in Rodenhouse et al. (2003) and Sillett and Holmes (2005).

Population regulation does not occur in the absence of density dependence. Thus a population showing pure exponential growth or decline would be unregulated. If population density had no effect on the per capita growth rate, there could be no range of population densities to which the population would return. In this context, Turchin (1995) uses the Murdoch and Wade (1989) definition of density dependence as "a dependence of per capita population growth on present and/or past population densities." While density dependence is a property of the overall population dynamics, which may involve time lags, no specific mechanism is necessarily responsible (Turchin 2003). That is, one aspect of the life history, such as birth rates may not be density dependent.

Although density dependence is necessary for population regulation, it is not a sufficient condition. For regulation to occur the following must also be true: (i) density dependence must be of the right sign (there must be a tendency to return to a carrying capacity), (ii) the return tendency must be strong enough to counteract potential disruptive effects of density-independent or stochastic factors, and (iii) the lag time over which the return tendency operates must not be too long (Turchin 1995).

Given more precise definitions of population regulation, and density dependence, Turchin (1995) found that the other remaining pieces needed are: (i) an acceptable statistical test of density dependence, and (ii) time-series data long enough to detect density dependence. From a review of the literature, Turchin (1995) concluded that whenever adequate data have been gathered for an appropriate period of time and tested with an appropriate method, density-dependence has been found. This does not mean that all populations are regulated at all times. Lack of regulation doubtless occurs temporarily, but the probability of detectable density dependence will increase with the length of time that data are collected on a population.

3.3 Combining density-dependent and density-independent factors

In Section 2.7 we added stochasticity to density-dependent models. To simulate density-independent effects we simply added environmental stochasticity by allowing the carrying capacity to randomly vary. We can more explicitly add a stochastic density-independent factor by modifying the Ricker equation (2.12). Let us assume that a certain level of density-independent mortality happens every year, but it varies randomly. We can modify the Ricker equation as shown below (3.1) by simply subtracting the number of deaths from density-independent factors.

$$N_{t+1} = N_t e^{r\left[\frac{K-N_t-D}{K}\right]} \tag{3.1}$$

We can also select different levels of density-independent death rates. As shown in Figure 3.1, the population growth is increasingly affected as we move from zero to increasingly higher amounts of density-independent growth.

Instead of assuming that density-independent effects are uniformly negative, let us now assume that some years are good years (bonanza years) and others are bad years.

Fig. 3.1 Effect of different levels of stochastic density-independent mortality on population growth, based on the Ricker equation (2.12). In all cases $r = 0.25$, $N_0 = 10$, and $K = 1000$. Population growth is increasingly affected as the strength of stochastic density-independent mortality increases.

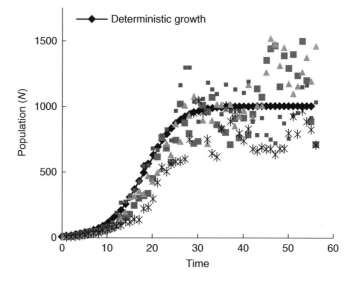

Fig. 3.2 Effect of stochastic "environmental noise" on population growth. The deterministic growth curve has no density-independent effects. In all cases $r = 0.25$, $N_0 = 10$, and $K = 1000$.

That is, environmental stochasticity does not always have a negative effect on our population. We can simply introduce a certain amount of "environmental noise." That is, density-independent effects that can be either beneficial or detrimental. To simulate this we can still use equation 3.1, but we now introduce stochastic variations that can be positive or negative, but with an average value of $D = 0$. In Figure 3.2 compare the four stochastic simulations with the deterministic growth curve with no density-independent mortality. Obviously, if we increase the variations around D, the "cloud" of points around K gets larger. The density-independent factors would get increasingly important and obscure the density dependence of the population.

3.4 **Tests of density dependence**

How can we detect density dependence in the field? For a density-independent population, Tanner (1966) proposed that we can simply use the equation for discrete growth, $N_{t+1} = \lambda N_t$. After taking natural logs of both sides of the equation we can write:

$$\ln N_{t+1} = \ln \lambda + \ln N_t = (1.0) \ln N_t + \ln \lambda \tag{3.2}$$

When we plot $\ln N_{t+1}$ versus $\ln N_t$, if λ is a constant, we should have a straight line with the slope of 1.0 and a y-intercept equal to $\ln \lambda = r$. But if there is density dependence and the growth rate slows with population size, when $\ln N_{t+1}$ is graphed against $\ln N_t$, a linear regression through the data should yield a slope less than 1.0.

For example, consider Figures 3.3 and 3.4. In Figure 3.3 we have an exponentially growing population with $\ln \lambda = r = 0.50$ and an initial population size of ten. In Figure 3.4 we have graphed $\ln N_{t+1}$ versus $\ln N_t$. Since this population is not showing

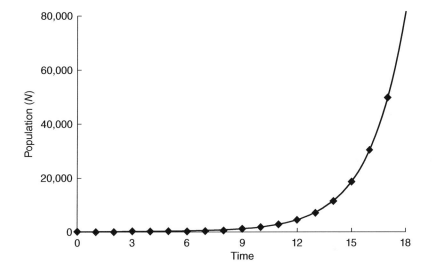

Fig. 3.3 Exponential growth. Initial population size = 10 and $r = 0.50$.

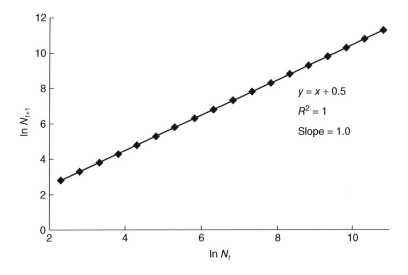

Fig. 3.4 Density-dependent test for data from Figure 3.3.

density dependence we get what we expect, a positive slope equal to 1.0. Now contrast this with a hypothetical yeast population, showing density-dependent growth (Figure 3.5 based on Pearl 1927). If we take natural logs and graph the data as we did in Figure 3.4, we find that the slope of the line is indeed less than one (Figure 3.6). The question remains, however, how far less than one should a regression slope be before we consider it significantly different?

Tanner (1966) examined 70 data sets for animal populations and found slopes significantly different from one in 63 of them. However, this method for detecting density dependence is fundamentally flawed. First, a linear regression assumes data points are independent. In this analysis the x-value in one time-series becomes the y-value in the next time series. Second, measurement error in the population estimates inevitably leads to a slope of less than one. Therefore a slope of less than one is often just an artifact of measurement error and not evidence for density dependence. Finally, the expected relationship between N_{t+1} and N_t is not necessarily linear if there is environmental

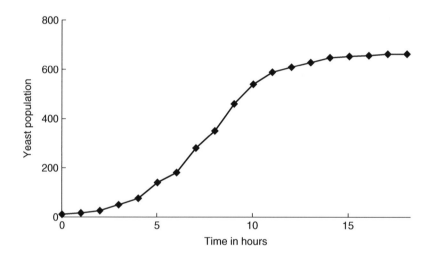

Fig. 3.5 Hypothetical yeast population growing in the laboratory. (*Source:* Adapted from Pearl (1927).)

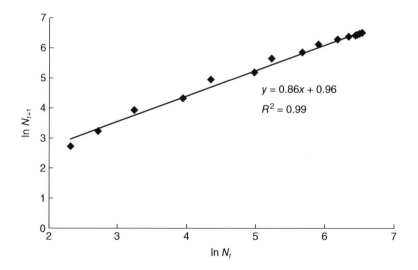

Fig. 3.6 Density dependence test for a yeast population.

variability or if the population has such a high growth rate that it approximates chaos (as described in Chapter 2).

A better test for density dependence is to examine the per capita growth rate versus population size (Turchin 1995; Case 2000). In the logistic equation we expect the per capita growth rate to have a negative slope when graphed against population number (Figure 2.2). By contrast, in exponential growth, the per capita rate remains more or less constant. For example, examine Figures 3.7 and 3.8 (based on exponential growth data). In Figure 3.7, (based on exponential growth data from Figure 3.3), the slope equals zero, indicating no change in the per capita growth rate with population density. By contrast, for our yeast population graphed in Figure 3.5 we find a significant negative slope for the same analysis (Figure 3.8). Examining plots of per capita growth versus N have many advantages including the detection of density dependence in populations with environmental noise (Case 2000).

Fig. 3.7 Per capita growth test for density dependence in a population with exponential growth and no carrying capacity.

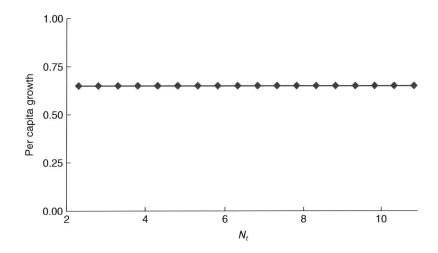

Fig. 3.8 Per capita growth test for density dependence in a yeast population.

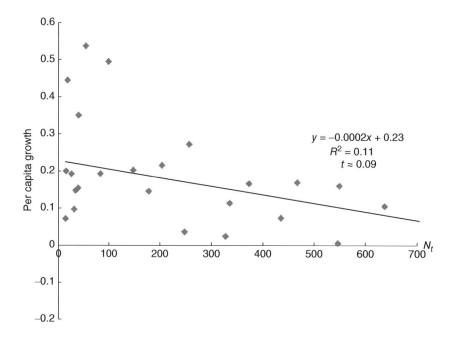

Fig. 3.9 Per capita growth versus population size in the Wisconsin Gray Wolf (*Canis lupus*) population from 1985 to 2011. (*Source*: Wydeven et al. 2012. Data from Wisconsin Department of Natural Resources.)

Let us now look at some field data. Recall the Wisconsin gray wolf population data set from Chapter 1 (Wydeven et al. 2012). From 1985 to 2011 this population increased exponentially with an $r = 0.16$ and no sign of a reduction in growth rate. Using these data, when we graph per capita grown versus current population size (Figure 3.9), the linear regression is not significant ($p > 0.05$) and the slope is not significantly different from zero. This is consistent with a population showing no density dependence.

By contrast in Figure 3.10, we present data on the Virginia Canada Goose (*Branta canadensis*) population based on Audubon Christmas Bird Counts from 1991 to 2011. *Christmas Bird Counts* (*CBC*) were initiated in the late 1800s as a replacement for the traditional Christmas hunt, and may be the oldest wildlife census in the world. The CBC, however, depends on volunteers, many of whom are not professional biologists, and the use of CBC data in scientific studies is somewhat problematic. Nevertheless, the CBC often represents the best long-term data on North American bird populations. Virginia waterfowl populations have been affected by habitat loss, hunting pressure, and environmental degradation. On the other hand, due to land use changes including the creation of new reservoirs and wetlands, waterfowl populations in general may be increasing in Virginia (Heath et al. 2012).

Indeed, if we examine CBC estimates of Canada Geese (*B. canadensis*) populations from 1991 to 2011 (see www.audubon.org and Anonymous 2012, www.christmasbirdcount.org), we see a definite increase for 10 years followed by 10 years of oscillations without any obvious further increase. Applying the same method to detect density dependence, that is, graphing per capita growth versus current population size, is illustrated in Figure 3.11. In this case we have a significant regression ($p < 0.02$) with a negative slope. Thus, based on this analysis, we can conclude that the Canada Goose population in Virginia is showing density dependence.

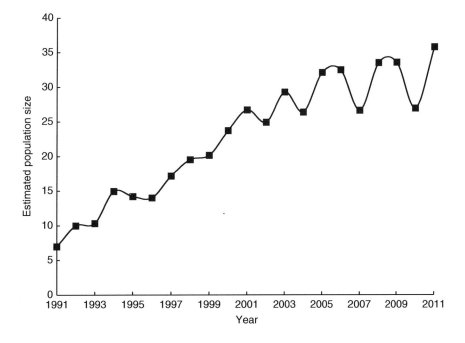

Fig. 3.10 Canada goose (*B. canadensis*) population in Virginia based on Audubon Christmas Bird Counts (www .christmasbirdcount.org) from 1991 to 2011. Population size is based on number per counting group per hour.

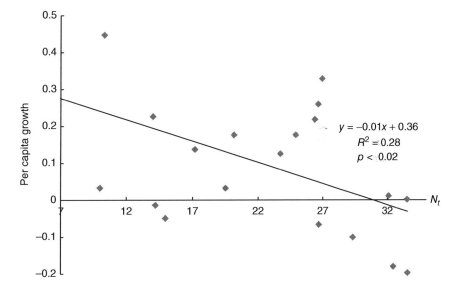

Fig. 3.11 Per capita growth versus population size in the Virginia Canada Goose population. (*Source:* Data from same source as Figure 3.10.)

$$y = -0.01x + 0.36$$
$$R^2 = 0.28$$
$$p < 0.02$$

3.5 Summary

Populations are affected by their own life histories, vital rates, and by feedback from the environment. Such feedback components, which may involve time lags, are termed *endogenous* by Turchin (2003). As we have reviewed before, demographic stochasticity can also seriously affect population behavior in spite of density-dependent feedback. By contrast, *exogenous* factors refer to environmentally related density-independent factors that affect population density, but are not, in turn, affected by it (Turchin 2003).

Our main conclusions are:

1. The per capita rate of change, r, is affected by both endogenous and exogenous factors, and both should be examined when attempting to understand population behavior (Turchin 2003).

2. As stated above, negative feedback between the realized growth rate of the population and its density is a necessary but not sufficient condition for population regulation.

3. Population dynamics are nonlinear and exogenous (density independent) factors may dominate population behavior.

4. A more reliable method for detecting density dependence is to plot per capita growth against N, as opposed to $\ln N_{t+1}$ versus $\ln N_t$.

Assuming we can reliably demonstrate population regulation, is it due primarily to competition or to predation/parasitism? Is regulation in many populations simply a reflection of habitat loss? How often is regulation due to local processes as opposed to metapopulation dynamics (Murdoch 1994)? Are populations under the influence of stochastic environmental factors to the extent that density-dependent population regulation is largely irrelevant?

Leirs et al. (1997) have shown that the population of the African rodent, *Mastomys natalensis* is regulated by an interaction between stochastic and deterministic seasonality and nonlinear density-dependent factors. Using a variety of statistical techniques, they analyzed population data based on a multiple mark-recapture method. The best fit to the data came from a model encompassing both density-independent factors (previous three month rainfall data), and density-dependent factors (a nonlinear demographic model). This example affirms that population regulation is often a combination of stochastic density-independent as well as density-dependent factors operating simultaneously.

References

Andrewartha, H.G. and L.C. Birch. 1954. *The Distribution and Abundance of Animals.* University of Chicago Press, Chicago, Illinois.

Anonymous. 2012. Audubon, Christmas Bird Counts. www.audubon.org and www.christmasbirdcount.org.

Case, T.J. 2000. *An Illustrated Guide to Theoretical Ecology.* Oxford University Press, Oxford, UK.

Davidson, J. and H.G. Andrewartha. 1948. The influence of rainfall, evaporation and atmospheric temperature on fluctuations in the size of a natural population of *Thrips imaginis* (Thysanoptera). *Journal of Animal Ecology* 17: 200–222.

Hairston, N.G., F.E. Smith, and L.B. Slobodkin. 1960. Community structure, population control and competition. *American Naturalist* 94: 421–425.

Heath, S.A., W.J.L. Sladen, and L.L. Rockwood. 2012. Wintering waterfowl on the Northern Virginia Piedmont. *The Raven* 83: 3–9.

Lack, D. 1954. *The Natural Regulation of Animal Numbers.* Oxford University Press, London, UK.

Lack, D. 1966. *Population Studies of Birds.* Clarendon Press, Oxford, UK.

Leirs, H., N.C. Stenseth, J.D. Nichols, J.E. Hines, R. Verhagen, and W. Verheyen. 1997. Stochastic seasonality and nonlinear density-dependent factors regulate population size in an African rodent. *Nature* 389: 176–180.

May, R.M. 1973. Stability in randomly fluctuating versus deterministic environments. *American Naturalist* 107: 621–650.

Pearl, P. 1927. The growth of populations. *Quarterly Review of Biology* 2: 532–548.

Murdoch, W.W. 1994. Population regulation in theory and practice. *Ecology* 75: 271–287.

Murdoch, W.W. and S.J. Wade. 1989. Analysis of insect population dynamics. Pages 113–140, *in* P.J. Grubb and J.B. Whiattaker, Editors. *Toward a More Exact Ecology*. Blackwell Scientific, Oxford.

Nicholson, A.J. 1957. The self-adjustment of populations to change. *Cold Spring Harbor Symposia on Quantitative Biology* 22: 153–172.

Ricklefs, R.E. 1990. *Ecology*, 3rd Edition. W.H. Freeman, New York, NY.

Rodenhouse, N.L., T. Scott Sillett, P.J. Doran, and R.T. Holmes. 2003. Multiple density-dependence mechanisms regulate a migratory bird population during the breeding season. *Proceedings of the Royal Society of London B* 270: 2105–2110.

Royama, T. 1977. Population persistence and density dependence. *Ecological Monographs* 47: 1–35.

Sillett, T. S. and Holmes, R.T. 2005. Long-term demographic trends, limiting factors, and the strength of density dependence in a breeding population of a migratory songbird. Pages 426–436, *in* R.S. Greenberg and P.P. Marra, Editors. *Birds of Two worlds*. Smithsonian Institution Press, Washington, DC.

Solomon, M.E. 1957. Dynamics of insect populations. *Annual Review of Entomology* 2: 121–142.

Strong, D.R. 1986. Density vagueness: abiding the variance in the demography of real populations. Pages 257–268, *in* J. Diamond and T.J. Case, Editors. *Community Ecology*. Harper and Row, New York, NY.

Tanner, J.T. 1966. Effects of population density on growth rates of animal populations. *Ecology* 47: 733–745.

Turchin, P. 1995. Population regulation: old arguments and a new synthesis. Pages 19–40, *in* N. Cappuccino and P.W. Price, Editors. *Population Dynamics: New Approaches and Synthesis*. Academic Press, San Diego, CA.

Turchin, P. 1999. Population regulation: a synthetic view. *Oikos* 84: 153–159.

Turchin, P. 2003. *Complex Population Dynamics: A Theoretical/Empirical Synthesis*. Princeton University Press, Princeton, NJ.

Wydeven, A., J. Wiedenhoeft, R. Schultz, J. Bruner, and S. Boles. 2012. Wisconsin endangered resources report #143. Status of the timber wolf in Wisconsin. Performance report 1 July2011 through 30 June 2012 (also progress reports for 15 April 2011–14, April 2012, and 2011 summaries. 37 pp.

4 Populations with age structures

4.1 Introduction

In the first three chapters, we examined either populations without distinct age classes or we specified that these populations had *stable age distributions* (SADs). We also assumed that models for continuous breeders would apply to seasonal breeders such as white-tailed deer. We danced around the problem of forecasting population growth for populations with complex age distributions and the fact that the age distribution itself can govern the behavior of the population, at least in the short term. Knowledge of the age-specific survivorship and fertility patterns of a population allows us to understand what age categories are most important to the future survival of the population. If you wanted to help conservation biologists ensure the long-term survival of a sea turtle population, for example, what recommendations would you make? Would it be most effective to protect the beaches where the females lay their eggs? Limit predation on the eggs? Gather up the hatchlings, sequester and feed them for a year before releasing them? Or take steps to reduce the mortality of adults through monitoring and regulating fishing fleets? To answer these questions, we will need to learn a number of skills that will allow us to extract growth rates from basic life history data and to project population growth using different assumptions.

When age-specific fertility and survivorship data are gathered, we put them together in the form of a life table. From the information in the life table, we are then able to calculate a variety of interesting statistics that acquaint us with the characteristics of the population. In addition to age-specific fertility and survivorship, we need information

Introduction to Population Ecology, Second Edition. Larry L. Rockwood.
© 2015 John Wiley & Sons, Ltd. Published 2015 by John Wiley & Sons, Ltd.
Companion Website: www.wiley.com/go/rockwood/populationecology

on the current number of individuals in each age class, because population growth in the short term is also strongly influenced by the latter.

The following example might fix in your mind the potential importance of age distribution on population growth. Imagine a group of several thousand young people attending a concert by the latest pop icon on an island off the coast of California. Suddenly a catastrophic series of earthquakes eliminates the entire local population while simultaneously California splits off from the rest of North America. Assume further that a few of the concertgoers are able to colonize this California island, and future population growth is now based on this group. Most of the concertgoers would obviously have been teenage girls (assume a few teenage boys also were dragged along). Further, assuming an abundance of food, this California island would rapidly be repopulated. Growth would, however, be irregular, since all of the girls would be the same age and reach menopause more or less simultaneously in the future. Growth would slow until their daughters began to reproduce. Now imagine the same scenario, except that survivors of the disaster were individuals attending an *AARP* (*American Association of Retired Persons*) convention. Presumably, all or almost all, of the females at the convention are over 55. What would the future of the California island population be in this case? Evidently age distribution can contribute to extinction of a population!

Many of the techniques we will examine in this chapter were developed for the life insurance industry and applied to human populations. For example, actuaries need to calculate the risk of insuring the life of their clients. Life tables were developed so that the probability that a 50-year-old pharmacist would live another 10 years could be determined; policy rates were then set accordingly. Such techniques were easily translated to animal populations, and the comparative study of survivorship among different groups of animals was initiated (Deevy 1947).

More problematic has been the application of life tables to plant populations. Rates of growth, reproduction, size, and mortality are not distinctly related to age in plants, as is the case for animals, but are highly variable and highly dependent on the local environment. This "phenotypic plasticity" can be demonstrated by growing genetically identical clones in different environments. Growth, size, and fertility, when measured against age, will be dramatically different (Silvertown and Doust 1993). The recommended solution is to develop a life history table in which stages are used instead of age classes (Hubbell and Werner 1979; Werner 1975; Werner and Caswell 1977). For example, in her study of teasel (*Dipsacus sylvestris*), Werner (1975) used the following stages instead of age classes: (i) first-year dormant seeds, (ii) second-year dormant seeds, (iii) small rosettes, (iv) medium rosettes, (v) large rosettes, and (vi) flowering plants. In teasel, plants die after flowering. Instead of calculating the probability that an individual would survive from one age class to the next, she calculated the probability that an individual of one life stage would survive to the next stage. We will also take a look at stage-based methods for animal populations later in this chapter.

Another problem in applying life tables to plants is that plant populations often spread through vegetative propagation. Grasses, for example, spread horizontally via rhizomes. New shoots arise from rhizomes and often separate from the original plant. A complex set of terminology has been developed to explain this phenomenon. For example, a genet is an individual that has arisen from a seed. A ramet is a new plant that is a clone but that has arisen through vegetative propagation and is now a completely independent plant with its own roots and shoots. Thus, a population of grasses may consist of several genets, each of which has several ramets. Clonal populations may proliferate indefinitely without flowering. This has led to some fascinating life cycles such as that of the giant bamboo (*Phyllostachys bambusoides*), in which clones of perennial ramets proliferate, forming large populations that flower only once every 120 years (Janzen 1976). All of the clonal ramets flower simultaneously and then the entire population dies, leaving behind only seeds with which to found the next population. Studies conducted since the great 1988 fire in Yellowstone National Park have forced biologists to revise the conventional wisdom that aspen (*Populus tremuloides*) does not reproduce by seed but spreads by cloning. Instead Turner et al. (2003) suggest that new genets of aspen and other perennial herbs of the forest floor are produced after fires, but recruitment of new individuals (ramets) during fire-free intervals is primarily through asexual reproduction.

In the following sections, as we refer to age-specific traits of survivorship or fertility, keep in mind that many plant and animal populations would often require a rather different approach in which we examine survivorship and fertility by size class or by stage in the life cycle.

4.2 Survivorship

The construction of a life table begins by gathering information on survivorship by age class. This sounds simple but is easier described than actually done. For example, one method is to study a cohort of individuals all born at the same time and follow the survivorship of these individuals until the last member of the cohort dies. At the beginning of such a study, it would be necessary to locate and mark all newborn individuals. Subsequently, one would need to verify when each individual died. Individuals that simply disappeared could not be assumed to have died; they might have emigrated. Studies of cohorts are obviously best done on small populations and on populations that move about in a predictable way. The advantage of studying a cohort is that one knows the exact age of each individual. The disadvantage is that such a study lacks generality, in that cohorts born in different years may have different survivorship schedules. In addition to the difficulties one might encounter in actually marking and gathering information on all members of a cohort, there are also practical problems. A cohort study on most species of turtles, for example, would require the entire professional life span of the investigator (picture a student waiting 50 years to finish her Ph.D. dissertation). A life table developed in this manner is known as a fixed-cohort, dynamic, or horizontal life table.

A second approach is to locate and examine all of the dead individuals in a population during some defined period. We would need a method for estimating the age of the animals or plants at death. This approach to the construction of a life table assumes that the rates of survival in the population are fairly constant. If this is not the case, age-specific mortality rates will be confused with year-to-year variation in mortality of the overall population. Data gathered in this manner produce a static, vertical, or time-specific life table.

A third approach is to collect life history data for several cohorts over as long a period as possible. In most populations, there is a large difference between juvenile and adult survivorship. Therefore, although survivorship data on adults are often relatively easy to gather, such data do not apply to the juvenile age classes. Depending on the age when reproduction begins, it is possible to find the growth rate of the population without specific data on juvenile survivorship, as described later in the chapter.

No matter how the data are gathered, the objective is to produce an estimate of age-specific survivorship and fertility. In human demography, age-specific survivorship is based on a theoretical cohort of 1000 individuals. If we let S_x equal the number of individuals surviving to age x, we set S_0 equal to 1000. Then, $S_1 = $ the number of individual surviving to age 1, $S_2 = $ the number surviving to age 2, and so on. Table 4.1 is based on data gathered by an ecology laboratory from the Fairfax City (Virginia) cemetery. In this case, 207 male grave stones were examined and the ages at death calculated. (Data were gathered only from graves in which the birth date was between 1800 and 1900. [Why?]). Because humans are so long lived, the data were placed into 5-year age intervals. The first step is to use the number who died in each age interval (column 2) to produce column 3, the number of survivors by age class. Next, we normalize the population to a theoretical cohort of 1000 by dividing each number by 207 and multiplying by 1000. This produces the S_x column. However, in most ecological studies, we do not, in fact, use the S_x data. Instead, each number in the S_x column is divided by 1000 to produce the survivorship function, l_x. Each value in the l_x column stands for the proportion of the population that survives to a given age, x. It is measured from birth until the last or oldest member of the population dies. l_x is known as age-specific survivorship and can be thought of as the probability, at birth, of living to a specific age class. By definition, $l_0 = 1.00$.

Table 4.1 Survivorship data for males born between 1800 and 1900.

Age category	Number who died in the age category	Number alive at the beginning of the age class	S_x based on a cohort of 1000	Survivorship, l_x Proportion of original cohort alive at the beginning of the age category
0–1	0	207	1000	1.000
1–4	1	207	1000	1.000
5–9	0	206	995	0.995
10–14	2	206	995	0.995
15–19	2	204	986	0.986
20–24	4	202	976	0.976
25–29	0	198	957	0.957
30–34	1	198	957	0.957
35–39	0	197	952	0.952
40–44	5	197	952	0.952
45–49	2	192	928	0.928
50–54	12	190	918	0.918
55–59	9	178	860	0.860
60–64	18	169	816	0.816
65–69	24	151	729	0.729
70–74	33	127	614	0.614
75–79	33	94	454	0.454
80–84	35	61	295	0.295
85–89	20	26	126	0.126
90–94	3	6	29	0.029
95–99	2	3	14	0.014
100–104	1	1	5	0.005
105–109	0	0	0	0.000
Total population	207			

Source: Data from the Fairfax City Cemetery, Fairfax, Virginia.

Fig. 4.1 (a) Survivorship "type" curves on an arithmetic scale. (b) Survivorship "type" curves on a log scale.

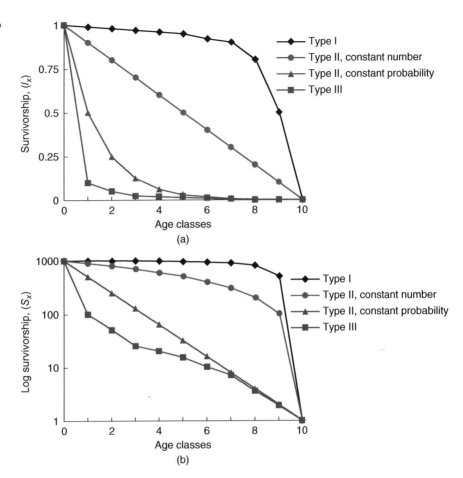

The survivorship table is used to construct the survivorship curves found in all ecology textbooks. In a survivorship curve, age (x), the independent variable, is graphed against survivorship. The y-axis may be on a straight arithmetic scale; however, many authors prefer a log (base 10) scale for survivorship. Pearl (1927) introduced the idea that biological populations routinely fit one of the three "types" of survivorship curves (Figure 4.1a). The type I curve, known as the "death at senescence" curve, is characterized by excellent survivorship at all ages from birth until "old age," at which time the death rate rapidly accelerates and survivorship plummets. The type II curve is linear and assumes that either a constant number or a constant proportion of the population dies in each age interval. Examine Figure 4.1a and b. When survivorship is expressed on an arithmetic scale, a constant number of deaths per age interval produces a linear curve. When log to the base 10 (Figure 4.1b) of survivorship is used, the constant probability of death per age interval produces a straight line. Finally, the type III curve applies to the majority of biological populations. In this curve, there is very high mortality among the juvenile age classes while adult survivorship is relatively high. This is illustrated most dramatically in Figure 4.1a, using the arithmetic scale for survivorship.

How realistic are these three survivorship "types?" Probably few populations exactly match any particular one. Furthermore, as found by Pefranka and Sih (1986) for the

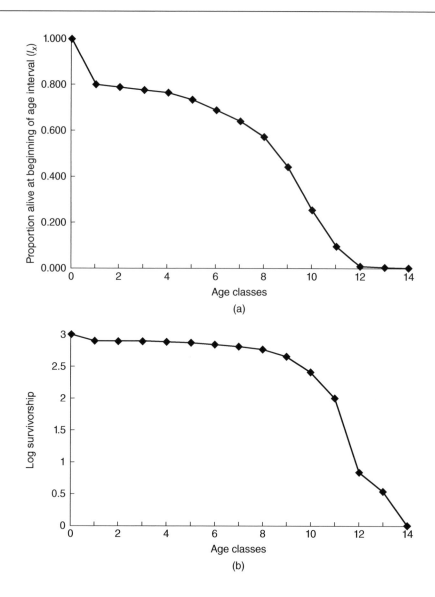

Fig. 4.2 (a) Survivorship (l_x) for Dall Sheep in Denali National Park, Alaska. (b) Log of survivorship for Dall Sheep in Denali National Park. (Source: Adapted from Deevy 1947).

salamander species *Ambostyma texanum*, survivorship curves may vary from year to year and place to place for the same species. (Recall our discussion in Chapter 1 of Population Viability Analysis, in which we emphasized that demographic traits are subject to both temporal and spatial variability.) However, we can make some general comments. The least realistic of the three types is type I. A type I curve applies to laboratory populations of animals such as *Drosophila*. If provided ample food, the population has a high rate of survivorship until the end of its maximum life span when individuals die more or less simultaneously (Hutchinson 1978).

Natural populations of mammals such as Dall mountain sheep (Deevy 1947), and many African ungulates (Caughey, 1966) have a type I survivorship curve although notice that 20% of the Dall sheep die in the first year of life (Figure 4.2a and b).

Fig. 4.3 (a) Human survivorship curves from two 18th Century English and two modern United States populations. (b) Human survivorship curves from two 18th Century English and two United States populations on a log (base 10) scale. (Source: Adapted from Lotka 1956 and Peters and Larkin 1989).

(a)

(b)

While modern human populations have a type I survivorship curve, in the not-so-distant past, living to a ripe old age was not assured (Figure 4.3a and b). By looking for the age where 500 of the original 1000 in a population are still alive, we have an idea of the average life expectancy at birth (Figure 4.3a). For a modern US population (1985), this figure is after the age of 80 (Peters and Larkin 1989). By contrast, for US males living early in the 20th century, this figure was less than 60 (Lotka 1956). For 18th-century English populations, average life expectancy was less than 10 years in Northampton and around 40 years in Carlisle (Lotka 1956)! When these same values are plotted on a log scale, however, they all approximate a type I survivorship curve (Figure 4.3b).

In order for an organism to have a type II survivorship curve, all stages of the life history must be more or less equally vulnerable to predation or other causes of death. Birds, especially the adult stages, are most commonly cited as having a type II survivorship curve. For example, when Gibbons (1987) examined longevity records of vertebrates in captivity, only birds displayed a type II survivorship curve on an arithmetic scale. In a study of white-crowned sparrows (*Zonotrichia leucophrys*) (Figure 4.4), Baker et al. (1981) found a type II survivorship curve on a log scale, which indicates a

Fig. 4.4 Log
survivorship for two
cohorts of
white-crowned sparrows.
(Source: Data from Baker
et al. 1981).

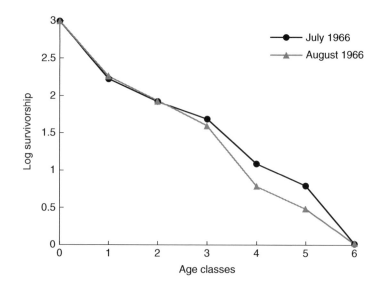

more or less constant probability of death, irrespective of age. The maximum life span
was 49 months in this species. Botkin and Miller (1974), however, argued that birds
do not, in fact, have an age-independent mortality rate. The survivorship curve for the
sooty shearwater (*Puffinus griseus*), based on an arithmetic scale, appears to be type II.
However, Botkin and Miller showed that, on closer examination, while the mortality
rate in the early age classes was 0.07 per year, there was an increase in the mortality
rate of 0.01 per year. They concluded that mortality was age-independent neither in
the sooty shearwater nor in most species of birds.

The type III survivorship curve, which features heavy mortality among young
age classes followed by good to excellent adult survivorship, applies to most biological populations from barnacles to sea turtles to plants (Hutchinson 1978). Even
medium-sized mammals, such as gray squirrels (Figure 4.5; Barkalow et al. 1970) and
golden lion tamarins (Figure 4.6, Jonathan Ballou, personal communication) have
type III survivorship curves, as do most amphibians such as *Ambystoma tigrinum* (e.g.,
Anderson et al. 1971). When an arithmetic scale is used, survivorship in Belding's
ground squirrel can be viewed as a type III. However, when a log scale is used,
both male and female (especially female) survivorship show linear, type II, curves
(Figure 4.5a and b).

Actually, most species do not follow any one of the type curves precisely, especially
when an arithmetic scale is used. For example, in *Phlox drummondii* (Leverlich and
Levin, 1979), survivorship drops from 1.00 to 0.67 in the first 63 days after germination. After 124 days, survivorship is down to less than 0.30. Mortality is minimal
thereafter until the plants are almost a year old (Figure 4.7). Yet this plant, when its
survivorship is plotted on a logarithmic scale, has been used as an example of a type I
curve (Smith 1996). Therefore, although survivorship curves are extremely useful in
order to visualize the large amount of data in a life table, there is little agreement as to
what constitutes a survivorship "type," and nothing is gained by attempting to fit a life
table to any of the three "type" curves.

Fig. 4.5 (a) Survivorship curve for a population of Belding's ground squirrel. (b) Log of survivorship for a population of Belding's ground squirrel. (Source: Data from Sherman and Morton 1984).

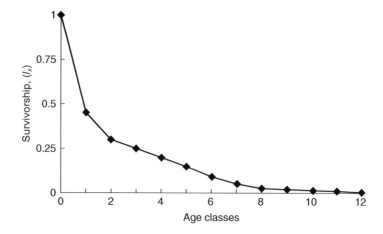

Fig. 4.6 Survivorship in captive female golden lion tamarins. Jonathan Ballou, personal communication.

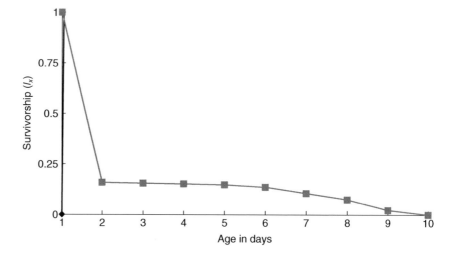

Fig. 4.7 Survivorship schedule for *Phlox drummondii* on an arithmetic scale. (Source: Data from Leverlich and Levin 1979).

4.3 Fertility

The other half of the life table is the **fertility column**, m_x. Here, each value represents the average number of female offspring produced per female of a given age. Again, gathering accurate data on fertility in the field is problematic for many populations. In order to simplify calculations, we count only the number of females. That is, the values are mean numbers of females by age class. Fertility, similar to survivorship, can be graphed as a function of age, and the resultant fertility curve is usually triangular or rectangular in shape. For example, Table 4.2 illustrates human fertility based on the 2010 US Census Bureau. Age classes until age 15 are usually termed "pre-reproductive." Ages 15–45 are considered the reproductive age classes. Figure 4.8 illustrates the usual triangular shape, with maximum reproduction occurring in the 25- to 30-year age classes. After age 45, fertility falls back to zero. These are the postreproductive age classes (Modern medical science, however, is pushing the normal boundary of reproduction past 45 years). The fertility schedule of Belding's ground squirrel also (Table 4.3a) has a more or less triangular shape (Figure 4.9a).

As usual, when populations are sampled, data do not necessarily follow generalized trends. For example, in a study by Grant and Grant (1992) on the cactus ground finch (*Geospiza scandens*), the fertility schedule had a very irregular shape, dropping

Table 4.2 Fertility data from US Census Bureau.

Age class	2010 US Census Bureau
0–1	0
1–4	0
5–9	0
10–14	0
15–19	0.073
20–24	0.218
25–29	0.242
30–34	0.207
35–39	0.127
40–44	0.032
45–49	0
50–54	0
55–59	0
60–64	0
65–69	0
70–74	0
75–79	0
80–84	0
85–89	0
Gross reproductive rate	0.899

Fertility is based on the average number of daughters born in 5-year age intervals.
Source: Anonymous 2010. Data from US Census Bureau

Fig. 4.8 Human fertility of the United States, 2010. (Source: Data from U.S. Census Bureau, http://www.census.gov /hhes/fertility/data/cps /2010.html).

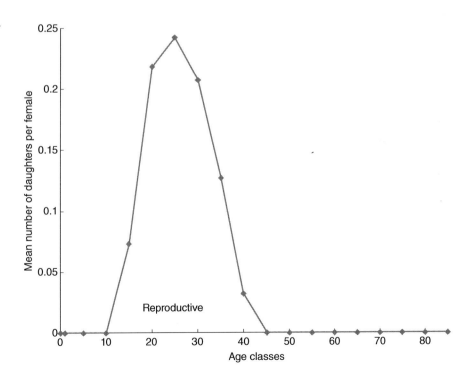

Age	l_x for females	m_x	p_x for females	q_x for females	$l_x m_x$
0	1.000	0	0.386	0.641	0
1	0.386	1.040	0.510	0.490	0.401
2	0.197	2.171	0.538	0.462	0.428
3	0.106	2.450	0.509	0.491	0.260
4	0.054	3.113	0.537	0.463	0.168
5	0.029	1.875	0.483	0.517	0.054
6	0.014	1.575	0.571	0.429	0.022
7	0.008	2.000	0.750	0.250	0.016
8	0.006	1.440	0.333	0.667	0.009
9	0.002	1.632	0	1.000	0.003
10		17.3 = GRR	xx	xx	$R_0 = 1.361$

Source: Data from Sherman and Morton (1984)

Age	l_x for males	p_x for males	q_x for males
0	1.000	0.350	0.650
1	0.350	0.434	0.566
2	0.152	0.316	0.684
3	0.048	0.313	0.688
4	0.015	0.200	0.800
5	0.003	0	1.000
6	0	0	1.000
7	0	0	1.000
8	0	0	1.000
9	0	0	1.000
10	0	0	1.000

Source: Data from Sherman and Morton (1984)

radically in the seventh year and showing an unexpected spike in the twelfth year (Figure 4.9b). In this case, environmental variation dominated the fertility schedule. Amboseli baboons also have an irregular fertility schedule (Alberts and Altemann 2003).

The sum of the m_x column defines the *gross reproductive rate* (*GRR*). This number is the average number of female offspring produced by a female that survives at least through the last reproductive age class (equation 4.1). For Belding's ground squirrel (Table 4.3a), this figure is 17.3.

$$GRR = \sum m_x \tag{4.1}$$

The two pillars of a life table are the survivorship (l_x) and fertility (m_x) columns. Long hours of fieldwork are necessary to gather the data in order to produce such a life table. Once the table is produced, many other calculations and projections are possible although all of them assume that the survivorship and fertility columns remain constant.

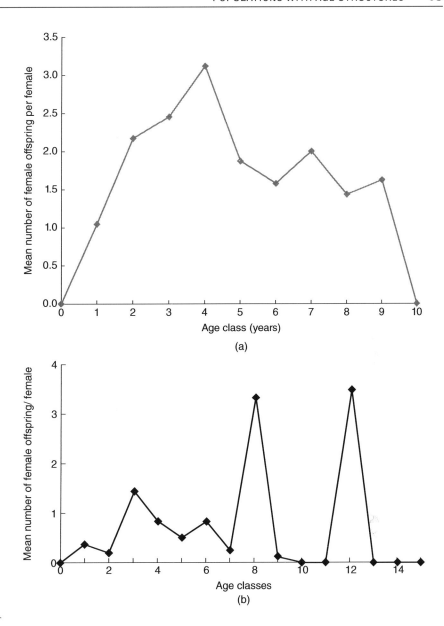

Fig. 4.9 (a) Fertility schedule of the Belding's ground squirrel. (b) Fertility schedule of the cactus ground finch. (Source: Data from Sherman and Morton 1984).

From the l_x column, we can develop two parallel columns, which provide informa-tion on how survivorship and mortality rates change with age. Consider Tables 4.3a and b. The l_x column is based on the probability, at birth, of surviving to a given age class. The p_x column, by contrast, is the age-specific probability of surviving to the next age class. That is, p_2 tells us the probability that an individual who has survived to the age of two will survive to be 3 years old. Similarly, p_4 would tell us the probability that a 4-year-old lives to age 5. These p_x values are critically important when we want to project future population growth as will become clear later in this chapter. The value of p_x is calculated according to the formula:

$$p_x = \frac{l_{x+1}}{l_x} \tag{4.2}$$

For example, in Table 4.3a, we see that $p_0 = l_1$, since $p_0 = 0.386/1.00$, $p_1 = 0.197/0.386 = 0.510$, and so on. Notice that $p_9 = 0$ for females, since no 9-year-old female squirrel lives to be 10 years old. P_{10} is undefined and is the equivalent of dividing zero by zero.

From Table 4.3b, we see that male Belding's ground squirrels have a much shorter life expectancy and $p_5 = 0$.

The companion value to p_x is q_x, which is the proportion of the population that has survived to a given age, x, but which will die in the next time or age interval. q_x is simple to calculate since it is equal to $1 - p_x$ (equation 4.3). This is based on the idea that $p_x + q_x = 1.0$. We recognize only two states of being, alive or dead:

$$q_x = 1 - p_x \tag{4.3}$$

4.4 Mortality curves

Caughey (1966, 1977) and others found that mortality curves (q_x) for female mammals, such as Orkney voles (Leslie et al. 1955), toque monkeys (Dittus 1977), buffalo (Sinclair 1977), Himalayan tahr, domestic sheep, Dall mountain sheep, and elk all follow a U-shaped pattern (Figure 4.10). The U-shape is the result of the fact that juvenile phases usually have high mortality, but they are coupled with adult phases that have low mortality. In the final phases of life, mortality increases (the senescent phase). Ralls et al. (1980) suggested that in polygynous species, although females show the U-shaped pattern, males have a spike of mortality in subadult to young adult

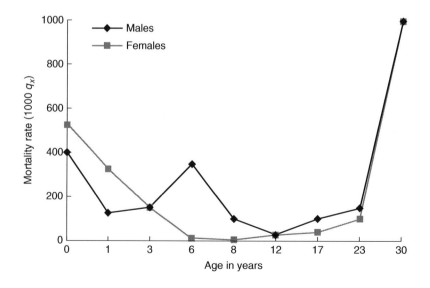

Fig. 4.10 Mortality curves for male and female impala. (Source: Data from Jarman and Jarman 1973; analysis from Ralls et al. 1980.)

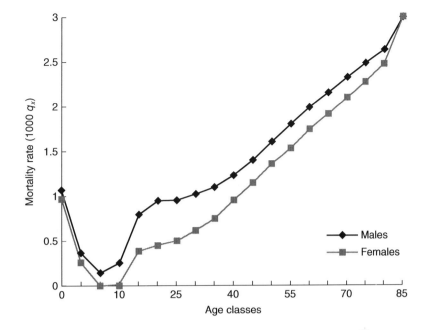

Fig. 4.11 Mortality rate for United States population based on 1986 vital statistics. (Source: Anonymous 1988.)

age classes (Figure 4.10). Impala (*Aepyceros melampus*) and toque monkeys (*Macaca sinica*), for example, illustrate this pattern. These higher male mortality rates are due to male–male competition for mates and the tendency for males to leave the natal group in many species (Ralls et al. 1980).

Interestingly, a similar pattern can be found for the US human population (Figure 4.11). United States Vita Statistics for 1986 (Anonymous 1988) have a U-shape for females, with high mortality in the first year of life, followed by a low rate of mortality until age 15. Thereafter, the mortality increases throughout life. Males show higher mortality rates from the age of one onwards. More striking, however, is that the largest separation in mortality rates between the sexes occurs from the ages of 15–30. These age classes are exactly those discussed by Ralls et al. (1980), in which male mammals suffer greater mortality in the subadult to early-adult age classes. Do human males, similar to other mammals, suffer these higher mortality rates due to male–male competition for females and/or due to dispersal away from the parental home, as Ralls et al. (1980) have suggested? Or are there other reasons such as the suggestion that human male brains mature at a slower rate than those of females (Thompson et al. 2000)?

Belding's ground squirrels also show dichotomy between male and female mortality curves (Figure 4.12). Female mortality is high in the first year and then changes little until the last two age classes, with a small number of females living up to 9 years. Male mortality is similar to that of females in the first year and then increases every year with no males living past 5 years. We do not have time to discuss the life history of Belding's ground squirrels here, but the high mortality rates and short lifetimes appear to be related to the breeding system in this species (Sherman and Morton 1984).

Fig. 4.12 Mortality rates
for Female and Male
Belding's ground
squirrel. (Source: Data
from Sherman and
Morton 1984.)

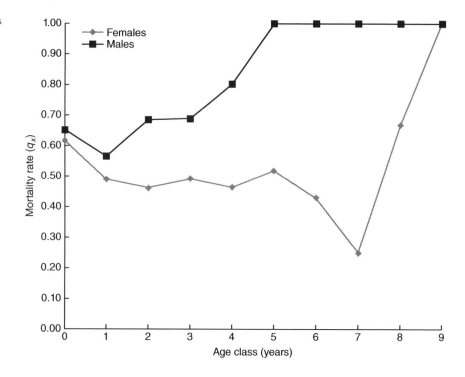

Fig. 4.12 Mortality rates for Female and Male Belding's ground squirrel. (Source: Data from Sherman and Morton 1984.)

4.5 Expectation of life

Another statistic of interest to demographers, to ecologists, and even to nonscientists is the age-specific expectation of life. The question to be addressed is, what is the average life expectancy of an individual of a given age, x? The survivorship table allows us to directly read the average life span at birth. We simply look for the age at which l_x equals 0.50. However, expectation of life is age specific. Obviously, if life expectancy at birth for a human male is 75 years, a male who has survived to be 65 years old should, on average, live well beyond 75. In fact, recent data suggest that a 65-year-old human male can look forward to an average of at least 15 more years of life.

Age-specific life expectancy is measured by taking the area under the survivorship curve beyond an age, x, and dividing by the number or proportion of survivors of that age, x. In theory, using integral calculus, expectation of life, e_x, to the last age class, (w), is found using the formula:

$$e_x = \frac{\int_x^w l_x d_x}{l_x} \tag{4.4}$$

In practice, we use discrete age classes. The survivorship between ages x and $x + 1$, if the age interval is reasonably small, is estimated as follows (equation 4.5):

$$L_x = \frac{l_x + l_{x+1}}{2} \tag{4.5}$$

L_x is the mean survivorship for any particular age interval and assumes that, on average, an organism dies half way between two age classes. The area under the

Table 4.4 Calculation of average, age-specific life expectancy for Belding's ground squirrel females.

Age, x	l_x	L_x	T_x	Expectation of life, e_x
0	1.000	0.693	1.302	1.30
1	0.386	0.292	0.609	1.58
2	0.197	0.152	0.318	1.61
3	0.106	0.080	0.166	1.57
4	0.054	0.042	0.086	1.59
5	0.029	0.022	0.045	1.53
6	0.014	0.011	0.023	1.64
7	0.008	0.007	0.012	1.50
8	0.006	0.004	0.005	0.83
9	0.002	0.001	0.001	0.50
10	0	0.000	0.000	0.00
		$\Sigma = 1.301$		

Source: Data from Sherman and Morton (1984)

survivorship curve for an individual of a given age, x, to the age, w, at which the oldest individual dies, is estimated as follows:

$$T_x = \sum_x^w L_x = \sum_x^w \frac{l_x + l_{x+1}}{2} \qquad (4.6)$$

Therefore, expectation of life is estimated by equation 4.7 (see Table 4.4):

$$e_x = \frac{T_x}{l_x} \qquad (4.7)$$

In this example, we see that if we use the l_x table and look for 0.50, we would assume that life expectancy at birth is less than a year. Using this method, we find that average life expectancy at birth is 1.30 years. For all age classes from 1 to 7, life expectancy is around an additional 1.5 years. As seen from Figure 4.12, mortality rate increases rapidly after age 7, and average life expectancy drops to less than a year.

4.6 Net reproductive rate, generation time, and the intrinsic rate of increase

Once we have the basic life table, we are in a position to calculate the same types of growth rate statistics that we discussed in the first two chapters. The first of these is the net reproductive rate, R_0. This is an equivalent to the R (equation 1.4) we developed for populations with discrete generations. The net reproductive rate represents the increase in the population per generation and is defined as the mean number of female offspring produced per female in the population per generation. This value is found by incorporating both the fertility and survivorship functions of the life table. For each age class, we calculate the product of $l_x m_x$. This product is the contribution a particular age class is making toward population growth per generation. The net reproductive rate for the population as a whole is the sum of these products for all age classes:

$$R_0 = \sum l_x m_x \qquad (4.8)$$

In Table 4.3a, the calculated net reproductive rate for Belding's ground squirrel population is 1.361. This means that the average female squirrel replaces herself with 1.361

female squirrels per generation. As in the case of the net reproductive rate for nonoverlapping populations, an $R_0 > 1$ means that the population, according to the life table, has the potential to increase every generation. The opposite is also true; an $R_0 < 1$ means that the population is decreasing every generation.

Although the net reproductive rate is an important statistic, we usually want to know the growth rate per year (or some other defined period). When we compare growth rates among different types of populations, the usual currency is r, the intrinsic rate of increase, or the finite rate of increase (λ), since both are measured for a specific unit of time. The intrinsic rate of increase can be extracted from life history data using an equation developed by Euler although some authors give credit to Lotka (see Mertz 1970, or Case 2000 for its derivation). It is most often known as the Euler equation; but in any event, it is considered to be a "characteristic equation" of demography (Dingle 1990):

$$\sum l_x m_x e^{-rx} = 1 \tag{4.9}$$

This equation is useful because it allows us to determine the intrinsic rate of increase from the life table. However, because r is an exponent in a summation, it cannot be explicitly solved for if there are more than two age classes. Values of r must be estimated and tried in the Euler equation until a value is found that satisfies it. However, Laughlin (1965) and May (1976a) showed that there exists an excellent approximation for r. Assuming an SAD, the approximation is based on the following:

If G = generation time, we can write $N_G/N_0 = R_0$. It is also true (equation 1.8) that $N_G/N_0 = e^{rG}$. Therefore, we can set $R_0 = e^{rG}$. Taking natural logs of both sides of the equation gives us $\ln R_0 = rG$ and therefore,

$$r = \frac{\ln R_0}{G} \tag{4.10}$$

This tells us that the intrinsic rate of increase can be found by dividing the natural log of the increase per generation by the generation time. We now have an approximation for r, but we must calculate G, the mean generation time. Mean generation time is actually a somewhat slippery concept and can be defined in various ways. Here we will use the definition, the mean age of the mothers at the time of their daughter's birth. This is the same definition as "the average interval between reproductive onset in two successive generations" (Dingle 1990). Generation time is estimated according to the following equations in which the age, x, is weighted by its realized fecundity, $l_x m_x$. In the second equation, discrete age intervals are used:

$$G = \int x l_x m_x d_x \tag{4.11a}$$

$$G = \sum x l_x m_x \tag{4.11b}$$

These equations, however, can only be used if the population is not growing. In order to account for the number of offspring being produced per individual female, the right side of the equation must be divided by the net reproductive rate, R_0. Therefore, we use equation 4.12 in estimating G:

$$G = \frac{\sum x l_x m_x}{R_0} \tag{4.12}$$

Once we have approximated the value of G, r can be estimated using equation 4.10. Note that all approximations of r gained using equation 4.10 must be verified by the Euler equation! Since equation 4.10 simply approximates r, the value of r must be verified, or the approximation adjusted using equation 4.9, the Euler equation. The way r is estimated and confirmed is shown in Example 4.1.

4.7 Age structure and the stable age distribution

In the next sections, we are ready to begin examining the interactions between the age distribution of a population and its life table. As stated earlier, the actual age distribution of a population has potentially dramatic effects on population growth in the short term. The age distribution of a population is defined as the proportions of the population belonging to various age categories at a given point in time. The proportion belonging to a given age category, x, is calculated by dividing the number of individuals in that age category by the total population size, N, producing c_x (4.13):

$$c_x = \frac{n_x}{N} = \frac{n_x}{\sum n_x} \tag{4.13}$$

The variable c_x is the proportion of the population belonging to an age category, x, and n_x equals the number of individuals in that age category.

Whenever survivorship and fertility remain constant for long enough, a population will converge on a particular age distribution, known as the stable age distribution, which is unique for each combination of survivorship and fertility. Once this SAD is achieved, the age distribution no longer changes unless and until survivorship or fertility change in the life table. Furthermore, the population will grow or decline at the steady rate, λ (unless the r − value = 0, in which case the population is unchanging and $\lambda = 1$), and each age class will change at the same rate as the population as a whole. If a population has an SAD, λ is easy to calculate, since $N_{t+1}/N_t = \lambda$. Since $r = \ln \lambda$ (equation 1.13), it is also simple to calculate r.

The SAD itself can be calculated from the survivorship column of the life table. In order to predict the SAD, however, it is also necessary to know the value of r as well as the survivorship function, l_x. Since the Euler equation requires the knowledge of fertility (m_x), we actually must know both survivorship and fertility. The formula for predicting the SAD is as follows (equation 4.14):

$$c_x = \frac{e^{-rx} l_x}{\sum e^{-rx} l_x} \tag{4.14}$$

4.8 Projecting population growth in age-structured populations

Examine Table 4.5. The basic information on survivorship and fertility by age class would have been gathered through field work. In order to project population growth into the future, we also must know the actual number of individuals belonging to each age class, again based on data we have obtained in the field. Once these data are available, we can do a year-by-year projection of not only the population size as a whole but also the expected number of individuals in each age class. We must assume, however, that the survivorship and fertility functions do not change.

In Table 4.5, we have done some basic calculations that will tell us generally what to expect from this population. The net reproductive rate, (R_0), tells us that we expect this population to grow every generation since $R_0 > 1$. In the ensuing projection, we will need the age-specific probability of surviving to the next age class (p_x); accordingly, we

Table 4.5 Hypothetical life table for a population.

Age, x	l_x	m_x	p_x	q_x	$l_x m_x$
0	1.00	0	0.50	0.50	0
1	0.50	2.0	0.40	0.60	1.0
2	0.20	1.0	0.50	0.50	0.2
3	0.10	1.0	0	1.00	0.1
4	0	0	–	–	–
Sums		GRR = 4.0			$R_0 = 1.3$

This table will be used to show a simple population projection (Table 4.6).

Table 4.6 Projected population growth based on life history from Table 4.5 and starting with 200 individuals in age class zero (newborn).

Age, x	n_x at $t = 0$	c_x at $t = 0$	n_x at $t = 1$	c_x at $t = 1$	n_x at $t = 2$	c_x at $t = 2$	n_x at $t = 3$	c_x at $t = 3$	Calculated stable age distribution
0	200	1.00	200	0.67	240	0.63	300	0.625	0.625
1	0	0	100	0.33	100	0.26	120	0.250	0.258
2	0	0	0	0	40	0.11	40	0.083	0.082
3	0	0	0	0	0	0	20	0.042	0.035
4	0	0	0	0	0	0	0	0	0
Sums	200	1.00	300	1.00	380	1.00	480	1.00	1.00

have used a column to p_x as well as to its opposite (q_x), the age-specific probability of dying in the next age class. The last column allows us to calculate R_0.

We begin this projection by assuming that we have founded this population with 200 new born females at time $= 0$ (Table 4.6). We have completed the first two columns based on that assumption (n_x = the number in an age category, c_x = the proportion of the population in that age category). To project the population, the number of individuals in each of the nonzero age categories is found by multiplying the number of individuals in the appropriate age category by the suitable p_x value. For example, to find the number of 1-year-old individuals at time $t = 1$, we multiply the number of newborn individuals at time $t = 0$ by p_0. The number of 2 year olds at time $t = 1$ is found by multiplying the number of 1 year olds at time $t = 0$ by p_1, and so on. In this case, to find the number of 1-year-old females at time $= 1$, we simply multiply 200 by p_0 (0.50), giving us 100. Since we lack 1-, 2-, or 3-year-old individuals at time $= 0$, there can be no 2-, 3-, or 4-year-old individuals at time $= 1$.

To find the number of newborn individuals at time $t = 1$, multiply the appropriate m_x values by the number of individuals in each age class at time $= 1$. In this example, to find the number of newborn individuals, we multiply the number of 1 year olds by the appropriate m_x value. In this case, $m_1 = 2.0$. That is, each of the 100 females who has survived has 2.0 female offspring, on average. Therefore, there are 200 new born females at time $= 1$. The total population is 300, and our c_x values are based on that number (Table 4.6). To find the number of 1-year-old females at time $= 2$, we repeat the procedure outlined earlier, giving us 100 again. The number of 2-year-old females at time $= 2$ is found by multiplying p_1 (0.40) by 100, giving us 40 2-year-old individuals.

The number of newborn individuals is found by multiplying 100 1-year-old females by 2.0 and 40 2-year-old females by 1.0. That is, multiply by the appropriate m_x values. The number of newborn individuals is, therefore, 240, and the total population is 380. Confirm the values for $t = 3$. The last column in Table 4.6 is the calculated SAD based on equation 4.14. Notice that after only three time periods, the population proportions (c_x) have moved to within a few tenths of the SAD, although we started at time $t = 0$ with only one age class (200, $x = 0$).

In the earlier exercise, the finite rate of increase can be calculated for each of the time periods t to $t + 1$ as N_{t+1}/N_t. The results are as follows:

λ from $t = 0$ to $t = 1$: $300/200 = 1.50$
λ from $t = 1$ to $t = 2$: $380/300 = 1.27$
λ from $t = 2$ to $t = 3$: $480/380 = 1.26$
λ from $t = 3$ to $t = 4$: $586/480 = 1.22$

Notice how quickly λ moves to 1.22. The calculated r from the Euler equation for this life table is 0.206, and therefore the predicted value of λ at the stable age (e^r) distribution is 1.23. Exercises similar to this show us three things: (1) in the short term, population growth is heavily influenced by its age distribution; (2) nevertheless, a population with a short generation time can move rapidly to its SAD; and (3) the finite rate of increase is influenced by the actual age distribution, but it settles in at the predicted value from the life history table ($= e^r$) as the population reaches the SAD.

In Example 4.1, find the necessary information to project this population. As you do that, test yourself to see whether you understand how to calculate generation time, G, the intrinsic rate of increase, and the SAD from the life history table.

Example 4.1 Find GRR, R_0, and G. Estimate r and then find its true value with the Euler equation. Verify the predicted stable age distribution (SAD). Project this population as described earlier. After reading the next section, use the Leslie matrix to project the population.

Age	l_x	m_x	p_x	q_x	$l_x * m_x$	$x * l_x * m_x$	Euler based on $r = 0.152$	Euler based on $r = 0.154$	$l_x * e^{-rx}$	c_x of SAD
0	1.00	0	0.250	0.750	0.00	0.00	0	0	1.000	0.733
1	0.25	0	0.400	0.600	0.00	0.00	0	0	0.214	0.157
2	0.10	7.0	0.800	0.200	0.70	1.40	0.517	0.514	0.073	0.054
3	0.08	7.5	0.500	0.500	0.60	1.80	0.380	0.378	0.050	0.037
4	0.04	5.0	0.250	0.750	0.20	0.80	0.109	0.108	0.022	0.016
5	0.01	0	0.000	1.000	0.00	0	0	0	0.005	0.003
6	0	0	–	–	0.00	0	0	0	0.000	0.000
Σ		GRR = 19.5			$R_0 = 1.50$	4.00	1.006	1.000	1.364	1.000

$G = 4.00/1.50 = 2.67$ Estimated value of $r = 0.152$ Predicted value of $\lambda = e^r = 1.17$

Population Projection

Age	n_x at $t=0$	c_x at $t=0$	n_x at $t=1$	c_x at $t=1$	n_x at $t=2$	c_x at $t=2$	n_x at $t=3$	c_x at $t=3$	n_x at $t=4$	c_x at $t=4$
0	250	0.714	318	0.743	359	0.729	420.6	0.733	492.10	0.733
1	60	0.171	62.5	0.146	79.5	0.162	89.75	0.156	105.15	0.157
2	20	0.057	24	0.056	25	0.051	31.8	0.055	35.90	0.054
3	12	0.034	16	0.037	19.2	0.039	20	0.035	25.44	0.038
4	6	0.017	6	0.014	8	0.016	9.6	0.017	10.00	0.015
5	2	0.006	1.5	0.004	1.5	0.003	2	0.003	2.40	0.003
6	0	0.000	0	0.000	0	0.000	0	0.000	0	0
	$N=350$	1.000	$N=428$	1.000	$N=492.2$	1.000	$N=573.75$	1.000	$N=670.99$	1.000
			$\lambda=428/350$ $=1.22$		$\lambda=492/428$ $=1.15$		$\lambda=574/492$ $=1.17$		$\lambda=671/574$ $=1.17$	

4.9 The Leslie or population projection matrix

This process of projecting the population one age class and 1 year at a time, as done earlier, is time consuming and tedious. Leslie (1945) showed that populations could easily be projected through the use of matrix algebra. If you are not familiar with matrix algebra, see the primer in Appendix 2. The matrix approach allows quick calculations of changes in the age structure and total population size as well as a quick method for finding λ when there is an SAD. The survivorship and fertility columns are placed in the matrix form $|A|$. The population itself is considered a column vector that, when multiplied by the matrix, produces a new column vector, which represents the population at time $t+1$ (equation 4.15):

$$N_{t+1} = |A|N_t \qquad (4.15)$$

The format for the matrix is as shown in Table 4.7. The p_x values (the probabilities of surviving from age x to age $x+1$) appear in the matrix in the off diagonal. The first row consists of the products $p_x * m_{x+1}$. The matrix must be a square matrix, with the final column consisting of zeros. In this example, with five age classes, we have a 5×5

Table 4.7 General matrix format for projecting a population with five age classes for one time period ($t=0$ to $t=1$).

$|AgeClasses|$			Matrix				$|t=0|$	$|t=1|$
0	$p_0 m_1$	$p_1 m_2$	$p_2 m_3$	$p_3 m_4$	0		n_0	n_0
1	p_0	0	0	0	0		n_1	n_1
2	0	p_1	0	0	0	*	n_2	= n_2
3	0	0	p_2	0	0		n_3	n_3
4	0	0	0	p_3	0		n_4	n_4

Column Vectors

For age class 0, the number of individuals is based
on $p_0 m_1 n_0 + p_1 m_2 n_1 + p_2 m_3 n_2 + p_3 m_4 n_3 + 0$.
For age class 1, the number of individuals is calculated as $p_0 n_0 + 0 + 0 + 0 + 0$.
For age class 2, the number of individuals is calculated as $0 + p_1 n_1 + 0 + 0 + 0$.
For age class 3, the number of individuals is calculated as $0 + 0 + p_2 n_2 + 0 + 0$.
For age class 4, the number of individuals is calculated as $0 + 0 + 0 + p_3 n_3 + 0$.

Table 4.8 Matrix projection based on life-history data found in Table 4.5.

AgeClasses	Matrix				Columns Vectors	
					$t = 2$	$t = 3$
0	$0.5*2 = 1.0$	$0.4*1 = 0.4$	$0.5*1 = 0.5$	0	240	300
1	0.5	0	0	0	100	120
2	0	0.4	0	0	40	40
3	0	0	0.5	0	0	20
Total	0	0	0	0	380	480

Age, x	Calculations for Column Vector, time = 3	Resultant Column Vector for time = 3
0	$(1.0*240) + (0.40*100) + (0.50*40)$	300
1	$0.50*240$	120
2	$0.40*100$	40
3	$0.50*40$	20
Total		480

Column vectors are based on time periods $t = 2$ and $t = 3$ in Table 4.6.

matrix. Given the rules of matrix multiplication, the product of the matrix times the column vector (representing the population by age classes at $t = 0$) results in a column vector at time $= 1$.

If we use the life table from Table 4.5, the resultant matrix is as shown in Table 4.8. If we then multiply this matrix by the column vector for time $= 2$ from Table 4.6, we can then calculate the column vector for time $= 3$. The result is identical to the projection we did earlier in Table 4.6.

4.10 A second version of the Leslie matrix

To this point, in constructing our life tables, we have assumed that the year begins with the reproductive season. For example, if we are developing a life table for white-crowned sparrows, we might assess the number of eggs that have hatched and assign the hatchlings to age class zero. However, in some studies, this approach is not practical, and the study begins with counts of animals that have completed at least 1 year of life. The count is done before the production of newborn individuals, but there is no count of the number of newborn individuals. Assume it is possible, however, to estimate both fertility and survivorship for age classes from year 1 onward. For example, see Table 4.9, which is based on Table 4.5. The life table simply begins with age class 1. Note that in this version of the matrix, all of the fertility values in the first row are simply multiplied by p_0, since in order to find the number of 1-year-old individuals, we are multiplying the total number of newborn by the probability of surviving to age class 1.

Table 4.9 A population projection matrix, based on Table 4.5, in which the year begins before reproduction and no count is made of the zero-year age class.

$$\begin{vmatrix} m_1 p_0 & m_2 p_0 & m_3 p_0 & m_4 p_0 \\ p_1 & 0 & 0 & 0 \\ 0 & p_2 & 0 & 0 \\ 0 & 0 & p_3 & 0 \end{vmatrix} * \begin{vmatrix} n_1 \\ n_2 \\ n_3 \\ n_4 \end{vmatrix}^{t=2} = \begin{vmatrix} n_1 \\ n_2 \\ n_3 \\ n_4 \end{vmatrix}^{t=3} \Rightarrow \begin{vmatrix} 1.0 & 0.5 & 0.5 & 0 \\ 0.4 & 0 & 0 & 0 \\ 0 & 0.5 & 0 & 0 \\ 0 & 0 & 0 & 0 \end{vmatrix} \begin{vmatrix} 100 \\ 40 \\ 0 \\ 0 \end{vmatrix}^{t=2} = \begin{vmatrix} 120 \\ 40 \\ 20 \\ 0 \end{vmatrix}^{t=3}$$

The population is projected from time $= 2$ to 3 as in Section 4.9

We end up with the same result by either method, but we do not have exact data on age class zero in this second method. Interestingly enough, since age class zero individuals do not reproduce, the identical finite rate of increase (λ) can be derived from either method. And since the second method requires a smaller matrix, calculations can be vastly simplified, as shown in the next section.

4.11 The Lefkovitch modification of the Leslie matrix

Lefkovitch (1965) noticed that for many organisms the yearly fertility and survivorship functions remain relatively constant once adulthood was reached. Instead of specific age classes, he proposed using "stage classes" based on life stages such as juvenile, young adult, and adult. Recall our discussion in the first part of this chapter of the necessity to use such a method for plant populations. Now review Table 4.3a in Section 4.3 for Belding's ground squirrel population. Notice that by using method two (see Section 4.10), the gray squirrel life table can be simplified to two stages as shown in Table 4.10. To do this, we have eliminated some of the variability in year-to-year adult survivorship, using 0.53 for p_A (adult survivorship). Further, we have estimated survivorship in the first year of life as $p_Y = 0.39$. Fertility is estimated as 1.0 for the young age class and 2.0 for mature adults.

Lefkovitch showed that in spite of this lumping, the growth rate λ is conserved. In setting up the projection matrix based on this lumping, we must realize that adults from many of the age classes that we have placed together are simply recycled back into the same stage class from which they came for many years. That is, a 3-year-old that survives is placed back in the "mature adult" stage class. Therefore, when placing p_A (mature adult survivorship) into the matrix, it ends up in the bottom right-hand corner of this 2×2 matrix.

Using $Y =$ "young adult" and $A =$ "mature adult" stages, the gray squirrel matrix can be simplified as follows:

$$\begin{vmatrix} m_Y p_Y & m_A p_Y \\ p_A & p_A \end{vmatrix} = \begin{vmatrix} 1.0 * 0.39 & 2.0 * 0.39 \\ 0.53 & 0.53 \end{vmatrix} = \begin{vmatrix} 0.39 & 0.78 \\ 0.53 & 0.53 \end{vmatrix}$$

Let us now project this population and determine its growth rate, λ. We will start with 340 "young adults" and 492 "mature adults." We project this population for 3 years in Table 4.11. After the first year, the calculated λ is $957/832 = 1.15$. Thereafter,

Table 4.10 Simplified life table for the Belding's ground squirrel.

Age or stage class	m_x	p_x
$Y =$ young adult	1.0	0.39
$A =$ mature adult	2.0	0.53

Table 4.11 Projection of Belding's ground squirrel population using the 2×2 simplified matrix.

Stage Class	Matrix		$t = 0$		$t = 1$		$t = 2$		$t = 3$
Y	0.39	0.78	340		516		545		608
A	0.53	0.53	492	=	441	=	507	=	558
			$\sum = 832$		$\sum = 957$		$\sum = 1052$		$\sum = 1166$

the $\lambda = 1.10$ and 1.11. By contrast, if you were to use all age classes and project the population until it reached the SAD, the λ value is 1.14. (You can prove this to yourself using Table 4.3a and projecting the population.) Given that we assumed one p_x for all adult age classes and the uncertainties of actual survivorship and fertility data we might gather in the field, these two estimates of the finite rate of increase are surprisingly close. As Lefkovitch (1965) emphasized, the reduced matrix has the same λ as its larger counterpart using all age classes.

4.12 Dominant latent roots and the characteristic equation

Following the rules of matrix multiplication, if the matrix does not change (meaning that l_x and m_x remain constant), we can write:

$$N_1 = |A|N_0$$

$$N_2 = |A|N_1 = |A||A|N_0$$

$$N_3 = |A|N_2 = |A||A||A|N_0$$

This generalizes to the following:

$$N_t = |A|^t N_0 \tag{4.16}$$

Therefore, a population can be projected to any time in the future. We can also project a population backward in time. The advantage of this is that we can examine properties a population might have had in the past, assuming the present life table. The backward projection requires the "identity matrix." The identity matrix is equivalent to the number one in algebra and is such that one can write the following:

$$I|A| = |A|I = |A| \tag{4.17}$$

Each number has an inverse in algebra (except zero) such that $(x) * (1/x) = 1$. The inverse of a matrix is such that $|A||A|^{-1} = I$ (the identify matrix). If the inverse of $|A|$ is $|B|$, then $|A||B| = I$.

For example, if $|A| = \begin{vmatrix} 2 & 5 \\ 3 & 8 \end{vmatrix}$; then, $|B| = \begin{vmatrix} 8 & -5 \\ -3 & 2 \end{vmatrix}$. The result of multiplying $|A||B|$ is the 2×2 identity matrix: $\begin{vmatrix} 1 & 0 \\ 0 & 1 \end{vmatrix}$ (see Appendix 2).

The inverse of a matrix only exists when the matrix is square and when the matrix has a "determinant." A determinant for a square matrix is a particular scalar number (see Appendix 2) that is easy to calculate for a 2×2 matrix but becomes increasingly complicated for larger matrices. The projection matrix has an inverse if $[A][B] = I$. We can then use $[B]$ to project the matrix backward from time $= 0$ to $t = -1, -2, -3$, and so on as in equation 4.18:

$$N_t = |B|^t N_0 \tag{4.18}$$

This allows us to compare the actual population of 1955, for example, with the potential population in that year with present day survivorship and fertility values.

As discussed earlier, when a population has an SAD, each age group grows as the same rate as the population as a whole; that is, at the finite rate of increase, λ.

Thus, $n_{x(t+1)} = n_{x(t)}\lambda$ for all age classes (x). And $N_{t+1} = N_t\lambda$. In matrix form, we have the following:

$$N_{t+1} = \lambda N_t = \lambda(IN_t) = \lambda I(N_t)$$

Since $N_{t+1} = |A|N_t = \lambda I N_t$, we can write (equation 4.19a):

$$|A| - \lambda I = 0 \qquad\qquad (4.19a)$$

Zero is a column vector consisting of all zeros.

Equation 4.19a is known as the characteristic equation for a matrix. It only exists for square matrices, and λ is known as the "latent root," "the characteristic root," or the Eigen value of the characteristic equation. When the matrix $|A|$ is of the order n (i.e., a 4×4 matrix is of the order 4), the characteristic equation is a polynomial of degree n and has n solutions. But for the Leslie matrix, there is only one positive root (or "dominant latent root"). This dominant latent root $= \lambda$, and Leslie has shown that $\lambda = e^r$. Thus, matrix algebra can be used to solve for r from basic life history data.

As explained in detail by Case (2000), one way of solving for λ is to find the determinant of the expression $|A| - \lambda I$ and setting it equal to zero. In other words, instead of projecting the matrix to find λ as we did earlier, we can find λ if we solve the following expression:

$$\det(|A| - \lambda I) = 0 \qquad\qquad (4.19b)$$

For more information on determinants and basic matrix operations, see Appendix 2. For a 2 × 2 matrix, we start by multiplying λ by the identity matrix, giving us $\begin{vmatrix} 1 & 0 \\ 0 & 1 \end{vmatrix}$ *

$$\lambda = \begin{vmatrix} \lambda & 0 \\ 0 & \lambda \end{vmatrix}$$

Now we subtract this from our 2×2 matrix (see rules of subtraction, Appendix 2) yielding the following:

$$\begin{vmatrix} a_{11} & a_{12} \\ a_{21} & a_{22} \end{vmatrix} - \begin{vmatrix} \lambda & 0 \\ 0 & \lambda \end{vmatrix} = \begin{vmatrix} a_{11} - \lambda & a_{12} \\ a_{21} & a_{22} - \lambda \end{vmatrix}$$

The determinant of a simple 2×2 matrix is found by taking the difference between the cross products resulting in the following (equation 4.20):

$$(a_{11} - \lambda)(a_{22} - \lambda) - (a_{21})(a_{12}) = \lambda^2 - \lambda(a_{11} + a_{22}) + (a_{11}a_{22}) - (a_{21}a_{12}) = 0$$

$$(4.20)$$

Solving for λ looks difficult, and finding the determinant for more complex matrices is extremely tedious without computer software. As discussed earlier, thanks to Lefkovitch (1965), many matrices can be simplified to this 2×2 form.

Now let us return to the 2×2 matrix for Belding's ground squirrel population we found in the previous section: $\begin{vmatrix} 0.39 & 1.06 \\ 0.53 & 0.53 \end{vmatrix}$.

Solving for λ using the method outlined in Equation 4.20, we have:

$$(0.39 - \lambda)(0.53 - \lambda) - (0.53 * 01.06) = 0$$

This gives us $\lambda^2 - 0.92\lambda + 0.21 = 0$.

We can solve this equation using the formula for the solution of a quadratic equation ($ax^2 + bx + c = 0$, recall your high school algebra!):

$$\lambda = \frac{-b \pm \sqrt{b^2 - 4ac}}{2a}$$

$$\lambda = (0.92 \pm \sqrt{0.85 + 0.84})/2$$

$$= \frac{0.92 \pm 1.30}{2} = 1.11$$

Although there is another solution, the positive or dominant root for λ is 1.11, which is very close to the 1.14 we find for λ at the SAD.

4.13 Reproductive value

As biologists and conservation biologists evaluate life histories, they are often interested in which age classes contribute most heavily to present and/or future population growth. For this analysis, they calculate a parameter known as the reproductive value. One application of the reproductive value focuses on whether natural selection can regulate events late in the life span of an organism when reproductive value is very low. Behavioral biologists have suggested that in dominance hierarchies, individuals with the greatest reproductive potential or value will be supported by their mothers or others in the population (Alberts and Altmann 2003). Conservation biologists attempting to evaluate what intervention strategy will give them the biggest bang for their buck (in terms of long-term survival of the population) may use reproductive value in determining which age classes are likely to produce the desired increase in growth rate.

To evaluate reproductive potential, simply examining the fertility column is usually misleading. For example, if 1-year-old females have, on average, two female offspring, but 2-year-old females have four female offspring, one might conclude that the 2-year-old individuals contribute more to population growth. However, suppose that survivorship values for $l_1 = 0.1$ and for $l_2 = 0.01$. The products of $l_x m_x$ tell us that 1-year-old females produce 0.2 females and 2 year olds only 0.04 females per generation. Evidently an evaluation of reproductive potential of a given age class must take into account both survivorship and fertility. Furthermore, the value of a given female depends not only on her present reproduction but also the potential for future reproduction. For example, in many species, females become more fecund and more adept at successfully raising offspring as they mature, leading to an increase in m_x with age, followed by a decrease as they senesce. On the other hand, in mammals that suckle their young for more than a year, reproductive rates may be suppressed in the year following a birth, leading to oscillations in birth rates by age class across the population. For example, in Amboseli baboons, $m_{10} = 0.816$, $m_{11} = 0.649$, $m_{12} = 0.818$ and $m_{13} = 0.554$ (Alberts and Altmann 2003).

Reproductive value is a number that measures the relative reproductive potential of an individual of a given age. It can also be thought of as the weighted average of present

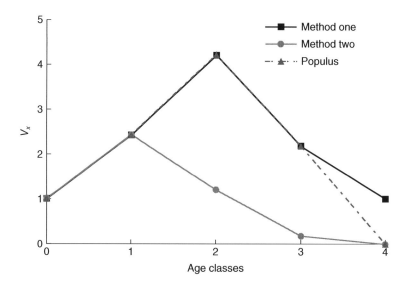

Fig. 4.13 Three methods of computing reproductive value.

and future reproduction by a female of age x. It is the relative value of a daughter born i time units in the future when the population size will be N_{t+i}. Reproductive value is scaled so that the value for the first age class equals 1.0 (Figure 4.13). There are several different ways of computing reproductive value (Lanciani 1998). In the following text, I present the traditional equation for reproductive value as method one (Lanciani 1998). A second method produces reproductive values that are closer to the original concept of Fisher (1930) and to that derived from the matrix algebra approach (Caswell 1989). Conceptually, the second value is defined as "the present value of future offspring" of a female of age x and ignores reproduction by the present age class (see method two). In other words, values from age $x + 1$ to the end of life are used. Reproductive value can also be found in the Populus computer simulations developed by the University of Minnesota ecologist Don Alstad (Alstad 2001). In method one, as in Populus, values from age x to the end of life are used.

The differences between the two approaches can be gleaned by going through the examples mentioned later. In the formula for reproductive value, the survivorship functions l_x, m_x, and r are as usual. In equation 4.21, the numerator is simply the Euler equation. In method two, the summation is from the age $x + 1$ to the end of life. The summation in the numerator is from the age class in question (x) to the end of the life span (z):

$$V_x = \frac{\displaystyle\sum_{x}^{z} e^{-rx} l_x m_x}{e^{-rx} l_x} \tag{4.21}$$

Sample calculations are found in Example 4.2. Note that you should be able to confirm that $r = 0.371$ and $\lambda = 1.449$, assuming an SAD.

Example 4.2 To calculate reproductive value, we need the following information.

Age, x	l_x	m_x	$e^{-rx}l_x m_x$	$e^{-rx}l_x$
0	1.00	0	0	1.000
1	0.60	0	0	0.414
2	0.50	3.0	0.714	0.238
3	0.40	2.0	0.263	0.131
4	0.10	1.0	0.023	0.023
5	0	0	0	0
Sum			1.000	

Calculation of reproductive value, V_x.

Age, x	Method One	Method Two
0	$(0 + 0 + 0.714 + 0.263 + 0.023)/1.000 = 1.000$	$(0 + 0.714 + 0.263 + 0.023)/1.000 = 1.000$
1	$(0 + 0.714 + 0.263 + 0.023)/0.414 = 2.42$	$(0.714 + 0.263 + 0.023)/0.414 = 2.42$
2	$(0.714 + 0.263 + 0.023)/0.238 = 4.20$	$(0.263 + 0.023)/0.238 = 1.20$
3	$(0.263 + 0.023)/0.132 = 2.17$	$(0.023)/0.132 = 0.17$
4	$(0.023)/0.023 = 1.000$	$0/0.023 = 0$

All methods produce the typical triangular shape when reproductive value, V_x, is graphed against age (Figure 4.13). In Figure 4.14, you will find the reproductive values for Belding's ground squirrel (again, after Sherman and Morton 1984). See Figure 4.8 of Alberts and Altmann (2003) for another example of triangular reproductive values for male and female Ambolseli baboons.

4.14 Conclusions: sensitivity analysis

We have learned that age- or stage-structured growth is common in most plant and animal populations, but the details of both survivorship and fertility differ greatly across species. For example, natural populations display many variations in survivorship, although a type III survivorship schedule is most common. Most of the "charismatic megafauna" of interest to conservation biologists and the general public shows age-structured growth. Thus, the techniques outlined in this chapter are necessary before we can make predictions about the potential for growth and recovery of an endangered population. We must understand the effects that age-specific survivorship and fertility have on the behavior of a population before we are in a position to implement a management plan that would actually be effective in promoting its long-term survival.

For example, examine Table 4.12. These data were gathered by Schmutz et al. (1997) on a population of the emperor goose (analysis from Morris and Doak 2002). Given this information, is this population growing or declining? What aspects of its life history are most important to its long-term growth rate? To answer these questions, we will use the matrix format from Section 4.10. It turns out that the long-term λ for this population is 0.989.

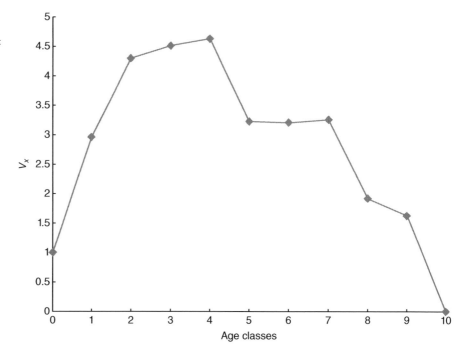

Fig. 4.14 Reproductive value, V_x of Belding's ground squirrel. (Source: Data from Sherman and Morton 1984.)

Table 4.12 Survivorship and fertility of an emperor goose population.

Survival of hatchlings $= S_0$	Survival of 1-year-old and older birds $= S \geq 1$	Fertility of 2-year-old birds $= F_2$	Fertility of 3-year-old and older birds $= F \geq 3$
0.136	0.893	0.639	0.894

$$
\begin{vmatrix} & & \text{Matrix} & \\ 0 & 0 & S_{\geq 1}F_2 & S_{\geq 1}F_{\geq 1} \\ S_0 & 0 & 0 & 0 \\ 0 & S_{\geq 1} & 0 & 0 \\ 0 & 0 & S_{\geq 1} & S_{\geq 1} \end{vmatrix} = \begin{vmatrix} & & \text{Matrix} & \\ 0 & 0 & 0.571 & 0.798 \\ 0.136 & 0 & 0 & 0 \\ 0 & 0.893 & 0 & 0 \\ 0 & 0 & 0.893 & 0.893 \end{vmatrix}
$$

Source: Data from Schmutz et al. (1997).

We now ask, "How sensitive is population growth (or extinction risk) to particular demographic changes?" Specifically, will a particular change in survivorship or fertility have a large impact on the growth rate λ? Using the above matrix, we have substituted survivorship values from 0.136 to 1.00 for both hatchlings (S_0) and older birds ($S_{\geq 1}$), and fertility rates of 0.136–2.000 for 2-year-old (F_2) and 3-year-old and older birds ($F_{\geq 3}$). Figure 4.15 summarizes this analysis. Obviously, survival, especially that of the older birds, has the greatest effect on the growth rate (λ). Increases in fertility have virtually no effect. A related point is that, since adult survival is so critical to population growth, errors in our estimates would have a very large impact on our conclusions about this population. As it stands now, from this matrix, our estimate of λ is less than 1.000, and we expect this population to slowly decrease. A small change in our estimation of survivorship, however, would lead us to believe that this population is

Fig. 4.15 The effect of changing vital rates on the value of lambda in an emperor goose population. An increase in survivorship of adults (1-year-old and older birds) has the greatest effect. Increases in fertility have a negligible effect on λ.

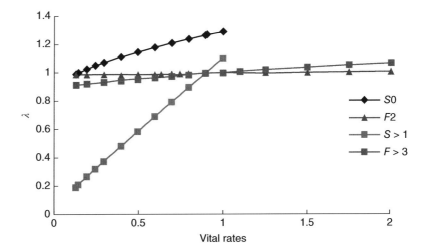

stable or growing. For example, a change in adult survival from 0.893 to 0.905 changes λ from 0.989 to 1.001. A change in juvenile survivorship from 0.136 to 0.155 changes the expected λ to 1.000.

An examination of this sort is known as a perturbation or "sensitivity analysis." Although Figure 4.15 is convincing, conservation biologists have sought to summarize the kind of analysis we have done earlier into a single number that would summarize the sensitivity of λ to particular vital rates. The most common basic measure of sensitivity is the slope of the tangent taken on the curve of λ as a function of each vital rate. Problems with this approach include the possibility of nonlinearity in the relationship between λ and a particular vital (survivorship or fertility) rate. A second issue is the scaling of sensitivity values. Obviously survivorship scales on a strict 0–1.0 scale, while reproduction can scale to very large numbers (number of acorns produced by an oak tree). These comparisons can be made more meaningful by examining the proportional change in λ as a proportion of change in the vital rates. These calculations result in a measure known as **elasticity**. Elasticity, then, is a standardized sensitivity that measures the effects of proportional changes in vital rates. That is, elasticities tell us the effect of perturbations in vital rates that are all of the same relative magnitude. Elasticities are standardized to sum to 100%.

For the emperor goose population discussed earlier, the majority (92% of elasticity) of sensitivity was in the survival of the 1-year-old and older birds (Morris and Doak 2002). An analysis of the Amboseli baboon population by Alberts and Altmann (2003) led to similar conclusions. That is, fertility represented just 9% of the total elasticity for both males and females. Survival of the prereproductive age classes accounted for 37% of the total elasticity for females and 62% for males. Details on the calculation of both sensitivity and elasticity values can be found in Morris and Doak (2002) or in Alberts and Altmann (2003).

What should be done to promote the long-term survival of these two populations or of other populations described in the first paragraph of this chapter? What evolutionary forces have led to a particular life history in the first place? These are just two of many questions for which there are no easy answers. Still, the analyses outlined in this chapter should have given you the tools necessary to at least begin to address these issues.

References

Alberts, S.C. and J. Altmann. 2003. Matrix models for primate life history analysis. Pages 66-102 in P.M. Kappeler and M.E. Pereira, Editors. *Primate Life Histories and Socioecology*. University of Chicago Press, Chicago.

Alstad, D. 2001. *Basic Populus Models of Ecology*. Prentice Hall, Upper Saddle River, NJ.

Anderson, J.D., D.D. Hassinger, and G.H. Dalrymple. 1971. Natural mortality of eggs and larvae of *Ambystoma t. tigrinum*. *Ecology* 52: 1107–1112.

Anonymous. 1988. *National Center for Health Statistics: Vital Statistics of the United States,1986*, Vol. 2, Sec.6, Life Tables. DHHS Pub. No. 88-1147. Public Health Service, Washington DC

Anonymous. 2010. U.S. Census Bureau http://www.census.gov/hhes/fertility/data/cps/2010.html

Baker, M.C., L.M. Mewaldt, and R.M. Stewart. 1981. Demography of white-crowned sparrows (*Zonotrichia leucophrys nuttalli*). *Ecology* 62: 636–644.

Barkalow, F.S., R.B. Hamilton, and R.F. Soots. 1970. The vital statistics of an unexploited gray squirrel population. *Journal of Wildlife Management* 34: 489–500.

Botkin, D.B. and R.S. Miller. 1974. Mortality rates and survival of birds. *The American Naturalist* 108: 181–192.

Case, T.J. 2000. *An Illustrated Guide to Theoretical Ecology*. Oxford University Press, Oxford, UK.

Caswell, H. 1989. *Matrix Population Models: Construction, Analysis and Interpretation*. Sinauer Associates, Sunderland, MA.

Caughey, G. 1966. Morality patterns in mammals. *Ecology* 47: 9067–9018.

Caughey, G. 1977. *Analysis of Vertebrate Populations*. Wiley, New York, NY.

Deevy, E.S. 1947. Life tables for natural populations of animals. *The Quarterly Review of Biology* 22: 283–314.

Dingle, H. 1990. The evolution of life histories. Pages 267–289 in K. Wohrmann and S.K. Jain, Editors. *Population Biology: Ecological and Evolutionary Viewpoints*. Springer-Verlag, Berlin.

Dittus, W.P.J. 1977. The social regulation of population density and age-sex distribution in the toque monkey. *Behavior* 63: 281–322.

Fisher, R.A. 1930. *The genetical theory of natural selection*. Clarendon Press, Oxford, UK.

Gibbons, J.W. 1987. Why do turtles live so long? *Bioscience* 73: 262–269.

Grant, P.R. and B.R. Grant. 1992. Demography and genetically effective sizes of two populations of Darwin's finches. *Ecology* 73: 766–784.

Hubbell, S.P. and P.A. Werner. 1979. On measuring the intrinsic rate of increase of populations with heterogeneous life histories. *The American Naturalist* 113: 277–293.

Hutchinson, G.E. 1978. *An Introduction to Population Ecology*. Yale University Press, New Haven, CT

Janzen, D.H. 1976. Why bamboos wait so long to flower. *Annual Review of Ecology and Systematics* 7: 347–391.

Jarman, M.V. and P.J. Jarman. 1973. Daily activity of impala. *East African Wildlife Journal* 11: 75–92.

Lanciani, C.A. 1998. A simple equation for presenting reproductive value to introductory biology and ecology classes. *ESA Bulletin* 79: 192–193.

Laughlin, R. 1965. Capacity for increase: a useful population statistic. *Journal of Animal Ecology* 34: 77–91.

Lefkovitch, L.P. 1965. The study of population growth in organisms grouped by stages. *Biometrics* 21: 1–18.

Leslie, P.H. 1945. On the use of matrices in certain population mathematics. *Biometrika* 35: 183–212.

Leslie, P.H., J.S. Tener, M. Vizoso, and H. Chitty. 1955. The longevity and fertility of the Orkney vole, *Microtus orcadensis*, as observed in the laboratory. *Proceedings of the Zoological Society of London* 125: 115–125.

Leverlich, W.J. and D.A. Levin. 1979. Age specific survivorship and reproduction in *Phlox drummondii*. *The American Naturalist* 113: 881–903.

Lotka, A.J. 1956. *Elements of Physical Biology*. Dover, New York, NY.

May, R.M. 1976. Estimating r: a pedagogical note. *The American Naturalist* 110: 496–499.

Mertz, D.B. 1970. Notes on methods used in life history studies. Pages 4–17 *in* J.H. Connell, D.B. Mertz, and W.W. Murdoch, Editors. *Readings in Ecology and Ecological Genetics.* Harper and Row, New York, NY.

Morris, W.F. and D.F. Doak. 2002. *Quantitative Conservation Biology, Theory and Practice of Population Viability Analysis.* Sinauer Associates, Sunderland, MA.

Pearl, P. 1927. The growth of populations. *Quarterly Review of Biology* 2: 532–548.

Pefranka, J.W. and A. Sih. 1986. Environmental instability, competition, and density-dependent growth and survivorship of a stream-dwelling salamander. *Ecology* 67: 729–736.

Peters, G.L. and R.P. Larkin. 1989. *Population geography*, 3rd Edition. Kendall/Hunt, Dubuque, IA.

Ralls, K, R.L. Brownell, Jr., and J. Ballou. 1980. Differential mortality by sex and age in mammals, with specific reference to the sperm whale. Rep. Int. Whaling Comm. (Special Issue 2): 233–243.

Schmutz, J.A., R.F. Rockwell, and M.R. Peterson. 1997. Relative effects of survival and reproduction on the population dynamics of emperor geese. *Journal of Wildlife Management* 61: 191–201.

Sherman, P.W. and M.L. Morton. 1984. Demography of Belding's ground squirrels. *Ecology* 65: 1617–1628.

Silvertown, J.W. and J.L. Doust. 1993. *An Introduction to Plant Population Biology.* Blackwell Scientific, Oxford, UK.

Sinclair, A.R.E. 1977. *The African Buffalo: A Study of Resource Limitation of Populations.* University of Chicago Press, Chicago, IL.

Smith, R.L. 1996. *Ecology and Field Biology*, 5th Edition. Harper Collins, New York, NY.

Thompson, P.M., J.N. Giedd, and R.P. Woods. 2000. Growth patterns in the developing brain detected by using continuum mechanical tensor maps. *Nature* 404: 190–3.

Turner, Monica G., William H. Romme, and Daniel B. Tinker. 2003. Surprises and lessons from the 1988 Yellowstone fires. *Frontiers in Ecology and the Environment* 1: 351–358.

Werner, P.A. 1975. Predictions of fate from rosette size in teasel (*Dipsacus fullonum* L.). *Oecologia* 20: 197–201.

Werner, P.A. and H. Caswell. 1977. Population growth rates and age vs. stage-dependent models for teasel (*Dipsacus sylvestris* Huds.). *Ecology* 58: 1103–1111.

5 Metapopulation ecology

5.1 Introduction

On August 27, 1883, Krakatau, an island about the size of Manhattan located between Sumatra and Java, underwent a series of volcanic eruptions releasing as much energy as 100 megatons of TNT (Wilson 1992). Magma, ash, and rock flew 5 km into the air and fell back into the sea, creating a tsunami 40 m in height, washing away villages in Java and Sumatra, killing 40 000 people. Waves were still a meter high when they came ashore in Sri Lanka. A total of over 18 km^3 of rock and ash were thrown into the air with dust and sulfuric acid aerosol reaching 50 km in the stratosphere, where its effects were seen as brilliant sunsets for several years thereafter. All of this airborne material produced "darkness at noon" in areas near the former Krakatau.

Only the southern end of Krakatau remained. This island, which became known as Rakata, was covered by pumice 40 m thick. The pumice had been heated to between 300 and 850 °C, and all living things had been destroyed; Rakata was a sterile island. Yet living things soon began colonizing this lifeless rock. Nine months after the explosion, a visitor found a small spider spinning its web. In the fall of 1884, a year after the eruption, biologists found a few shoots of grass. By 1886, there were 15 species of grasses and shrubs; by 1897, there were 49 species; and by 1928, around 270 species of plants were found, about 200 of which were seed plants and 50 were ferns (MacArthur and Wilson 1967). According to Schoener's 2010 review, subsequent surveys found that by 1983 the number of species of both seed plants (Bush and Whittaker 1991) and ferns (Thornton et al. 1993) had slowly moved toward a dynamic equilibrium at similar numbers of species although the cumulative number of species recorded over time was over 300 (Bush and Whittaker 1991). However, immigration and extinction curves

Introduction to Population Ecology, Second Edition. Larry L. Rockwood.
© 2015 John Wiley & Sons, Ltd. Published 2015 by John Wiley & Sons, Ltd.
Companion Website: www.wiley.com/go/rockwood/populationecology

estimate about two new species of higher plants (including ferns) arriving per annum with an estimated one species going extinct per year. It should be noted, however, that neither the immigration nor the extinction curves are monotonic as predicted by MacArthur and Wilson (1967) (see Bush and Whittaker 1991, Schoener 2010).

What Wilson (1992) described in the earlier paragraphs sounds like a typical successional sequence, proceeding from a community of colonizing species to a mature or "climax" forest. What we want to emphasize, as we consider the topic of metapopulations, is the two processes at work on Rakata. Obviously, one of those processes is colonization. New species continuously arrive on this island from the nearby mainlands. The other process is local extinction. Some species that arrived on this island, and were recorded early in the 20th century, are no longer present. Again, this may not be surprising to students of succession. So-called climax species are supposed to outcompete and eliminate earlier successional species. But is that what happened? MacArthur and Wilson's theory of island biogeography proposed a different model, which we explore later. We should note, however, that a species-by-species analysis by Whittaker et al. (2000) concluded that successional theory accounts for most of the extinctions.

The bird community is, perhaps, more to the point. In 1908, 13 species of birds were recorded in Rakata; by 1920, there were 31 species; and by 1933, 30 species were found. Wilson (1992) believed that an "equilibrium" number for Krakatau was approximately 30 species of birds, and that number had been reached in about 36 years. More important, however, is that the actual composition of the bird community has not remained stable. During the interval between the 1920 and 1933 surveys, five species of birds went extinct on Krakatau, to be replaced by four new species (MacArthur and Wilson 1967). For example, the bulbul (*Pycnonotus aurigaster*) and the gray-backed shrike (*Lanius schach*) had become extinct in Rakata between 1920 and 1933.

Surveys of Rakata in 1963 and 1983 show an actual decline in bird species richness of about 20% since 1951 (Bush and Whittaker 1991), with extinction rates hovering around 0.25%–0.42% species per annum (Schoener 2010). Again, Bush and Whittaker (1991) assert that these changes in the avifanua are largely attributable to successional change, not the processes described in the MacArthur and Wilson (1967) theory.

Nevertheless, the history of Krakatau illustrates two major points: (1) local populations are continuously subject to the twin processes of colonization and extinction and (2) communities are continuously changing. Even when the number of species in the community is static, the composition of the community is not. For example, butterfly counts have been carried out at the Environmental Studies of the Piedmont field site near Warrenton, Virginia, for 16 years (Wood 2012). Over that time period, 68 species have been recorded, but no more than 43 and no less than 21 in any given year. The mean number of species recorded is 31. Only three species have been recorded in every year of the survey. Some of these results may be due to sampling error or inconsistent sampling, yet they illustrate the point that community composition is not necessarily a constant at a given site (Table 5.1).

5.2
Metapopulations and spatial ecology

Many of the population models we have examined, particularly the deterministic models, have underlying assumptions that natural populations are numerous, widespread, and occupy contiguous habitats. The reality is that these assumptions may never have been true for certain populations, and that the natural world is now increasingly fragmented due to human activities. Wildlife populations are now more likely to be

Table 5.1 Results from 16 annual butterfly counts at Environmental Studies of the Piedmont near Warrenton, Virginia.

Year	Total number of individuals sampled	Total number of species found
1997	609	34
1998	674	28
1999	546	24
2000	958	26
2001	632	31
2002	2245	38
2003	1387	36
2004	713	28
2005	874	30
2006	248	25
2007	1122	24
2008	341	21
2009	1448	37
2010	1287	43
2011	930	37
2012	1114	36
Mean	945.5	31.1 ± 6.3
	Total number of species over 17-year period	68

small, restricted in distribution, and increasingly isolated from each other. Partially as a reaction to these realities and partially because of increasingly sophisticated theoretical developments, ecologists have begun stressing the importance of the spatial context in populations, communities, and ecosystems. What we can broadly call spatial ecology is the progressive introduction of spatial variation and complexity into ecological analysis, including changes in spatial patterns over time. Krakatau/Rakata is an example of both temporal and spatial complexity. Not only has its communities and populations changed over time, but the community found in present-day Rakata is different from the nearby mainland forests of Java and Sumatra. The local community of Rakata remains distinct from that of the mainland (Wilson 1992) and from the nearby islands of Panaitan and Peucang (Bush and Whittaker 1991).

Spatial ecology is distinguished by two different approaches: (1) landscape ecology and (2) metapopulation ecology.

Landscape ecology usually focuses on a larger geographic scale than traditional ecological studies; it was founded largely by community- or ecosystem-oriented ecologists, geographers, and landscape planners. Landscape ecology explicitly recognizes the heterogeneity or "patchiness" of the environment both spatially and over time. It provides a large-scale perspective that describes the physical structures of patchy environments as well as the movements of both individuals and resources among them. Generally, landscape ecologists focus at the community level, as opposed to the population level. Furthermore, landscapes have a more complex structure than usually allowed in simple metapopulation models, with habitat suitability being on a continuous scale, rather than simply "suitable" or "unsuitable" (Hanski 1999). Landscape ecologists do not usually work on population dynamics (Turner et al. 2001), but much of their work is relevant to metapopulations. For example, both landscape and metapopulation models often attempt to incorporate the roles of edge habitats, movements of individuals between patches via habitat corridors, spatial location of the patches, habitat fragmentation, landscape disturbance, and spatial and temporal variation in the quality of the habitat.

The metapopulation approach begins by stressing that local populations are influenced by immigration/emigration and extinction, as well as by birth and death processes. Until the 1960s, the idea that populations might routinely go locally extinct was rarely discussed in the literature. However, the population geneticist, Wright (1940), as well as ecologists such as Andrewartha and Birch (1954), introduced the notions that populations are connected by migration, and that local extinctions might be commonplace (Hanski 1999). The importance of immigration and emigration to the long-term persistence of a local population, however, was first emphasized by Levins (1970), who coined the term "metapopulation." For Levins, a metapopulation was a population consisting of many local populations. He asserted that all local populations have a finite probability of extinction, and long-term survival of a species was at the regional or metapopulation level (Hanski 1999). Beginning in the 1990s, as it became obvious that the natural world was becoming increasingly fragmented, the metapopulation approach became standard in the world of conservation biology. An understanding of metapopulations, the probabilities of local extinctions in different-sized natural reserves, and the rates of immigration and emigration between these preserves, became one of the fastest growing research areas in population, community, landscape, and conservation biology. As currently defined, metapopulations are regional assemblages of plant and animal species, with the long-term survival of the species depending on a shifting balance between local extinctions and re-colonizations in the patchwork of fragmented landscapes.

In landscape ecology, we begin with the assumption of complex environmental heterogeneity, whereas in a simple metapopulation analysis, the landscape, from the perspective of a given species, is assumed to consist of discrete patches of suitable habitat, surrounded by unsuitable areas. All of these patches are of the same quality and size, and while these patches are isolated from each other, there is no special recognition given to how far apart they are or any measure of ease of movement from one patch to another (connectivity). Nor are local population dynamics emphasized. The concern is simply whether a patch is occupied or not, the extinction rate on patches, and the overall colonization rate among patches. From this simple approach, more complex spatially realistic models have been developed to recognize differences in habitat quality and size, local population dynamics, and differences in connectivity among local patches.

Spatial ecology includes a variety of approaches, including the so-called lattice models. In patch or lattice models, the habitats occupied by local populations are represented in continuous space or made up of a series of spatially subdivided cells or "lattice" segments. These models include explicit spatial locations for the habitats. In patch models, the habitats have just two states, occupied or empty, and no effort is made to estimate the size of the population in the occupied patches. Cells change states according to simple rules, and stochastic extinctions and colonizations take place. Many of these simplifying rules can be relaxed in a spatially explicit variation of this approach, known as the *incidence function model* (IFM), examined later in this chapter (Hanski 1999).

The spatial distribution of a species is based on environmental patchiness. However, the recognized patterns of spatial distribution (clumped, random, or regular) only describe the current spatial pattern; they do not address the underlying causation of the pattern, or the long-term persistence of the population at that site.

The term "metapopulation" has been used for any spatially structured population, and "metapopulation dynamics" has been used to refer to any population dynamics involving spatial patterns (Hanski 1998). Furthermore, as Harrison (1994) pointed out several different types of metapopulations exist: classical metapopulations, mainland-island metapopulations, nonequilibrium metapopulations, and patchy metapopulations. We will explore these later in this chapter.

As elucidated by Tilman and Kareiva (1997), we must recognize that an individual organism only interacts with its local environment and with the competitors and/or predators present in that local environment. Such realities have been largely ignored by ecologists in the past, particularly when deriving theoretical models for competition or predation. Both classical and modern studies of competition and predator–prey interactions can be easily integrated into a simple metapopulation (spatial) context. For example, envision the environment as a series of patches and apply the competitive exclusion principle (to be elaborated in a later chapter). If we have a superior competitor that drives other species extinct on a given patch, the inferior competitors are only driven locally extinct. They can remain in the region as long as they can have a higher dispersal rate than extinction rate, and there are empty habitat patches. Similarly, a predator may drive a prey species locally extinct, but the prey population remains regionally present as long as it continues to colonize empty habitat patches faster than its predator. Although this is an oversimplification of models presented later, consider the following two examples.

Huffaker (1958) worked with orange mites and their predators in the laboratory. He found that coexistence of the two species was impossible on any given orange. Here one orange represents a small homogeneous habitat or patch. Through the addition of spatial complexity, however, as well as barriers to limit the rates of movements between patches, the orange mite and its predator co-existed in the laboratory for many months. A higher dispersal rate and environmental complexity allowed the prey species to remain regionally present, although it was continually driven locally extinct on a given orange once the predator arrived. Notice the parallels to epidemiological and host–parasite interactions. An unoccupied, but suitable, habitat patch is the equivalent to an individual susceptible to a parasite, yet not infected. An infected individual is the equivalent of an occupied habitat patch.

Spatial complexity can also have important effects on competitive interactions. For example, spatial complexity can help explain the coexistence of more species than expected based on the theory that the number of coexisting species should not exceed the number of limiting resources (Hutchinson 1961). As pointed out by Lehman and Tilman (1997), usually there is a trade-off, often expressed in terms of energetic investments, between competitive ability and colonization ability. While a superior competitor may take over a given site, if the less competitive species is a better colonizer, it may simply escape to a new site before competitive exclusion can occur. For example, Hubbell et al. (1999) proposed that high tree diversity on Barro Colorado Island in Panama is due, at least in part, to the low dispersal ability of competitively dominant species. Through a combination of low local abundance, low dispersal, and chance events, many plant species are absent from the local neighborhood in which a tree is located. Many sites are colonized by "default" species that were not the best competitor for the site. For example, an individual tree sapling competes with only 6.3 neighbors on average. Thus, plants compete only with those individuals sufficiently nearby to shade them or whose roots overlap with theirs in the soil. In order to "win" locally, a tree must only compete with those species that have "shown up" in the local neighborhood. Inferior competitors "win" by default. Because winners are only the best competitors that happen to have colonized a specific site, this process can lead to an almost unlimited diversity (Tilman 1994, Hubbell et al. 1999).

While landscape ecology and metapopulation ecology have started at different scales and with different assumptions and traditions, these two branches ask similar questions. Many landscape ecology courses include sections on metapopulations. An important future task will be the reconciliation of these two approaches and the establishment of common methodologies and principles.

5.3 MacArthur and Wilson and the equilibrium theory

Spatial ecology has its roots in the MacArthur and Wilson equilibrium (or dynamic) theory of island biogeography. MacArthur and Wilson (1963, 1967) brought a quantitative theoretical framework to the study of biogeography. As stated by Losos and Ricklefs (2010, p. xi), "*The Theory of Island Biogeography* was a compelling call for a comprehensive re-fashioning of biogeographical thinking."

Even before Darwin carried out his pioneering work on the Galapagos, island examples have been of great importance in biology and islands have been analyzed as natural laboratories and experimental systems. Islands are small, contained ecosystems in which certain species found in continental ecosystems may be missing. The lessons learned from examining islands can also be applied to those continental areas that are comparable to islands. That is, streams, lakes, tidal pools, caves, and mountaintops can be thought of as habitat islands in a "terrestrial sea." The approach of island biogeography has also been applied to host animals as habitat patches for parasites. Finally, as noted earlier, the natural world is increasingly fragmented, surrounded by roads, agricultural crops, shopping malls, industrial sites, and urban development. As conservation biologists became increasingly aware that wildlife preserves were essentially islands, a set of rules for the design of natural areas was inferred from the MacArthur and Wilson theory (Diamond 1975; Terborgh 1975; Wilson and Willis 1975; Willis 1984).

In his recent book, Letters to a Young Scientist, Wilson (2013) describes how he and MacArthur developed the theory of island biogeography. He quotes a letter he received from MacArthur as follows (now it would be an email, I assume!).

"Start with an empty island. As it fills up with species, there are fewer species available from other islands to arrive as immigrants, so the rate of immigration falls. Also, as the island fills up with species, it becomes more crowded and the average population size of each species decreases. As a result, the rate of species extinction rises. Therefore, as the island fills up, the immigration rate falls, and the extinction rate of the species already present rises. Where the two curves cross the extinction rate equals the immigration rate and the number of species is at equilibrium" (Wilson 2013, pp. 226–227). There, in one paragraph, is the essence of the MacArthur and Wilson (1963, 1967) theory.

The basic principles derived from the MacArthur and Wilson theory are as follows:
1. There is a relationship between habitat island area and the number of species found there (the species-area curve).

2. Local extinction is a normal, common occurrence, particularly on small islands with small populations.
3. Local diversity is based on interplay between colonization from a "mainland" source of species and local extinction, resulting in an "equilibrium" number of species.
4. Island size and distance from the source of species will affect the "equilibrium" number of species. That is, large islands that are close to the mainland will have more species than small islands, far from the mainland.

A simple differential equation captures the essence of the model (Hanski 2010):

$$\frac{dS}{dt} = I(P - S) - ES \tag{5.1}$$

where S = the number of species on the island, P = the pool of species on the mainland, I is a colonization rate parameter, and E is the extinction rate parameter. This equation assumes that all species have the same colonization and extinction rates. At equilibrium, found by setting $dS/dt = 0$, the result is as follows:

$$\hat{S} = \frac{IP}{(I + E)} = P\left\{\frac{I}{(I + E)}\right\} \tag{5.2}$$

The implication here is that unless the extinction rate is zero, the number of species on the island will always be less than the number of species on the mainland. Ricklefs (2010) points out that if the species pool were much larger than the number of species on the island, the steady state diversity $\hat{S}$ simplifies to I/E times P.

The relationship between number of species on an island and the area of the island is one of cornerstones of island biogeography theory. The species-area relationship has been discussed since the 19th century, and MacArthur and Wilson (1967) proposed that the number of species on an island could be approximated by the equation:

$$S = CA^z$$

where S = the number of species on the island, A = the area of the island, C = a constant (the y-intercept, see later), and z = a constant that remains fairly consistent within a taxonomic group and/or the types of islands being considered.

The above equation can be log transformed as follows:

$$\log S = \log C + z \log A \tag{5.3}$$

This is an equation for a straight line with a slope = z and with $\log C$ as the y-intercept. Thus, if data are gathered on the area of islands of different sizes and on the number of species on each island, a regression of the log-transformed data will produce a linear equation with slope z. The slope is relatively consistent within a taxonomic group but also depends on the type of island system. That is, the z-value depends on whether we are dealing with true oceanic islands, recently isolated islands ("land-bridge" islands), or habitat islands. According to MacArthur and Wilson (1967), z-values range from 0.20 to 0.40 for oceanic islands, 0.12 to 0.25 for arbitrary portions of the mainland, and greater than 0.26 for habitat islands (Gould 1979; Quinn and Harrison 1988). Preston (1962) showed that a z-value of 0.26 is expected when the log of species abundance versus the number of species has a normal distribution.

Gould (1979) pointed out that a slope of 0.25 is extremely common for species area curves. What is of interest are those z-values differing significantly from 0.25. When we simply sample larger and larger areas of habitats not isolated from each other, the z-values are theorized to be smaller than the expected value of 0.25. When small areas are sampled, they include a number of transient species passing through the area, raising the number of species. The result is a smaller than expected rise in the number of species with increasingly large sample areas. Thus, ants from nonisolated continental areas in New Guinea have a z-value of 0.17 (Wilson 1961), mammals from the Sierra Nevada in California have a z-value of 0.12 (Brown 1971), and birds from the Great Basin of the United States have a z-value of 0.17 (Brown 1978).

By contrast, larger than expected z-values arise when islands contain great habitat diversity, with semi-isolated, unique biota encountered as sample areas are increased. Examples include terrestrial invertebrates found in caves ($z = 0.72$, Culver et al. 1973), mites on cushion plants ($z = 0.42 - 0.69$; Tepedino and Stanton 1976), and mammals on isolated mountain tops ($z = 0.43$, Brown 1971; $z = 0.33$, Brown 1978). Lawrey (1991, 1992) has suggested that pollution, by reducing interspecific competition, produces larger than expected z-values for lichens species on rocks of differing sizes. While z-values varied from 0.16–0.21 for six undisturbed sites, a site disturbed by air pollution near the Capital Beltway in Maryland yielded a species area curve with a z-value of 0.28. For more discussion on the slope of the species-area relationships, see Ricklefs (2010) and Schoener (2010).

Some scientists have asserted that as islands get larger the topography becomes more complex, there are more habitats, and therefore, we have more species. In their study of red mangrove islands, however, Simberloff and Wilson (1969, 1970) found that species number increased with island size alone and was unrelated to habitat diversity. For an overview of the strength of island area versus habitat diversity in promoting species diversity, see Ricklefs and Lovette (1999) or Schoener (2010).

The number of species found on an island, according to MacArthur and Wilson, was due to two contrasting processes of (1) **immigration** and (2) **extinction**. Extinction was envisioned as a normal, locally common event, while new species were added through immigration form the mainland. Diversity was the result of the equilibrium between immigration and extinction. Furthermore, the theory indicated that once the "equilibrium" number of species was reached, the only constant was the **number of species** in the community, not the identity of the species involved (Figure 5.1). As extinction is a locally common process, there should be a regular "turnover" in the species found on the island.

The expected number of species on an island is affected not only by the area of the island but also by the distance of the island from the source of species. Immigration rates are lower on smaller islands and on islands farther from the "mainland" source of species. By contrast, immigration rates are higher on larger islands and on islands closer to the "mainland." Extinction rates are expected to be higher on small islands, because average population sizes are smaller (Wilson 1992).

The rate at which new immigrant species establish themselves on the island falls as the number of species on the island increases. As more species become established on the island, fewer individual immigrants will belong to a species not already present; moreover, it will be harder for a new species to successfully colonize due to competition with the already established species. Species with high dispersal rates are those that

Fig. 5.1 Immigration and extinction curves from the island biogeography model of MacArthur and Wilson (1967). S is the equilibrium number of species, where the two curves intersect.

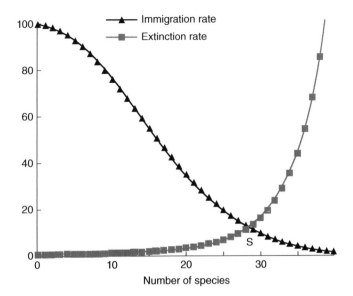

arrive quickly, while those with lower dispersal rates arrive more slowly. Owing to the proposed colonization-competition tradeoff, the species with lower dispersal rates are likely more competitively dominant.

The extinction curve rises as more species arrive on the island. The more species that are present, the more that can become extinct. But, again, as more species are present, competition increases and the average population size per species declines, leading to an increased probability of extinction. Finally, if succession proceeds to a "climax" stage, the community will be saturated with species. At equilibrium, the number of species will be constant on the island, although some new species will continue to arrive while others will go extinct. Again, see Losos and Ricklefs (2010) for updated information on the preceding paragraphs.

The MacArthur and Wilson equilibrium theory captured the imagination of ecologists, conservation biologists, and biogeographers, making it the leading paradigm for the spatial dynamics of species during the 1980s. It shares much of the conceptual framework of metapopulation biology. Both view nature as subdivided into discrete fragments of suitable habitat, both view local populations as subject to stochastic processes and prone to extinction, and both stress the importance of movements of individuals between habitats (or islands). A key difference is that island biogeography stressed the community property of diversity rather than focusing on the dynamics of individual populations. Furthermore, island theory was developed to explain patterns at large spatial scales as opposed to fragmentation of landscapes at small scales (Hanski 2002).

The MacArthur and Wilson model is now categorized as a **mainland-island metapopulation**. There is a constant source of species, the mainland. The mainland population is seen as permanent, with no chance of extinction. Furthermore, dispersal is one-way. Species move from the mainland to the island; the reverse is not significant. Finally, no movement from one island to another is included in this type of metapopulation.

5.4 The Levins or classical metapopulation

According to the Levins model, metapopulation persistence is due to a stochastic balance between local extinction and recolonization of empty habitat patches. The rate of change in occupied habitat patches is a function of colonization rates (c) and extinction rates (ε) as shown in equation 5.4 (Levins 1969). P is the proportion of patches occupied by the population under consideration:

$$\frac{dP}{dt} = cP(1 - P) - \varepsilon P \tag{5.4}$$

As described by Hanski (2001), if we define P' as the number of habitat fragments occupied by the species (rather than the proportion), and define T as the total number of habitat patches available, the equation can be modified as follows (equation 5.5):

$$\frac{dP'}{dt} = cP'(T - P') - \varepsilon P' \tag{5.5}$$

Both of these models are deterministic descriptions of the rate of change of metapopulation size, even though the models are based on stochastic events. Assumptions include the following:
1. The local populations are identical and have the same behavior.
2. Extinctions occur independently in different patches and therefore local dynamics are asynchronous.
3. Colonization spreads across the entire patch network and all patches are equally likely to be "encountered."
4. Furthermore, all patches are equally connected to all other patches.

In the Levins model, we are not concerned with population dynamics within each population. We do not attempt to assess the number of individuals in each patch; we simply record a patch as occupied or not occupied. For this reason, we also do not assess the size or quality of the patches.

The equilibrium value of P can be obtained by setting $dP/dt = 0$. This produces the expected proportion of patches to be occupied and amounts to a carrying capacity term such as is found in the logistic equation:

$$0 = cP(1 - P) - \varepsilon P = P(c - cP - \varepsilon)$$

Since $P = 0$ is not an interesting solution, we have the following:

$$0 = c - cP - \varepsilon, \text{ and } \varepsilon = c - cP$$

The equilibrium value of P, defined as $\hat{P}$ and found by solving the above for P, is shown in equation 5.6.

$$\hat{P} = \frac{c - \varepsilon}{c} = 1 - \frac{\varepsilon}{c} \tag{5.6}$$

The implication here is that colonization must be greater than extinction or the equilibrium proportion of patches occupied will be zero, and the colonization rate must be greater than the extinction rate for persistence of the metapopulation. If we consider colonization a "birth" event and extinction a "death" event (thereby using $c - e$ as the equivalent of the growth rate, r, in the logistic equation) and we use $1 - e/c$

Fig. 5.2 Expected proportion of habitat patches occupied based on equation 5.7. In this example $p_0 = 0.50$ and 0.15, the colonization rate, $c = 0.75$, and the extinction rate, $e = 0.55$. The expected proportion of patches occupied at equilibrium $= 1 - \varepsilon/c = 0.27$.

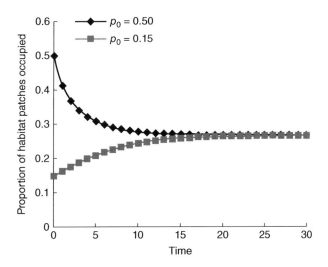

to represent a "carrying capacity" term (equivalent to K in the logistic equation) as mentioned earlier, we can model metapopulation dynamics as a modification of the logistic (equation 5.7). No matter what the starting patch frequency is (assuming $1 \geq P > 0$), over time it moves to the expected value based on $\hat{P} = 1 - \varepsilon/c$ (see Figure 5.2):

$$\frac{dP}{dt} = (c - \varepsilon)P\left(1 - \frac{P}{1 - \frac{\varepsilon}{c}}\right) \tag{5.7}$$

This simple model has helped ecologists develop insights into the consequences of habitat destruction and fragmentation. For example, imagine a fragmented landscape in which a fraction of the habitat patches is destroyed. The extinction rate is not affected, but the colonization rate is. This is because there are fewer local populations and fewer empty patches. If the patch connectivity is reduced, it can be modeled by reducing the value of c. Habitat destruction can therefore lead to a reduction in the proportion of patches that are occupied. Alternatively, if no patches are destroyed but they are reduced in area, this would result in lower average population sizes, which would increase the extinction rate. Simultaneously, colonization rate would be reduced due to the smaller population sizes in the occupied patches. The net result again is a reduction in the fraction of occupied patches (Figure 5.3a and b).

5.5 Lande's extension of the Levins model

Lande (1987, 1988) proposed an extension of the Levins model to account for habitat loss in which a fraction $(1-h)$ of habitat patches are permanently destroyed, and thus only the fraction h is suitable for colonization. If colonists are unable to discriminate between suitable and unsuitable patches, they will attempt to colonize both types. The colonization rate of empty patches $(1 - P)$ is diminished to cPh and the model becomes

$$\frac{dP}{dt} = cPh(1 - P) - \varepsilon P \tag{5.8}$$

P is now the fraction of occupied patches among the suitable patches.

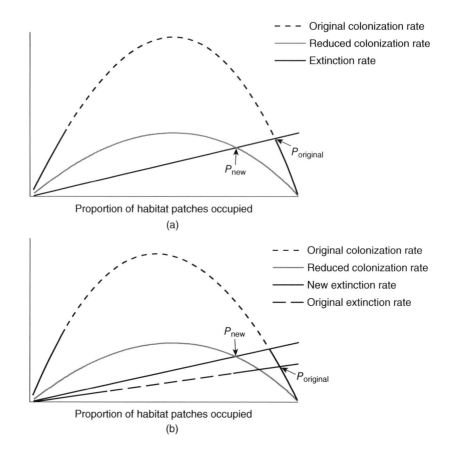

Fig. 5.3 (a) Graphical illustration of the effect on patch occupancy of a lowered colonization rate due to a reduction in the number of habitat patches. (b) Expected changes in patch occupancy with lower colonization rate and increased extinction rate. Colonization rate is reduced by loss of habitat number; extinction rate is increased by reduction in patch area. (Source: Data from Hanski 1999.)

As in the above equation, we find the equilibrium value, $\hat{P}$ by setting equation 5.8 equal to zero we have

$$0 = cPh(1 - P) - \varepsilon P = ch(1 - P) - \varepsilon \tag{5.9}$$

Solving for $\hat{P}$ we have

$$\hat{P} = \frac{ch - \varepsilon}{ch} = 1 - \frac{\varepsilon}{ch} \tag{5.10}$$

The result of interest here is that $\hat{P} > 0$ when $1 > \varepsilon/ch$ and $h > \varepsilon/c$.

The key prediction here is that metapopulation equilibrium is greater than zero if and only if the amount of habitat (h) exceeds the threshold value of ε/c.

In the Levins model, when $h = 1$, we had $1 > \varepsilon/c$, which means the condition for long-term persistence was $c > \varepsilon$. Here we conclude that long-term persistence in a fragmented landscape is facilitated by a large amount of suitable habitat, h, as well as by low extinction and good colonization.

If we designate P^* as the fraction of suitable habitat originally occupied and h_0 is the amount of original suitable habitat in the environment, we can substitute these values into equation 5.9 we have $ch_0(1 - P^*) - \varepsilon$. Rearranging, we have $\varepsilon/c = h_0(1 - P^*)$.

For persistence of the metapopulation as described earlier, h must be greater than ε/c. Thus, for long-term persistence, we have equation 5.11:

$$h > h_0(1-P^*) \tag{5.11}$$

This has led to what has been called Levins rule (Hanski et al. 1996). "The minimum amount of habitat necessary for long-term metapopulation persistence is the amount of non-occupied habitat available when the species is still common."

Carlson (2000) applied this rule to the woodpecker, *Dendrocopos leucotos*, in Sweden and Finland. Using data from Poland where it is still common, Carlson found that h_0 was 0.66 and P^* was 0.81. Recall that h_0 is proportion of suitable habitat in the landscape, and P^* is the proportion of suitable habitats that are occupied. The result, using equation 5.11, is that h must be greater than 0.125. But, in Sweden and Finland, $h < 0.12$ and indeed, the populations have been severely declining. This is obviously a very simplistic analysis "but the results are of pedagogic value" (Hanski and Gaggiotti 2004, p. 81).

5.6 Extinction in metapopulations

In metapopulation dynamics, as well as in the MacArthur and Wilson theory, the extinction of a local population is not uncommon. But we are more interested here in the extinction of the metapopulation. Table 5.2 (Hanski 1998) summarizes the potential causes of both local and metapopulation extinctions.

We need only comment on the comparison of stochastic processes in local versus metapopulation extinctions. One of the assumptions for long-term metapopulation persistence is that the expected number of new populations generated by one existing population during its lifetime must be greater than one. That is, the replacement rate must be greater than one, as is true for a local population to persist. In a small metapopulation, however, all local populations may go extinct by chance. This is known as "extinction-colonization stochasticity" (Hanski 1998). This is an exact analogue to demographic, stochastic extinction of a local population (Chapter 1). This may happen even if the replacement rate is greater than one in both local populations and metapopulations. Regional stochasticity is due to processes such as large-scale weather patterns, which produce synchrony among the independent local populations. This effectively reduces the number of independent populations and makes metapopulation level persistence less likely.

5.7 Metapopulation dynamics of two local populations

Recall that in the discrete-time population model, when r was set >2.69 (Figure 2.19) the population underwent chaotic behavior (May 1974, 1976). However, if two such populations are connected to each other by migration, a number of interesting and unexpected changes occur (Hanski 1999).

Table 5.2 Potential causes of local and metapopulation extinctions.

	Local extinction	Metapopulation extinction
Stochastic processes	**(a)** Demographic **(b)** Environmental	**(a)** Extinction–colonization Interaction **(b)** Regional processes
Extrinsic causes	Habitat loss	Habitat loss and fragmentation

Source: Data from Hanski (1998).

Fig. 5.4 Population one. $N_0 = 1$, $r = 3$, $K = 3$; connected to population two at $t = 49$.

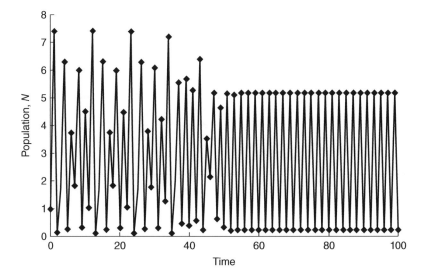

Fig. 5.5 Population two. $N_0 = 2$, $r = 3$, $K = 3$; connected to population one at $t = 49$.

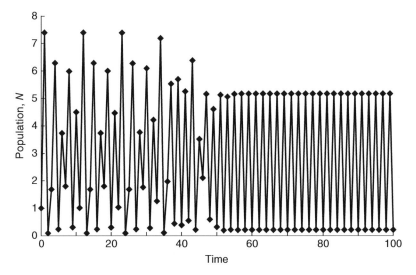

In this example, the Ricker model is used:

$$N_{t+1} = N_t e^{r\left(1 - \frac{N_t}{K}\right)}$$

(5.12)

In populations one and two, shown in the accompanying figures, $r = 3$ and $K = 3$. Population one (Figure 5.4) is initiated with one individual ($N_0 = 1$), while for population two (Figure 5.5) $N_0 = 2$. Each population and the metapopulation (Figure 5.6) behave chaotically. The metapopulation is simply the sum of populations one and two. At time = 49, the two populations are connected by allowing 30% of the individuals to emigrate. The emigrants are divided equally between the two populations. The migrations calm the chaotic behavior, and by time = 55 the two populations (Figures 5.4 and 5.5) have moved to a two-point limit cycle. The two populations go through cycles out

Fig. 5.6 Metapopulation of populations one and two. The two populations are connected at $t = 49$. The metapopulation is stabilized at $t = 55$.

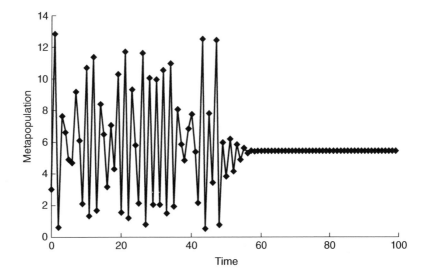

of phase with each other, and the metapopulation (Figure 5.6) is completely stabilized (Hanski 1999, Chapter 3).

Gyllenberg et al. (1993) have confirmed that migration can help stabilize local population dynamics although some mortality must occur during migration to have a stabilizing effect. Similarly, migration from a permanent (mainland) population or from a population with a low growth rate also has a stabilizing effect. Movement of individuals between local populations has, at least theoretically, a stabilizing effect on the local populations themselves as well as on the metapopulation. We are well advised, however, not to push this theoretical point too far in field populations (Hanski 1999).

5.8 Source-sink metapopulations and the rescue effect

The concept of source-sink metapopulations was put into the literature by Pulliam (1988) and is based on the fact that habitats are not uniform but differ in quality. High-quality patches produce large populations with positive growth rates and are likely to be a source of emigrants. These high-quality areas, where $r > 0$, are known as source patches. Other habitat patches are of low quality, have small populations, and consistently have negative growth rates. That is, populations in sink patches have a negative r in the absence of immigration (Hanski 1999). If and when migrants from source populations arrive at sink patches, they are either founders of new populations or they become new members of established populations.

The rescue effect is based on the idea that emigrants from source areas regularly supplement these small, extinction prone populations. If the expected size of the small population is increased through this supplementation, it becomes less prone to, or is rescued from, extinction (Brown and Kodric-Brown 1977). In a true sink habitat, a population would decline to extinction if cut off from its source population. A pseudo-sink population is one in which the population would decline to a lower equilibrium, but not go extinct, if cut off from its source population (Watkinson and Sutherland 1995). In practice, it is difficult to distinguish between a true sink and a pseudo-sink population in the field.

The MacArthur and Wilson (1967) island biogeography theory, a mainland-island metapopulation, has much in common with source-sink metapopulations in that mainlands are large compared to islands and, in theory, not prone to extinction, while islands are small and there is always a finite probability of extinction. The mainland is the source, while the islands are, collectively, sink populations. While MacArthur and Wilson (1967) emphasized that extinction is a normal process in a community, they envisioned the mainland as a more or less uniform patch when compared to the island populations, and that the mainland would function as a source indefinitely. Moreover, while extinction was assumed to occur on islands, the island populations were also assumed to have positive growth rates in a normal year (Elmhagen and Angerbjörn 2001).

One of the best illustrations of the source-sink concept comes from the study by Hubbell and Foster (1986) of a 50-hectare plot of Tropical Moist Forest on Barro Colorado Island in Panama. They mapped over 238,000 individual trees and shrubs of 314 species over a 13-year period. They found that at least one-third of the rare species were not self-maintaining populations. They were not reproducing effectively, and their presence in the plot was a result of immigration from outside of the 50-hectare plot.

Sink populations can be important to the long-term survival of a source population. Assume that the source population is prone to chaotic behavior or shows great fluctuations in size because of disease or sensitivity to environmental fluctuations such as drought or fire. In this case, the source population itself could be rescued if connected to a sink (Gyllenberg et al. 1996).

5.9 Nonequilibrium and patchy metapopulations

In examining the literature on metapopulation studies, Harrison (1991, 1994) found that the term has also been used to describe two other kinds of situations. In a nonequilibrium metapopulation, the rate of extinction among the populations exceeds the colonization rate. Without some change in the dynamics of the system, including perhaps restoration of functioning habitat patches, the ultimate fate of such a metapopulation is extinction.

By contrast, in a patchy population, the migration rate is so high that the so-called subpopulations function effectively as one single population. Hanski (1999) has asserted that patchy populations should be excluded from metapopulation theory since they lack distinct breeding subpopulations. The classical metapopulation of Levins is found between the extremes of patchy and nonequilibrium metapopulations (Elmhagen and Angerbjörn 2001).

5.10 Spatially realistic models

The Levins model assumed that local populations were identical; extinctions occurred independently in different patches; all patches were equally likely to be found; and patches were equally connected to each other. He did not try to describe how individuals move from one population to another, nor did he allow for differences in patch size or quality, in spite of the importance assigned to island size by MacArthur and Wilson.

Most metapopulation models also assume unconditional emigration. That is, there are no consequences to the source population from losing individuals. This seems to be a reasonable assumption in most populations since emigrants are often pictured as "extra" individuals when a population has reached or exceeded its carrying capacity. In many species, non-breeding males venture away from the family group looking for an empty territory or an existing group that they may join in order to become breeders. Females of butterflies and other insects emigrate in order to find newly available host plants for egg deposition. Male and female dung beetles emigrate when the dung pile they were born into has become depleted. If the mortality rate of dispersers, however, is much higher than that of individuals who remain "home," a very high population dispersal rate (large percentage dispersing) can lead to local, and sometimes metapopulation, extinction (Hanski 2002; Mickelberg, 2011).

The rate and the scale of recolonization of an empty habitat depend on the shape of the dispersal curve of the population. A simple model of animal movement is based on a random walk, using a coefficient of diffusion (D) and a normal distribution for movements, the variance of which increases with time ($2Dt$; Okubo 1980). Studies of animal emigration, however, indicate that more individuals move very short and very long distances than predicted by random walks (Johnson and Gaines 1990). In metapopulation models, emigration distances are usually modeled using a negative exponential function, which is a reasonable approximation of reality (Hanski 2002; equation 5.13; Figure 5.7). Two characteristics determine the dispersal efficiency of a species. One is the number of individuals dispersing (here equal to β) and the dispersal ability of each individual (α). Species differ a great deal in the amount of energy invested in reproduction each year. Obviously the more energy invested in reproduction, the greater the number of dispersal units. But, for a given reproductive effort, species also differ in whether they produce a smaller number of large offspring or a larger number of small offspring. The smaller the dispersing unit, the greater distance it is likely to travel. Many highly dispersed organisms are so small that the wind can carry them hundreds or even thousands of miles. On the other hand, acorns only move as far as gravity or squirrels will take them. In equation 5.10, α is directly associated with a greater colonizing ability per unit of dispersal, while β is associated with the

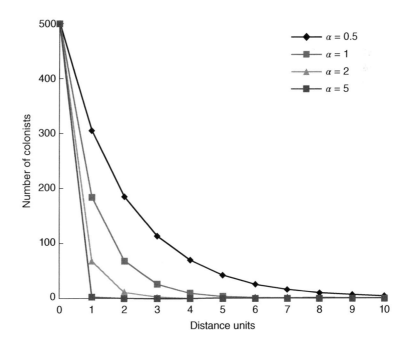

Fig. 5.7 Number of colonists arriving at different distances based on a negative exponential model, $C_i = \beta e^{-\alpha d_i}$, and using different alpha values. C_i = the number of colonists arriving at a distance d_i from the source population. β = 500 in all cases.

number of colonists produced per individual from the source population. C_i is the colonization probability per unit time for population i, and d_i is the distance from the source population.

$$C_i = \beta e^{(-\alpha d_i)} \tag{5.13}$$

where α and β are site and species specific parameters.

What is most important to a metapopulation is the proportion of individuals that move long distances. The exact shape of the distribution of migration distances among emigrants is therefore of great importance but is very hard to estimate in most populations (Hanski 2002).

We will discuss one approach that has been proposed to make metapopulation models more realistic. Such models describe spatially realistic metapopulation theory (SMT) (Baguette 2004). Hanski and Ovaskainen (2003) argue that this class of models is best represented by stochastic patch occupancy models (SPOMs), one example of which is the Incidence Function Model (IFM) championed by Hanski (1994a, 1994b, 1997, 1999). The idea of incidence function can be traced to Diamond (1975) according to Hanski (2010). The IFM is a stochastic patch model in which the population in each patch has one of two states, presence or absence. The IFM includes: (1) a finite number of habitat patches; (2) patches of different sizes (sometimes including differences in quality and shape); and (3) each patch has a unique spatial coordinate so that interactions among patches are localized in space. Since habitat patches are simply occupied or not, there is usually no estimation of population sizes or dynamics within patches. The major virtue of the IFM is that it is constructed so that parameters can be estimated from field data. This allows the application of this model to real populations.

The IFM begins with the assumption that for an empty habitat patch, i, there is a constant probability, C_i, of recolonization per unit time. If a patch is occupied, there is a constant probability, E_i, of extinction per unit time. One event, either colonization of extinction is allowed per time period. The long-term probability of the patch being occupied is called the "incidence" or J_i (Equation 5.14). The IFM is based on discrete time intervals and is a stochastic rather than a deterministic model (Hanski 2001):

$$J_i = \frac{C_i}{(C_i + E_i)} \tag{5.14}$$

There are a number of difficulties if we are modeling a true metapopulation with no "mainland" source of species. With no external mainland, metapopulation extinction is the only true steady state (Hanski 1999). However, a metapopulation may theoretically persist for very long periods.

Recall that in the rescue effect the probability of extinction on a habitat patch is reduced through immigration of individuals from other patches. In order to allow for the rescue effect, Hanski modified the probability of extinction between times t and $t + 1$ by substituting $(1 - C_i)E_i$ for E_i. This modified equation is shown as follows (equation 5.15):

$$J_i = \frac{C_i}{C_i + E_i - C_i E_i} \tag{5.15}$$

Hanski (1999) then derived a relationship between extinction probability, E_i, and the size of the patch area, A_i, using the basic reasoning of the species-area curve. That is, extinction probability depends on population size, which is a function of patch area. The general relationship is expressed in equation 5.16, where e and X are estimated from the data. The parameter X is a measure of environmental stochasticity (Hanski 1999):

$$E = \frac{e}{A^X} \tag{5.16}$$

Colonization probability, C_i, is a function of the number of immigrants, M_i. In a simple mainland-island metapopulation, the colonization probability is a function of distance (d_i) from the mainland, which was expressed as equation 5.13.

For a true metapopulation, however, M_i is the sum of individuals arriving from all of the surrounding habitat patches. M_i can be written as a summation for all patches (equation 5.17):

$$M_i = \beta S_i = \beta \sum_{j \neq i} e^{(-\alpha d_{ij}) p_j A_j} \tag{5.17}$$

In this equation, d_{ij} represents the distance between patches i and j; p_j is 0 for an unoccupied patch and 1 for an occupied patch; and A_j is the size of the patch. The summation term is represented by S_i, which becomes a measure of patch isolation or to put it positively, patch connectivity. If population sizes (N_j) are known for each patch, S_i can be written as 5.16:

$$S_i = \sum_{j \neq i} e^{(-\alpha d_{ij}) N_j} \tag{5.18}$$

As in equation 5.18, the term α describes how fast the immigration rate from patch j declines with distance. Hanski suggests that this term can be found through mark-recapture data.

If interactions among immigrants are negligible, C_i increases exponentially with M_i. However, there is often a sigmoid relationship between the number of immigrants and successful recolonization by a given species. Therefore, C_i can be written as in equation 5.19.

$$C_i = \frac{M_i^2}{M_i^2 + y^2} \tag{5.19}$$

where y is a parameter fitted from the data. In the following sections, the terms y and β are combined simply as y (Hanski 1999).

Once equations were developed for the dependence of extinction on patch size and colonization on patch connectivity, Hanski combined them into the usual form of the IFM (Equation 5.20):

$$J_i = \frac{1}{1 + \dfrac{ey}{S_i^2 A_i^X}} = \left[1 + \frac{ey}{S_i^2 A_i^X} \right]^{-1} \tag{5.20}$$

As before, J_i is the probability that a patch, i, is occupied. The value e is a parameter related to the probability of extinction per unit time in a patch of a given size, y is a parameter related to successful immigration, A_i represents the area of the patch i, X is the rate of change of extinction per unit time with increasing patch size (a measure of environmental stochasticity), and S_i describes the connectivity between patches. That is, the effect of distance on immigration rate. Equation 5.20 can be rewritten as follows:

$$J_i = \frac{1}{1 + e^{\ln(ey) - 2\ln S_i - X \ln A_i}} \tag{5.21}$$

If we manipulate and take natural logs of both sides of this equation, we come up with a linear relationship between the expected patch occupancy (J_i) and the two independent variables, connectivity (S_i) and size (A_i) of the habitat patches (equation 5.22):

$$\ln\left(\frac{J_i}{1 - J_i}\right) = -\ln(ey) + 2\ln S_i + X \ln A_i \tag{5.22}$$

Hanski (1999) has described how to estimate all of these parameters from field data and has applied the model to simulate metapopulation dynamics in butterflies (Hanski et al. 1995; Wahlberg et al. 1996), the American pika (*Ochotona princeps*; Moilanen et al. 1998), and a number of other species. The basic information needed in order to apply this model in the field is simply the area of each habitat patch and the interpatch distances (d_{ij}). Subsequently, model parameters must be estimated. First α is estimated from mark-recapture data or estimated from patch occupancy data. The parameters y and e are fitted to empirical data using nonlinear regression techniques. The value of X is fitted from the data and can be modified to include the rescue effect.

In Table 5.3, from Kindvall (2000), are the results of fitting the IFM to field data gathered from a fragmented population of the bush cricket (*Metrioptera bicolor*). Kindvall (1995) found that using occupancy data from a single year did not result in realistic predictions about the metapopulation. When parameters for the IFM were estimated from patch occupancy over a 5-year period better results were obtained.

Kindvall (2000) compared the IFM to three other spatially realistic models, including a logistic regression model by Sjögren-Gulve and Ray (1996) and found the logistic

Table 5.3 Results of fitting the incidence function model to occupancy data of the bush cricket (*Metrioptera bicolor*) for two areas of Sweden during the period 1989–1994.

Parameters	Western area	Eastern area
Number of available patches	66	50
Actual P	0.82	0.71
Predicted P	0.80	0.62
Multiplier of X For the Rescue Effect	0.05	0.001
α	2.0	6.0
X	0.876	0.514
Y	7.278	2.571
e	0.072	0.029

P = mean proportion of available habitats actually occupied from 1990 to 1994. Predicted mean proportion of patches occupied is based on 100 replicates
Source: Data from Kindvall (2000).

Table 5.4 The predicted and observed annual extinction and colonization rates for three species of shrews (*Sorex*) on small islands.

Species	Body size (g)	Parameter estimates		Predicted		Observed	
		X	E	C	E	C	E
S. araneus	9	2.30	0.20	0.26	0.04	0.20	0.04
S. caecutiens	5	0.91	00.53	0.03	0.28	0.05	0.33
S. minutus	3	0.46	0.73	0.18	0.53	0.13	0.46

Parameters for this mainland-island incidence function model were from 68 islands and applied to a different set of 17 islands. X and e are annual extinction parameters. C = annual colonization rate; E = annual extinction rate.

regression model performed best in its ability to predict regional occupancy, local occupancy, and the number of colonizations and extinctions. Nevertheless, he found the IFM did a reasonable job of predicting the actual outcomes for the bush cricket in Sweden.

The application of the IFM to mainland-island populations can be simpler since there are fewer parameters to estimate. One example is a study on the occurrence of small mammal populations on islands in lakes and in the sea (Peltonen and Hanski 1991; Hanski 1993). The IFM was based on the occurrence of three species of shrew (*Sorex*) on 68 islands. Using parameters estimated from this study, Hanski (1993) predicted the annual colonization and extinction probabilities on 17 additional islands. The observed and the predicted rates were well matched (Table 5.4). In this study, only the extinction parameters (X and **e**) were estimated since colonization was from the mainland and was assumed not to differ among islands (Hanski 1999). The value X is inversely related to the strength of environmental stochasticity. A large value of X means weaker environmental stochasticity. The value of X is directly correlated with body mass in the shrews described in Table 5.4, as well as in birds from four different areas (Cook and Hanski 1995). What this implies is that species with larger mass (such as *Sorex araneus*) are less affected by environmental stochasticity as compared with smaller massed species (Hanski 1999). Table 5.4 also suggests that only *S. araneus* has a long-term metapopulation survivorship, since it is the only species with a $C > E$.

5.11 Assumptions and evidence for the existence of metapopulations in nature

The different types of metapopulations described earlier (mainland/island, classical, source/sink, patchy, and nonequilibrium) are all variations on the same themes. Local extinctions are commonplace, there is an equilibrium involving colonization and extinction rates and so on. They differ in the levels of detail, whether they allow for patches to be of different quality, whether they allow for differing levels of connectivity between patches, and whether they include local population dynamics.

What are the general assumptions we are making in all of these models? How can we demonstrate that the long-term persistence of a species in a landscape is due to metapopulation, rather than local population processes? One difficulty is that long-term population data on patch occupancy are hard to gather. For this reason, many studies of population persistence in a fragmented landscape, meeting the criteria for a metapopulation, come from short-lived, easily monitored organisms. Accordingly, there has been an emphasis on butterfly populations such as the Glanville fritillary (*Melitaea cinxia*) and the Bay checkerspot (*Euphydryas editha*; Hanski 1999; Hanski et al. 2004; Ehrlich and Hanski 2004). On the other hand, metapopulation theory was famously, although perhaps inappropriately, used in designing a conservation plan for the northern spotted owl (Boyce 2002) and is being implemented in the management of both black and golden-lion tamarin (*Leontopithecus rosalia*) populations in South America (Mickelberg 2011).

Assumptions that metapopulation dynamics are decisive to the structure of the regional population include (Hanski and Kuussaari 1995, Hanski 1999) the following:

1. The species has local breeding populations in relatively discrete habitat patches. This condition stresses that the population is spatially structured, and therefore, most individuals interact with others only in the local habitat patch.
2. No single local population is large enough to have a longer expected lifetime than the expected lifetime of the metapopulation itself. This excludes mainland-island populations.
3. Empty habitat patches are common. In the Glanville fritillary butterfly study in Finland, for example, 70% of approximately 1600 habitat patches have been empty at a given time (Hanski et al. 1995).
4. The habitat patches are not too isolated to prevent recolonization. Long distance movements may be facilitated by habitat corridors or other mechanisms.
5. Local dynamics are sufficiently asynchronous to make simultaneous extinction of all local populations unlikely. With complete synchrony, the metapopulation only lasts as long as the local population with the lowest chance of extinction. The greater the asynchrony, the longer the metapopulation is likely to last. In a review of the literature, Elmhagen and Angerbjörn (2001) found eight studies (four insect species and four small mammal species) in which asynchrony of population dynamics among patches was confirmed.
6. Population turnover, local extinctions, and the establishment of new populations form the basis for metapopulation dynamics, and metapopulations persist despite population turnover. Elmhagen and Angerbjörn (2001), in a review of the literature, found 22 studies confirming turnover (see Section 5.11.1).
7. Population size or density is significantly affected by migration. This is the basis for source-sink populations and the rescue effect (Pulliam 1988, 1996).
8. Population density, colonization rate, and extinction rates are all affected by patch size and isolation.
9. Metapopulations can affect competitive, predator–prey, and parasite–host interactions. These ideas were discussed in the introduction and will be elaborated in future chapters.

Reviews of metapopulation studies by Elmhagen and Angerbjörn (2001) and Harrison (1991, 1994) found that many of these criteria are frequently not met in the published metapopulation literature. Nevertheless, there is an extensive literature, particularly from studies of butterflies, supporting most of these assumptions (Ehrlich and Hanski 2004).

5.11.1 *Supporting field studies*

1. Boycott (1930) studied fresh-water mollusk populations in 84 ponds in England. Over a 10-year period he recorded 64 extinctions and 93 colonizations of 18 species.
2. In their study of arthropod populations on red mangrove islands, Simberloff and Wilson (1969, 1970) removed all arthropod species by fumigation. After recolonization of the mangroves by the arthropods, they found an equilibrium of between 20 and 40 species per island, depending on island size. But, the turnover rate of species was approximately 2% of the species pool per day.

Table 5.5 Number of surviving local populations, the number of extinctions, and the number of colonizations per year for the metapopulation of *Melitaea cinxia* on Sottungia island in the Baltic Sea off the coast of Finland.

Year	Number of local populations	Number of extinctions	Number of colonizations	Total number of turnover events
1991	10	–	–	–
1992	5	5	0	5
1993	5	1	1	2
1994	6	2	3	5
1995	3	3	0	3
1996	6	0	3	3
1997	10	0	4	4
1998	14	0	4	4
1999	2	12	0	12
2000	11	0	9	9

Extinctions plus colonizations equal number of turnover events. Larvae were transported in August of 1991.

3. As described at the beginning of this chapter, after the explosion on Krakatau, the number of bird species on Rakata seemed to reach equilibrium at 30 by 1934. Yet a consistent turnover has continued.

4. Hanski et al. (2004) described an experiment with the Granville fritillary (*Melitaea cinxia*) in which 10 local populations, derived from 72 larval groups, were transported from the Finnish island of Åland to the island of Sottungia in August of 1991. Sottungia is a 4 km by 2 km island containing 20 small meadows suitable for this butterfly, although there were none at the time of the introduction. An examination of Table 5.5 shows that this metapopulation has persisted, in spite of the fact that none of the original 10 populations lasted for the entire 11 years of the study. These data confirm the persistence of the metapopulation as a stochastic balance between local extinctions and re-colonizations of available habitat patches. Note that we are only keeping track of presence or absence of the population in a habitat; we are not making an assessment of local population numbers or local population dynamics, and there is no assessment of habitat quality.

5. Long-term work on the same species of butterfly on the Åland islands (Nieminen et al. 2004) over an 8-year period found: (1) the number of extinct populations varied from 131 to 234 per year; (2) the number of colonizations ranged from 97 to 230 per year; (3) the total number of extant populations varied from 303 to 496 per year; and (4) the number of empty patches differed from 749 to 3507 per year.

6. Crone et al. (2001) examined the 6-year data gathered by Pokki (1981) on vole (*Microtus agrestis*) populations found on the Tvärminne archipelago in Finland. Extinction and recolonization of local island subpopulations were common and a spatially explicit model such as the IFM provided reasonable predictions of the structure and function of this metapopulation. However, contrary to assumptions, the mainland did not prove to be a significant source of dispersing animals; in fact, an important source of immigrants to the larger islands came from tiny, ephemeral populations found on small islands. While these small populations were unlikely to persist, they were an important source of immigrants to larger islands on the archipelago. Crone et al. (2001) also found that the parameters fitted to the IFMs varied dramatically from year to year. Most importantly, this study suggests that the

"rescue effect" can be turned around such that a small ephemeral population may help stabilize a larger, more permanent population.

In his 2010 essay, Schoener reviewed evidence for three important features of the MacArthur and Wilson theory, which are also relevant to metapopulation theory. These are: (1) evidence for species equilibrium, (2) evidence for species turnover, and (3) evidence related to species-distance and species-area relations. Some of this evidence has been discussed earlier and we will only summarize his remarks concerning the first two.

1. Schoener points out that species equilibriums can be steady as postulated by MacArthur and Wilson, but they can also be cyclical, featuring regular fluctuation in the number of species. This latter possibility is illustrated by Osman's (1978) work on invertebrate diversity of marine epifauna. These fluctuations are driven by seasonal environmental cycles. Next, Schoener points out that due to stochastic forces we should expect variance around any average equilibrium number of species. Third, frequent disturbances may disrupt the attainment of an anticipated equilibrium number of species, leaving the community with fewer species. Finally, equilibrium numbers of species will be attained more quickly by organisms that are good dispersers and have high immigration rates.

2. Schoener (2010) also points out that species turnover is likely to be underestimated unless very thorough surveys of ecosystems are made with short time intervals between surveys. Second, turnover is more common on small islands and on islands far from the mainland. Third, species turnover usually occurs among rare species, which may constitute a small fraction of the community. This results mostly from the relationship between extinction rate and population size. Small populations are extinction prone due to stochastic forces as we have noted earlier.

5.12   Summary

Metapopulation and spatial ecology have provided a new framework for both population and conservation biology. Spatial locations of populations and the interactions among local populations have as great an effect as the traditional parameters of birth and death rates, age structures, and interspecific interactions. Although a metapopulation was originally just a "population of populations" (Levins 1970), many of Levins' simplifying assumptions have been relaxed in modern models. As we have seen, the persistence of the metapopulation is highly influenced by: (1) the number of patches; (2) the size and quality of the patches; and (3) the connectivity among the patches. Spatially explicit models attempt to encompass these variables.

We have not specifically discussed the topic of corridors between patches. The usefulness of corridors to metapopulation persistence has been widely discussed. The general idea is that, since movements between populations have the potential to stabilize both the local and the metapopulation, a corridor between habitat patches would increase connectivity and facilitate these movements. Yet in spite of intuitive appeal and theoretical support, the benefits of corridors remain a controversial topic in metapopulation and conservation biology. Although corridors have not yet been proven irrefutably to be beneficial to wild populations, Laurance and Laurance (2003) concluded that the preponderance of available evidence is positive. They recommended that corridors be regarded as beneficial in fragmented landscapes unless specific local evidence suggests otherwise.

The metapopulation approach has challenged the dogma that populations only exist in locations where they are optimally adapted. Rather, we know that local populations go extinct on even the best quality habitats, and that so-called sink populations hang on in areas of marginal habitat. Furthermore, most natural populations are small enough to be subject to stochastic extinctions. Metapopulations, in which populations in different patches have independent growth and decline dynamics, may therefore be necessary for the long-term persistence of the regional population (Hanski 1999). Foppen et al. (2000) have even demonstrated that sink populations can be essential to the preservation of the populations found in larger patches.

Within a decade of the publication of MacArthur and Wilson's theory of island biogeography, it dominated conservation biology to the extent that "rules" of refuge design were based it. In the 1990s, there occurred a shift, and now metapopulation theory has replaced island biogeography as the major theoretical basis for conservation biology, although its applications to specific situations should not be undertaken lightly (Doak and Mills 1994). On the other hand, citations of Mac Arthur and Wilson's 1967 book, after reaching an "equilibrium" in the 1980s, have increased dramatically since 2000 (Schoener 2010). Island biogeography and metapopulation theory may be replaced by some other paradigms in the future. But, the overall message is clear. Spatial dynamics matter.

References

Andrewartha, H.G. and L.C. Birch. 1954. *The Distribution and Abundance of Animals*. University of Chicago Press, Chicago, IL.

Baguette, M. 2004. The classical metapopulation theory and the real, natural world: a critical appraisal. *Basic and Applied Ecology* 5: 213–224.

Boyce, M.S. 2002. Reconciling the small-population and declining-population paradigms. Pages 41–49 *in* S.R. Beissinger, and D.R. McCullough, Editors. *Population Viability Analysis*. University of Chicago Press, Chicago, IL

Boycott, A.E. 1930. A re-survey of the fresh-water mollusca of the parish of Aldenham after ten years with special reference to the effect of drought. *Transactions of the Hertforfshire Natural History Society* 19: 1–25.

Brown, J.H. 1971. Mammals on mountaintops: non-equilibrium insular biogeography. *The American Naturalist* 105: 467–478.

Brown, J.H. 1978. The theory of insular biogeography and the distribution of boreal birds and mammals. *Great Basin Naturalist memoirs* 2: 209–227.

Brown, J.H. and A. Kodric-Brown. 1977. Turnover rates in insular biogeography: effect of immigration on extinction. *Ecology* 58: 445–449.

Bush, M.B. and R.J. Whittaker. 1991. Krakatau: colonization patterns and hierarchies. *Journal of Biogeography* 18: 341–356.

Carlson, A. 2000. The effect of habitat loss on a deciduous forest specialist species: the White-backed Woodpecker (*Dendrocopos leucotos*). *Forest ecology and management* 131: 215–221.

Cook, R.R. and I. Hanski. 1995. On expected lifetimes of small and large species of birds on islands. *American Naturalist* 145: 307-315.

Crone, E.E., D. Doak, and J. Pokki. 2001. Ecological influences on the dynamics of a field vole metapopulation. *Ecology* 82: 831–843.

Culver, D.C., J.R. Holsinger, and R. Barogdy. 1973. Toward a predictive cave biogeography: the Greenbrier Valley as a case study. *Evolution* 27: 689–695.

Diamond, J.M. 1975. The island dilemma: lessons of modern biogeographic studies for the design of natural reserves. *Biological Conservation* 7: 129–146.

Doak, D.F. and L.S. Mills. 1994. A useful role for theory in conservation. *Ecology* 75: 615–626.

Ehrlich, P.H. and I. Hanski. 2004. *On the Wings of Checkerspots*. Oxford University Press, New York/Oxford.

Elmhagen, B. and A. Angerbjörn. 2001. The applicability of metapopulation theory to large mammals. *Oikos* 94: 94–100.

Foppen, R.P.B., J.P. Chardon, and W. Liefveld. 2000. Understanding the role of sinkpatches in source-sink metapopulations: reed warblers in an agricultural landscape. *Conservation Biology* 14: 1881–1892.

Gould, S.J. 1979. An allometric interpretation of species-area curves: the mean of the coefficient. *The American Naturalist* 114:335–343.

Gyllenberg, M., G. Soderbacka, and S. Ericsson. 1993. Does migration stabilize local population dynamics? Analysis of a discrete metapopulation model. *Mathematical Biosciences* 118: 25–49.

Gyllenberg M., I. Hanski, and A. Hastings. 1996. Structured metapopulation models. Pages 93–108 *in* I. Hanski and M. Gilpin, Editors. *Metapopulation Biology: Ecology, Genetics and Evolution*. Academic Press, London.

Hanski, I. 1993. Dynamics of small mammals on islands. *Ecography* 16: 372–375.

Hanski, I. 1994a. A practical model of metapopulation dynamics. *Journal of Animal Ecology* 63: 151–162.

Hanski, I. 1994b. Patch-occupancy dynamics in fragmented landscapes. *Trends in Ecology and Evolution* 9: 131–135.

Hanski, I., T. Pakkala, M. Kuussaari, and G. Lei. 1995. Metapopulation persistence of an endangered butterfly in a fragmented landscape. *Oikos* 72: 21–28.

Hanski, I., A. Moilanen, and M. Gyllenberg. 1996. Minimum viable population size. *American Naturalist* 147: 527–541.

Hanski, I. 1997. Predictive and practical metapopulation models: the incidence function approach. Pages 21–45 *in* D. Tilman and P Kareiva, Editors. *Spatial Ecology*. Princeton University Press, Princeton, N.J.

Hanski, I. 1998. Metapopulation dynamics. *Nature* 396: 41–49.

Hanski, I. 1999. *Metapopulation Ecology*. Oxford University Press, Oxford, UK.

Hanski, I. 2001. Spatially realistic theory of metapopulation ecology. *Naturwissenschaften* 88: 372–381.

Hanski, I. 2002. Metapopulations of animals in highly fragmented landscapes and population viability analysis. Pages 86–108 in S.R. Beissinger and D.R. McCullough, Editors. *Population Viability Analysis*. University of Chicago Press, Chicago, IL.

Hanski, I. and O. Ovaskainen. 2003. Metapopulation theory for heterogeneous fragmented landscapes. *Theoretical Population Biology* 64:119–127.

Hanski, I. & Gaggiotti, O.E. 2004. Metapopulation biology: past, present, and future: Ecology, genetics, and evolution of metapopulations. *In*, Hanski, I. and Gaggiotti, O.E., Editors, *Ecology, Genetics, and Evolution of Metapopulations*. Academic Press, San Diego.

Hanski, I., P.R. Ehrlich, M. Nieminen, D.D. Murphy, J.J. Hellmann, C.L. Boggs, and J.F. McLaughlin. 2004. Checkerspots and conservation biology. Pages 264–287 in P.R. Ehrlich and I. Hanski, Editors. *On the Wings of Checkerspots*. Oxford University Press, New York/Oxford.

Hanski, I. 2010. The theories of island biogeography and metapopulation dynamics. Pages 186–213 in J.B. Losos and R.E. Ricklefs, editors, *The Theory of Island Biogeography Revisited*. Princeton University Press, Princeton, NJ.

Hanski, I. and M. Kuussaari. 1995. Butterfly metapopulation dynamics. Pages 149–171 in N. Cappuccino and P.W. Price, Editors, *Population Dynamics: New Approaches and Synthesis*. Academic Press, San Diego, CA.

Harrison, S. 1991. Local extinction in a metapopulation context: an empirical evaluation. *Biological Journal of the Linnean Society* 42: 73–88.

Harrison, S. 1994. Metapopulations and conservation. Pages 111–128 *in* J.P. Edwards, R.M. May, and N.R. Webb, Editors. *Large-Scale Ecology and Conservation Biology*. Blackwell, London, UK.

Hubbell, S.P., R.B. Foster, S.T. O'Brien, K.E. Harms, R. Condit, B. Wechsler, S.J. Wright and S. Loo de Lao. 1999. Light-gap disturbances, recruitment limitation, and tree diversity in a Neotropical forest. *Science* 283: 554–557.

Hubbell, S.P. and R. Foster. 1986. Commonness and rarity in a Neotropical forest: implications for tropical tree conservation. Pages 205–231 *in* M.E. Soule, Editor, *Conservation Biology: The Science of Scarcity and Diversity*. Sinauer. Sunderland, MA.

Huffaker, C.B. 1958. Experimental studies on predation: dispersion factors and predator-prey oscillations. *Hilgardia* 27: 343–383.

Hutchinson, G.E. 1961. The paradox of the plankton. *The American Naturalist* 95: 137–147.

Johnson, N.C. and M.S. Gaines 1990. Evolution of dispersal: theoretical models and empirical tests using birds and mammals. *Annual Review of Ecology and Systematics* 21: 449-480.

Kindvall O. 1995. The impact of extreme weather on habitat preference and survival in a metapopulation of the bush cricket, *Metrioptera bicolor* in Sweden. *Biological Conservation* 73: 51–58.

Kindvall, O. 2000. Comparative precision of three spatially realistic simulation models of metapopulation dynamics. *Ecological Bulletins* 48: 101–110.

Lande, R. 1987. Extinction thresholds in demographic models of territorial populations. *American Naturalist* 130: 624–635.

Lande, R. 1988. Genetics and demography in biological conservation. *Science* 241: 1455–1460.

Laurance, S.G.W. and W.F. Laurance. 2003. Bandages for wounded landscapes: faunal corridors and their role in wildlife conservation in the Americas. Pages 313–325 in G.A. Bradshaw and P.A. Marquet, Editors. *How Landscapes Change*. Ecological Studies 162. Springer-Verlag, Berlin.

Lawrey, J.D. 1991. The species-area curve as an index of disturbance in Saxicolous Lichen communities. *The Bryologist* 94: 377–382.

Lawrey, J.D. 1992. Natural and randomly-assembled lichen communities compared using the species-area curve. *The Bryologist* 95: 137–141.

Lehman, C.L. and D. Tilman. 1997. Competition in spatial habitats. Pages 185–203 *in* D. Tilman and P. Kareiva, Editors. *Spatial Ecology*. Princeton University Press, Princeton, N.J.

Levins, R. 1969. Some demographic and genetic consequences of environmental heterogeneity for biological control. *Bulletin of the Entomological Society of America* 15: 237–240.

Levins, R. 1970. Extinction. Pages 77–104 *in* M. Gerstenhaver, Editor. *Some Mathematical Problems in Biology*. American Mathematical Society, Providence, RI.

Losos, J.B. and R.E. Ricklefs. 2010. *The Theory of Island Biogeography Revisited*. Princeton University Press, Princeton, NJ.

MacArthur, R.H., and E.O. Wilson. 1963. An equilibrium theory of insular zoogeography. *Evolution* 17: 373–387.

MacArthur, R.H. and E.O. Wilson. 1967. *The Theory of Island Biogeography*. Princeton University Press, Princeton, NJ.

May, R.M. 1974. Biological populations with non-overlapping generations: stable points, stable cycles and chaos. *Science* 186: 645–647.

May, R.M. 1976. Models for two interacting populations. Pages 49–70 *in* R.M. May, Editor. *Theoretical Ecology, Principles and Applications*. W.B. Saunders, Philadelphia, PA.

Mickelberg, J.L. 2011. Understanding and managing isolation in a fragmented population of the Golden Lion Tamarins, *Leontropithecus rosalia*. Ph.D. Thesis. George Mason University.

Moilanen, A., A.T. Smith, and I. Hanski. 1998. Long-term dynamics in a metapopulation of the American pika. *American Naturalist* 152: 530–542.

Nieminen, M., M. Siljander, and I. Hanski. 2004. Structure and dynamics of *Melitaea cinxia* metapopulations. Pages 63–91 in P.R. Ehrlich and I. Hanski, editors. *On the Wings of Checkerspots*. Oxford University Press, New York/Oxford.

Okubo, A. 1980. *Diffusion and Ecological Problems: Mathematical Models.* Springer-Verlag, Berlin, Germany.

Osman, R.W. 1978. The influence of seasonality and stability on the species equilibrium. *Ecology* 59: 383–399.

Peltonen, A. and I. Hanski. 1991. Patterns of island occupancy explained by colonization and extinction rates in shrews. *Ecology* 72: 1698–1708.

Pokki, J. 1981. Distribution, demography, and dispersal of the field vole, *Microtus agrestis* (L.), in the Tvärminne archipelago, Finland. Acta. *Zoologica Fennica* 164: 1–48.

Preston, F.W. 1962. The canonical distribution of commonness and rarity. *Ecology* 43: 185–215.

Pulliam, H.R. 1988. Sources, sinks, and population regulation. *The American Naturalist* 132: 652–661.

Pulliam, H.R. 1996. Sources and sinks: empirical evidence and population consequences. Pages 45–70 *in* O.E. Rhodes, Jr., R.K. Chester and M.H. Smith, Editors. *Population Dynamics in Ecological Space and Time.* University of Chicago Press, Chicago, IL.

Quinn, J.F. and S.P. Harrison. 1988. Effects of habitat fragmentation and isolation on species richness. Evidence from biogeographic patterns. *Oecologia* 75: 132–140.

Ricklefs, R.E. 2010. Dynamics of colonization and extinction on islands. Pages 388–414 *in* J.B. Losos and R.E. Ricklefs, Editors. *The Theory of Island Biogeography Revisited.* Princeton University Press, Princeton, NJ.

Ricklefs, R.E. and J. Lovette. 1999. The role of island area *per se* and habitat diversity in species-area relationships of four Lesser Antilliean faunal groups. *Journal of Animal Ecology* 68: 1142–1160.

Schoener, T.W. 2010. The MacArthur and Wilson equilibrium model. Pages 52–87 *in* J.B. Losos and R.E. Ricklefs, Editors. *The Theory of Island Biogeography Revisited.* Princeton University Press, Princeton, NJ.

Simberloff, D.S. and E.O. Wilson. 1969. Experimental zoogeography of islands. The colonization of empty islands. *Ecology* 50: 278–296.

Simberloff, D.S. and E.O. Wilson. 1970. Experimental zoogeography of islands: a two year record of recolonization. *Ecology* 51: 934–937.

Sjögren-Gulve, P. and C. Ray. 1996. Using logistic regression to model metapopulation dynamics: large-scale forestry extirpates the pool frog. Pages 111–137 in D.R. McCullough, editor. *Metapopulations and Wildlife Conservation.* Island Press, Washington, DC.

Tepedino, V.J. and N.L. Stanton. 1976. Cushion plants as islands. *Oecologia* 25: 243–256.

Terborgh, J. 1975. Faunal equilibrium and the design of wildlife preserves. Pages 369–380 *in* F. Golley, and E. Medina, Editors. *Tropical Ecological Systems: Trends in Terrestrial and Aquatic Research.* Springer, Berlin and New York, NY.

Thornton, I., R.A. Zann, and P.A. Rawlinson. 1993. Colonization of Rakata (Krakatau Is.) by non-migrant land birds from 1883 to 1992 and implications for the value of island biogeography theory. *Journal of Biogeography* 20: 441–452.

Tilman, D. 1994. Competition and biodiversity in spatially structured habitats. *Ecology* 75: 2–16.

Tilman, D. and P. Kareiva. 1997. *Spatial Ecology.* Princeton University Press, Princeton, NJ.

Turner, M.G., R.H. Gardner, and R.V. O'Neil. 2001. *Landscape Ecology in Theory and Practice.* Springer, New York.

Wahlberg, N., A. Moilanen, and I. Hanski. 1996. Predicting the occurrence of endangered species in fragmented landscapes. *Science* 273: 1536–1538.

Watkinson, A.R. and W.J. Sutherland. 1995. Sources, sinks, and pseudo-sinks. *Journal of Animal Ecology* 64: 126–130.

Whittaker, R.J., R. Field, and T. Partomihardjo. 2000. How to go extinct: Lessons from the lost plants of Krakatau. *Journal of Biogeography* 27: 1049–1064.

Willis, E.O. 1984. Conservation, subdivision of reserves, and the antidismemberment hypothesis. *Oikos* 42: 396–398.

Wilson, E.O. 1961. The nature of the taxon cycle in the Melanesian ant fauna. *The American Naturalist* 95: 169–193.

Wilson, E.O. 1992. *The Diversity of Life*. Norton, New York.

Wilson, E.O. 2013. *Letters to a Young Scientist*. Liveright Publishing Corporation. New York, NY.

Wilson, E.O. and E.O. Willis. 1975. Applied biogeography. Pages 523–534 *in* M.L. Cody and J.M. Diamond, Editors. *Ecology and Evolution of Communities*. Harvard University Press, Cambridge, MA.

Wood, T. 2012. Environmental Studies of the Piedmont. http://envstudies.org/

Wright, S. 1940. Breeding structure of populations in relation to speciation. *The American Naturalist* 74: 232–248.

6 Life history strategies

6.1   **Introduction**

To the uninitiated, nothing could be worse than accompanying a bunch of "birders" on a field trip. They keep stopping, peering through their binoculars, whispering to each other, and motioning you to keep quiet. Why are they so fascinated with birds? Aren't they all pretty much the same?

Of course, not. Even the most naïve nonbiologist knows that birds come in an amazing variety of colors and sizes; amateur birders are legion. What we are interested in exploring in this chapter, however, is the variety and potential adaptive value of life histories found in all groups of organisms. Because ornithologists such as David Lack have contributed so much to our understanding of life histories, we begin by using birds to illustrate the complexity and diversity of life histories. These accounts are mostly based on Janzen (1983).

1. The Groove-billed Ani (*Crotophaga sulcirostris*) is a common and conspicuous bird found in the lowlands and midelevations of Central America. Females are about 65 g in mass but lay extremely large 11 g eggs. As each female may deposit four to eight eggs in the nest, the combined total mass of her eggs may exceed her body weight. What is more extraordinary, however, is that this species engages in a communal breeding system. The birds live and breed in a group ranging from two to eight adults, with an equal number of males and females. The group defends a common territory, year-round. A single nest is constructed, and all females deposit their eggs in it, forming a communal clutch. All members of the group contribute to incubation and feeding of the nestlings. Anis are highly social, roosting and sleeping in close contact with each other while engaging in mutual grooming. However,

Introduction to Population Ecology, Second Edition. Larry L. Rockwood.
© 2015 John Wiley & Sons, Ltd. Published 2015 by John Wiley & Sons, Ltd.
Companion Website: www.wiley.com/go/rockwood/populationecology

Verhencamp (1977, 1978) found that there are specific costs and benefits to the individuals participating in this group endeavor. For example, there are individual differences in the number of eggs that get into the nest, in the amount of time and effort put into incubation, and in the care of the nestlings. Furthermore, the eggs and offspring of the dominant females and males benefit the most. Dominant females lay their eggs last and actually remove eggs laid by other females from the nest. These dominant females then behave like brood parasites in that they actually put less effort into incubation and feeding than do the subordinate females. On the other hand, so-called alpha males, who have the most eggs in the nest (and the most to lose), perform a large share of the incubation. What is the advantage of communal nesting, especially for the subordinate birds? How do the dominant females get away with dumping the subordinates' eggs while they do less of the work?

2. The Northern Jaccana (*Jacana spinosa*) is found from Costa Rica northward in Central America wherever there is floating aquatic vegetation. Jacanas have reversed the usual roles of the genders. Males build the nests and incubate and care for the young. Females lay one egg a day for 4 days in a typical clutch. Females are able to lay a second clutch of eggs elsewhere within 7–10 days, if necessary. The eggs are quite small (7.9 g) compared with the average weight of the females (160.9 g). Males are smaller (mean weight of 91.4 g) than females.

   The mating system is polyandrous. Each male defends a small territory while each female defends a territory containing one to four males. Once chicks reach 12–16 weeks of age, the females often provide a second clutch for the males to care for. The ratio of males to females varies seasonally and from place to place but is often skewed in favor of the males. For example, the long-term average at Turrialba, Costa Rica, was 2.3 males per female (Jenni 1983). Jacanas have a very high reproductive potential, but the hatching and fledging survivorship rates are very low.

3. The Frigatebird (*Fregata magnificens*) is a large (800–1700 g) seabird with a life history that is unusual because of its low reproductive potential. Both sexes do not become mature until 5–8 years of age. Females breed only every other year and lay one egg in a clutch. The egg is incubated for 55 days and the nestling grows very slowly. They are fed primarily by the females for as long as 14 months. Given a 50:50 sex ratio, a new female is produced, on average, only every four years! The potential *r*-value for this species is extremely low, but by contrast, survivorship of adults is very high. The life span is 40 or more years. What selective pressures resulted in a life history so radically different from that of most bird species?

4. Brown Pelicans (*Pelecanus occidentalis*) are one of the best known birds in the Western Hemisphere. They are found on both the Atlantic and Pacific coasts from North Carolina to Brazil and from British Columbia to Chile. Breeding colonies may contain as many as 500 pairs. An adult brown pelican weighs from 2 to 5 kg; it takes 3–5 years to attain adult plumage. Males and females share chick-raising duties equally, and the normal clutch size is three eggs. Incubation takes 30 days, and the nestlings need 10–12 weeks to fledge. Schreiber and McCoy visited a pelican colony four times during the breeding season of 1979 on Isla Guayabo in Costa Rica. Of 430 nests surveyed, most had three eggs, but the average was 2.42. By their fourth visit, the number of surviving fledglings was 506, which was an average 1.18 per nest. The Brown Pelican is much larger than the Frigatebird, has a much higher reproductive potential, and has a lower survival rate as well.

5. Oropendolas (*Zarhynchus wagleri*), which are related to blackbirds and orioles, nest in colonies. Males weigh twice as much as females (212 vs 110 g), and they have been shown to take twice as much energy to fledge as a female. As a result, male mortality among chicks is much higher during times of food scarcity. The sex ratio at colonies is normally 5:1 in favor of females. In Costa Rica, nesting begins with the dry season (December), and three complete breeding cycles are possible before the beginning of the rainy season in May. The normal clutch size is two, but breeding success is very low. The average number of chicks fledged per nest is 0.40. On the other hand, survivorship of adults is very high. Adults have been recorded living beyond the age of 26 in the field. By contrast to frigate birds, which also have very long adult life spans, this species has a much higher reproductive potential.

So what have we learned about life histories from these birds? Nesting ranges from communal to colonial to pairwise. Breeding systems varied from communal to polyandrous to simple pair bonds. Fecundity varied from one egg every other year to as many as eight in one clutch. Survivorship of the chicks is as low as only 0.40 per nest, but adult survivorship is as high as 40 years. What accounts for all this variation in life histories? Under what conditions do we find high versus low fecundity and/or survivorship? These are questions we want to approach in this chapter.

But, first, what is a life history? According to Sibly (2012), a life history includes data such as at what age an organism begins reproduction, the number of offspring produced at different ages, and the chances of survival at each age. Reznick (2011, p. 211) defines a life history as a composite of all of the variables that contribute to how a species reproduces itself. These variables include size at birth, age and size at maturity, and reproductive effort. Reproductive effort is a composite of all of the investments a parent makes in its offspring, while fitness is the relative success an individual has in contributing offspring to the next generation. Fitness is a composite of survival and the successful production of offspring (Reznick 2011, p. 214). The diversity of life histories is just as overwhelming as the diversity of species itself. Can we make sense of this diversity? That is, are there patterns that we can uncover such that we could predict that a particular life history will have evolved in a particular environment?

Another set of questions we wish to address concerns the relationship between the body size of an organism and its reproductive potential. Although body mass does not determine all aspects of life history, it has a powerful influence. For example, Figure 6.1 is based on data for 24 species of mammals found in Costa Rica. The log of the length of the prereproductive period was graphed against the log of adult body mass. The obvious conclusion is that there is a higher likelihood of delay in reproductive maturity in the larger animals. Similarly, Figure 6.2 demonstrates that animals with larger mass also have a longer interval between births. We will return to this theme.

Basic life history equations were available early in the twentieth century (Lotka 1925). But, a serious comparison of life history parameters across a large number of species did not begin until 1954. In that year, Frederick Smith and L.C. Cole published important papers that have become the foundation of life history analysis.

Smith (1954) surveyed the literature and published what was known about r, R_0, and G (generation time) at that time. Using the well-known relationship:

$$N_t = N_0 e^{rt}$$

Fig. 6.1 Log length of the pre-reproductive period versus log of mass in 24 species of Costa Rican mammals. Linear regression is significantly positive. (Source: Data from Janzen 1983.)

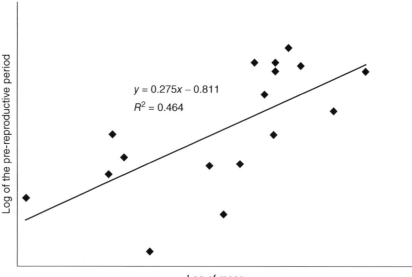

$y = 0.275x - 0.811$

$R^2 = 0.464$

Log of mass — Log of the pre-reproductive period

Fig. 6.2 Log interval between births in years as a function of the log of mass based on 24 species of Costa Rican mammals. Linear regression is significantly positive. (Source: Data from Janzen 1983.)

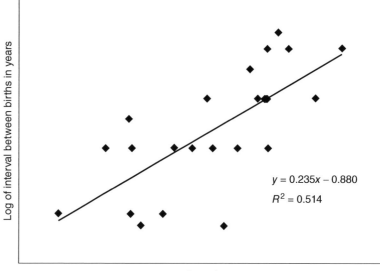

$y = 0.235x - 0.880$

$R^2 = 0.514$

Log of mass — Log of interval between births in years

and setting $t =$ the generation time, G, we have

$$N_G = N_0 e^{rG} \quad \text{and} \quad \frac{N_G}{N_0} = e^{rG}$$

Taking an equation from Chapter 1, we have

$$N_G = N_0 R_0 \quad \text{and} \quad \frac{N_G}{N_0} = R_0$$

The result is

$$R_0 = e^{rG} \tag{6.1}$$

or

$$\ln R_0 = rG \tag{6.2}$$

If two of these variables are known, the third can be determined, at least for a population with a stable age distribution. However, there is no necessary relationship among all three of these variables.

Smith (1954) and a number of more recent surveys of life histories have shown that:

1. r is inversely related to generation time, G. r-values respond very strongly to generation time as we will show later in this chapter.
2. Generation time is directly related to size. Generation time is highly influenced by the length of the prereproductive period and by intervals between births Figures 6.1 and 6.2 illustrate the relationship between these parameters and size (mass). Points 1 and 2 lead us to point three.
3. r is inversely related to size (mass).
4. Life span and growth rate are negatively associated (Stearns 1992).
5. High individual growth rates are positively correlated with r.
6. Growth rate is inversely related to body mass.

In summary, small size is associated with fast growth rates, high r-values, and short generation times, while organisms with large mass have slower growth rates, small r-values, and long generation times (Smith 1954; Enquist et al. 1999).

Smith also speculated that r is related to the harshness of an animal's environment, and that large organisms live in a more benign world. That is, small environmental changes are less catastrophic for large organisms. This is a point made by Hanski (1999) in the last chapter. He speculated that environmental stochasticity was reduced for larger organisms. By contrast, organisms with high r-values have the ability to recover more quickly from events that decimate their populations. Furthermore, over evolutionary time, the fossil record shows many lines of organisms continually increase in size until they go extinct. This is known as "Cope's rule" (Southwood 1976).

The general questions are as follows: Under what conditions is it advantageous to have a high r-value, combined with small size? Under what conditions will evolution select for large size, with its presumed lower r-value? Is it inevitable that large size must be combined with a low r-value?

6.2 Power laws

The relationship between size (mass) and a wide variety of metabolic, physiological, and ecological functions (everything from skeletal muscle contraction rate to incubation period to maximum life span) has been hypothesized to follow general power or scaling laws (Brown et al. 2012; equations 6.3a and 6.3b).

$$Y = Y_0 M^b \tag{6.3a}$$

M is mass; b is an allometric constant; Y is a physiological rate or some other variable dependent on mass; and Y_0 is the normalization constant.

$$\log Y = \log Y_0 + b \log M \tag{6.3b}$$

Taking the log of both sides of equation 6.3a leads to equation 6.3b, which shows us that on a log–log scale, we can expect a linear relationship between mass and a dependent variable, with the slope equal to *b*.

Despite the fact that organismal mass ranges over 21 orders of magnitude (from bacteria to whales), these scaling laws consistently apply, and it has been asserted that b is a multiple of $1/4$ (Brown et al. 2004). For example, heart rate versus mass scales as −0.25, life span as 0.25, and the length of both mammalian aortas and tree trunks as 0.25 (West et al. 2000). These are known as allometric relationships because the scaling factor, *b*, is not equal to one. If *b* = 1, the relationship would be isometric and plot as a straight line on both an arithmetic and a logarithmic axis. However, in these allometric relationships, *b* is not equal to one. Kleiber (1932) showed long ago that the relationship between mass and basal metabolic rate in mammals followed an allometric relationship with *b* = 3/4. Figure 6.3 shows us that, in fact, the 3/4 relationship between metabolic rate and mass appears to apply to everything from unicellular organisms to both poikilothermic and homeothermic vertebrates. More recently, West et al. (2000) claimed that $M^{3/4}$ also "coincides with the respiratory rate … of mammalian mitochondria … (and) … even with that of the molecular respiratory complex and terminal oxidase molecular units within mitochondrial membranes!" However, these generalizations have been questioned (White et al. 2007; Duncan et al. 2007; Packard and Birchard 2008), and a reanalysis of mammalian metabolic rate indicates the relationship of the log-transformed data is curvilinear (Kolokotrones et al. 2010). This suggests that claims about a particular slope value are biased, and that other relationships correlated with metabolic rate are also likely curvilinear. In a new book on Metabolic Ecology, Sibly et al. (2012) pull back from some of the universal claims but advocate a metabolic approach to ecology. We will return to this topic later.

Fig. 6.3 Allometric scaling of metabolic rate for organisms ranging from unicellular. Protista to homeothermic mammals. The regression line has a slope of 0.75. (Source: Data from Bonner (2006).)

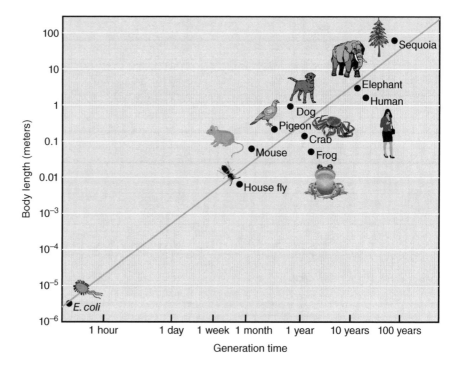

The 3/4 power scaling relationship for mammals was known as Kleiber's law. But a different scaling power was rooted in the botanical literature based on geometric or Euclidean principles. The basic idea was that "resources are acquired by surfaces and used by volumes" (Horn 2004). For example, a cell acquires nutrients across a two-dimensional surface area but must distribute those resources throughout its three-dimensional volume. For a forest, the area of ground that a tree covers scales as the square of the diameter of ground area covered or shaded (canopy "footprint"). The volume scales according to volume or the cube of this same diameter. The number of individuals that can fit into a given area is assumed to be the reciprocal of the area covered. Plant volume and biomass were therefore proposed to be proportional to negative 3/2 (equation 6.4a) of population density.

This is the basis, for example, of the "geometric model" of self-thinning.

Agronomists and plant ecologists had proposed that mortality during the intraspecific self-thinning process followed a −3/2 power law (Yoda et al. 1963). This law asserted that the mean weight per plant increased faster than density decreased. During the thinning process, if density (N) and mean weight per plant (ϖ) are points on a graph, over time these points will produce a linear relationship in which the slope will be −3/2 (equations 6.4a and 6.4b). Total mass continues to increase until the constant final yield (C) is reached.

$$\varpi = CN^{-3/2} \tag{6.4a}$$

$$\log \varpi = \log C - (3/2) \log N \tag{6.4b}$$

The −3/2 power law, however, began to be questioned late in the twentieth century (Lonsdale 1990), since thinning exponents were shown to be highly variable among species. In a series of papers (Enquist et al. 1998; West et al. 1997), a different model was proposed. This model assumes that growth continues until it is limited by a resource, and that resources are acquired through branching, linear or "fractal" networks that distribute resources throughout the organism. In plants, Enquist et al. (1998) used whole-plant xylem transport as an estimate of resource use. They found that resource use scales as the 3/4 power of mass, the same exponent that Kleiber found. West et al. (1997) and Enquist et al. (1998) therefore asserted that the real exponent in equations 6.4a and 6.4b should be −4/3, not −3/2.

Again, the quarter scaling rule has many applications in both animal and plant populations. Calder (1984) showed that maximum life span, reproductive maturity, incubation time for birds, and a large number of metabolic functions scaled to the $1/4$ power against body mass on a log–log scale. For example, a regression line through ln(density) versus ln(mass) in terrestrial mammals has a slope of −0.77 ($b \approx -3/4$) (see Figure 6 of Brown et al. 2004). Enquist et al. (1998) showed that the population density of both plant and animal populations scale as $M^{-0.75}$. From such similarities, they suggested that plants and animals share a similar scaling law that "reflects how resource requirements of individual organisms affect competition and spacing among individuals within ecological communities."

West et al. (1997) and Enquist et al. (1998, 2000) proposed the existence of a common mechanism underlying the scaling or power laws in both plants and animals. They asserted that living things are sustained by the transport of materials through linear networks that branch to supply all parts of the organism. These networks include mammalian blood vessels, bronchial trees, plant vascular systems (xylem and phloem), as well as tracheal tubes in insects. Most of these distribution systems can be described as a branching network in which the size of the tubes decreases. This "vessel-bundle" structure is not only characteristic of plant vascular systems but also of vertebrate and invertebrate circulatory systems. These networks vary in the properties of the tubes, in the type of fluid transported (liquid to gas), and in the nature of the pump. A pulsating compression pump is found in cardiovascular systems, a bellows in respiratory systems, diffusion in insect tracheae, and osmotic and vapor pressure in plant vascular systems. In spite of these differences, West et al. (1997) proposed that they all follow the same scaling laws. The scaling laws are based on the following principles, to which all biological networks should adhere. (i) The system must fill the volume of the organisms so as to distribute essential nutrients to all of its cells. (ii) The terminal branches of these networks should be roughly the same size, regardless of body size. Indeed, capillary size does not vary much among animals, and the same is true of petioles or terminal xylem in plants. (iii) Supply networks should be so efficient that fluids move through them with a minimal loss of energy. Models of fluids flowing through networks lead to the conclusion that the summed cross-sectional areas of the "daughter" branches at each level should be equal to that of the "parent" branch. Such an area-preserving pattern is found in trees and in artery-arteriole branches. Enquist et al. (2000) have modeled the physical structures of the networks and substituted them into equations describing fluid volume and rate of flow in a system. The end result is a formula for metabolic rate versus mass with a scaling exponent of 0.75. In other words, their model seemed to successfully confirm the 3/4-power scaling law.

In plants, Enquist et al. (1998, 2000) predicted that if the log of average mass per plant is graphed against density, the slope would be −0.75. However, since the independent variable is density, and mass is the dependent variable, the expected slope is the inverse of −3/4, or −4/3. Enquist et al. (1998) then gathered data on plants ranging of *Lemma* to *Sequoia* and showed that the slope is as predicted, −4/3 with an $R^2 = 0.963$ on the regression model. The slope of −3/2, predicted by the geometric model, was not found.

Enquist et al. (1998) concluded that both animals and plants share a common set of allometric relationships involving body mass, growth rates, life spans, maximum size, and densities. Like all grand theories, there are likely to be complications and exceptions. Darveau et al. (2002) raised doubts about any "single cause" explanation for scaling of basal metabolic rates, since metabolic rates are controlled by a number of steps and not by any single, rate-controlling factor. They believe that the scaling constant b in equations 6.3a and 6.3b is based on the sum of multiple metabolic factors, not on a limit based on any one metabolic rate. As Agutter and Wheatley (2004) point out, not everyone is convinced that a uniform scaling constant exists, and many other models of metabolic scaling have been proposed. According to Packard and Birchard (2008), traditional allometric analysis fails to provide a valid predictive model for mammalian metabolic rates. Recently, Enquist and Bentley (2012) have discussed some incorrect assumptions that had made about plant anatomy. Nevertheless, Brown, Enquist, West, and others have proposed a "*metabolic theory of ecology*" (*MTE*) (Brown et al. 2004; Sibly et al. 2012).

6.3 The metabolic theory of ecology

Brown et al. (2004) and Brown and Sibly (2012) begin by linking basic metabolic processes to body mass and temperature. Since metabolism determines the rate of acquisition of energy and nutrients by organisms, it determines the rate of resource use and sets constraints on the allocation of resources to growth, reproduction, and other components of fitness.

Given Kleiber's finding that metabolic rate scales according to $b = 3/4$, we can write an equation (6.5) for whole-organism metabolic rate (I) as a function of mass (M) and a normalization constant, I_0. Again, the justification for the $1/4$ power scaling is based on West et al. (1997, 1999a, b), in which "whole-organism metabolic rate is limited by rates of uptake of resources across surfaces and rates of distribution of materials through branching networks. The fractal-like designs of these surfaces and networks cause their properties to scale as $1/4$ powers of body mass or volume, rather than $1/3$ powers that would be expected from Euclidean geometric scaling" (Brown et al. 2004, p. 1773):

$$I = I_0 M^{3/4} \tag{6.5}$$

Next, Brown et al. (2004) introduced the effect of temperature on biological processes. Biological activity increases exponentially with temperature, and a general equation for the kinetics of this process is described as equation 6.6. In this equation, E is the activation energy, k is the Boltzmann constant, and T is absolute temperature in degrees Kelvin. The Boltzmann factor describes how temperature affects reaction rate by changing the proportion of molecules with sufficient kinetic energy:

$$e^{-E/kT} \tag{6.6}$$

What interests us here is that nearly all biological rates are temperature dependent, including population growth rates, development time, and life span (Brown et al. 2004).

The effect of mass and temperature on metabolic rates can be combined into one equation as shown in equation 6.7a, where i_0 is a normalization constant independent of both body size and temperature. After taking the natural log of both sides of equation 6.7a, we have 6.7b:

$$I = i_0 M^{3/4} e^{-E/kT} \tag{6.7a}$$

$$\ln\left(IM^{-3/4}\right) = -E\left(1/kT\right) + \ln(i_0) \tag{6.7b}$$

Equations 6.7a and 6.7b essentially predict that mass-corrected whole-organism metabolic rate is a linear function of the inverse of absolute temperature ($1/kT$). Metabolic rate, in turn, determines the rate of resource acquisition from the environment. Brown et al. (2004) then showed how these metabolic relationships can determine life history characteristics such as population growth rate, development rate, age at maturity and life span. For example, Brown et al. (2004), based on a variety of data, showed that maximum growth rate (r_{max}) is dependent on both temperature and mass. The slope ($b = -0.23$) seems to confirm the $-1/4$ allometric relationship between mass and r_{max} that has been reported earlier (Slobodkin 1962; Blueweiss et al. 1978).

The MTE has already generated a great deal of commentary and controversy (see Volume 85, issue 7 of Ecology). As expected, many scientists believe this theory is an overreach. One issue is the scaling exponent. For example, White et al. (2007)

and Duncan et al. (2007) asserted that the data do not support a universal metabolic allometric exponent of 0.25.

In the 2012 book, *Metabolic Ecology, a Scaling Approach,* Brown and Sibley retreat from some of the claims of the original 2004 paper but strongly advocate a metabolic approach to ecology. They point out that they attempted to make a particular idealized approach to a MTE. Their purpose now, however, is to advocate for a metabolic approach to ecology, rather than to promote the MTE, per se.

Brown et al. (2012) state that their overriding purpose in proposing this approach to ecology includes:

1. to highlight the fundamental role of energy metabolism;
2. to emphasize the role of metabolic processes in ecology. Interactions between organisms and their environment (the definition of ecology) all involve exchanges of energy and chemicals;
3. to use the principles of metabolism to develop a quantitative basis for many aspects of ecology, including population ecology.

Again, they emphasize that the metabolic rates of all organisms are governed by body size, temperature, and availability of resources such as water and nutrients. However, Brown and Sibly (2012) recognize the following: (i) There is considerable variation in the exponent *b*. Clearly, it was an overgeneralization to claim that the scaling exponent should be $3/4$ based on Kleiber's law. The power law does not necessarily have a single scaling exponent. The relationship between mass and metabolic rate in mammals, for instance, increases from small to larger species. (ii) Metabolic rates vary due to factors other than body size. Climate, habitat, diet, body temperature, and taxon can all affect metabolic rate; (iii) Temperature dependence of metabolic rate and other functions can be complicated and deviate substantially from pure exponential relationships.

Brown and Sibly (2012, p. 31) quote G.E.P. Box: "All models are wrong, but some are useful." Is MTE useful? It cannot capture many aspects of population and community ecology that focus on niche relationships, foraging behavior, and mutualism. But it may be able to: (i) explain major patterns of variation of organisms of different sizes, living in diverse environments of different temperatures; (ii) incorporate basic principles of biology, physics, and chemistry to provide a quantitative, mechanistic basis for how body size and temperature affect metabolic rate and various ecological relationships; and (iii) provide a framework for interpreting new data and for making and testing predictions. Brown et al. (2012) recognize the controversies surrounding their theories but assert that we can look to metabolism as a mechanism to provide for a unifying theory of ecology.

One question we posed in the introduction (is it inevitable that large size must be combined with a lower *r*-value?) has been largely answered in the affirmative. But other questions remain: Under what conditions is it advantageous to have a high *r*-value, combined with small size? Under what conditions will evolution select for large size and/or high adult survivorship combined with a low *r*-value? How do we explain all of those mating systems that have been evolved by the birds we looked at in the introduction?

6.4 Cole and Lewontin

Cole (1954) reviewed the basic equations of population ecology and noted that it was easy to visualize situations where natural selection would tend to increase the value of *r* (after all, fitness is measured in terms of differential production of offspring). However, there exist many life histories in which the value of *r* is quite low. For example, condors like frigatebirds, lay one egg every 2 years, and one female egg once every 4 years. What ecological or evolutionary conditions lead to these life histories?

Cole broke down life histories into components. He analyzed each separately in order to determine which were most crucial to the r-value and therefore to the life history of the organism. In his simple models, he analyzed three of them. What is remarkable about Cole's work is that he did it without the aid of modern calculators or computers. In 1954, computers were available to few individuals, and hand calculators, as we know them, did not exist.

Cole recognized that since $r = b - d$, a species may increase r by reducing mortality. However, he chose to limit his investigations to factors that might influence b. In most of his models, therefore, he assumes no mortality! Next, he broke his investigation of reproduction down into three areas:

1. What are the advantages of "repeated breeding" or iteroparity, as opposed to "one-time breeding" or semelparity? This basic dichotomy in life histories had neither been widely discussed nor previously investigated. Cole appears to have invented these two terms.
2. What is the effect of the length of the prereproductive period on r?
3. How does litter size (B) or clutch size (defined as the number of offspring produced at one time) affect r?

Cole made several simplifying assumptions, which he asserted did not seriously alter his general conclusions. Furthermore, he wanted to show the maximum gains a species might realize from altering a life history pattern. These assumptions were:

1. no mortality (indefinite survival);
2. the litter size does not change with age;
3. iteroparous species continue to breed indefinitely.

It is unfortunate that the first part of the paper became more famous than the remainder. What became known as "Cole's result," is really due to his first assumption, which was corrected in the literature by Charnov and Schaffer (1973). "Cole's result" centers on his discussion of semelparity versus iteroparity. In the following analysis, Cole is trying to determine what the advantage of iteroparity was in terms of its effect on r.

Consider an annual plant or animal. If it matures in late summer, dies in the fall at the time of reproduction, then, $B = \lambda = e^r$. For a semelparous organism, we have the following:

$$r = \ln B \tag{6.8}$$

where $B =$ the litter size in terms of the number of female offspring.

By how much would r be increased if the organism lived for another year? First of all, if the litter size is only 1, since $\ln 1 = 0$, no growth would be possible and the species would have to be iteroparous for the population to grow.

Cole next developed an equation involving $\alpha =$ the age at first reproduction, $\omega =$ the age at last reproduction, and $B =$ litter size.

Starting with the Euler equation ($1 = \Sigma(l_x m_x e^{-rx})$, assume that $l_x = 1$ for all ages, and m equals b and is a constant for all ages after α. Then the number of litters beginning at age α and including age ω is $n = (\omega-\alpha) + 1$. Cole showed that the Euler equation could be approximated under these conditions by

$$1 = e^{-r} + Be^{-r\alpha} - Be^{-r(n+\alpha)} \tag{6.9}$$

Now consider the extreme case of $\alpha = 1$ with n infinitely large. That is, reproduction begins in the first year after birth, and there is an infinite life span. The final term becomes zero, and we have the following:

$$1 = e^{-r} + Be^{-r} = e^{-r}(1 + B)$$

Taking natural logs:

$$0 = -r + \ln(1 + B)$$

which results for an iteroparous organism, in:

$$r = \ln(1 + B) \tag{6.10}$$

for an iteroparous organism.

Compare this with, $r = \ln B$ for the semelparous organism.

This is "Cole's result." "The absolute gain in intrinsic population growth which could be achieved by changing to the perennial habit is equivalent to adding one individual to the average litter size" (Cole 1954, p. 118).

In other words, a litter or clutch size of 101 in an annual organism would serve the same purpose as a clutch size of 100 forever in a perennial. Therefore, why would organisms evolve mechanisms for adults to survive winter if the same result is obtained through adding one individual to the litter size?

The answer, as was shown by Charnov and Schaffer (1973), is that there is great difference for most species between *juvenile and adult survivorship*. Organisms with high juvenile mortality, small litter sizes, or long prereproductive periods (slow maturity) have a great deal to gain by repeated reproduction (iteroparity). Similarly, if organisms live in an unpredictable environment, such that reproduction can fail completely in some years, selection would strongly favor adult survivorship and repeated reproduction. This is sometimes called a "bet-hedging" life history. Recall the life histories of the frigatebirds, pelicans, and oropendulas described in the introduction.

"Cole's result" is based on his assumption of complete survivorship of all life stages. If, on the other hand, juvenile survivorship is superior to adult survivorship, then semelparity should be favored. Examples include dragonflies, mayflies, and periodical cicadas.

Cole (1954) then analyzed the relative gains on r: (i) age of first reproduction, (ii) litter size, and (iii) iteroparity. As was hinted at in our analysis of mammal reproduction, it turns out that r is most sensitive to changes in the prereproduction period (maturation time). The r-value is somewhat sensitive to litter size and is least influenced by changes from semelparity to iteroparity (but see the earlier paragraph).

For example, examine Table 6.1. Presented are four life histories. In each case, the survivorship function is held constant. The fertility columns reflect differences in the either maturation time, or the amount of reproduction that is "front loaded" to an earlier age class. In scenario one, reproduction begins at age 2, and the majority of reproduction takes place in this age class. The result is an r of 0.102 and a generation time of only 2.2 years. In scenario two, the *gross reproductive rate* (GRR) is greater than that of scenario one, while R_0 remains at 1.25. The prereproductive period is also the same as in scenario two. However, because reproduction is mostly "backloaded" onto

Table 6.1 Four fertility scenarios demonstrating the importance of early reproduction on the value of r.

Age	l_x	Scenario one m_x	Scenario two m_x	Scenario three m_x	Scenario four m_x
0	1.00	0	0	0	0
1	0.100	0	0	0	0
2	0.050	20	10	0	0
3	0.025	10	30	40	50
4	0.010	0	0	0	0
5	0	0	0	0	0
		GRR = 30	GRR = 40	GRR = 40	GRR = 50
		$R_0 = 1.25$	$R_0 = 1.25$	$R_0 = 1.00$	$R = 1.25$
		$G = 2.2$	$G = 2.6$	$G = 3.0$	$G = 3.0$
		$r = 0.102$	$r = 0.086$	$r = 0.000$	$r = 0.074$
		$\lambda = 1.11$	$\lambda = 1.09$	$\lambda = 1.00$	$\lambda = 1.08$

l_x = survivorship, m_x = fertility, GRR = Gross Reproductive Rate, R_0 = Net Reproductive Rate, G = Generation Time, r = intrinsic rate of increase, and λ = finite rate of increase.

age class three, the generation time is 2.6 years, and the resulting r-value is only 0.086. In scenario three, by increasing the prereproductive period to age 3, even though we increase the GRR to 40, the net reproductive rate falls to 1.00 and $r = 0$. In scenario 4, we see that, because reproduction does not start until age 3, we must raise the GRR to 50 in order for the population to grow. This produces an R_0 of 1.25, but since generation time is 3 years, the r-value is only 0.074.

In Table 6.2, we apply Cole's ideas to a human population. The survivorship column is based on 1985 Vital Statistics from the United States (Peters and Larkin 1989). The four fertility scenarios are based on a population in which the average female has a GRR of 1.5 females. In scenario one, the average female has 1.0 female offspring between the ages of 15 and 20 and the remainder between 20 and 25. In each of the other scenarios, the females delay reproduction by 5 years, 10 years, and, finally, 15 years. The result is that when reproduction begins at the age of 15, the r-value is 0.091, compared to an r-value of 0.052 when reproduction begins at 30. Generation time is obviously almost doubled when reproduction begins at 30 compared to at 15 years. If a population had

Table 6.2 Four population scenarios for a hypothetical human population.

Age	l_x	Scenario one m_x	Scenario two m_x	Scenario three m_x	Scenario four m_x
0	1.00	0	0	0	0
1	0.991	0	0	0	0
5	0.989	0	0	0	0
10	0.988	0	0	0	0
15	0.987	1.000	0	0	0
20	0.985	0.500	1.000	0	0
25	0.982	0	0.500	1.000	0
30	0.979	0	0	0.500	1.000
35	0.975	0	0	0	0.500
40	0.970	0	0	0	0
45	0.961	0	0	0	0
		$r = 0.091$	$r = 0.073$	$r = 0.061$	$r = 0.052$
		$R_0 = 1.48$	$R_0 = 1.48$	$R_0 = 1.47$	$R_0 = 1.47$
		$G = 16.6$	$G = 21.6$	$G = 25.6$	$G = 31.6$
		$\lambda = 1.095$	$\lambda = 1.076$	$\lambda = 1.063$	$\lambda = 1.053$

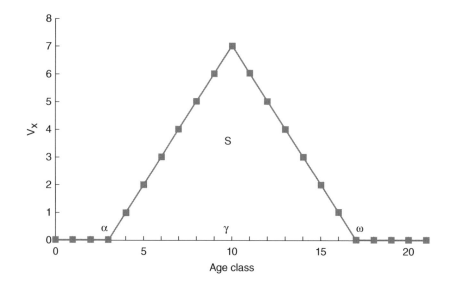

Fig. 6.4 Summary of the Lewontin analysis of the effect on r of changes in life history parameters. V_x = the product of l_x and m_x · α = age at which reproduction begins. γ = the age of maximum reproduction. ω = age at which reproduction ends. S = area under the triangle, which equals the total life time reproductive output. In the Lewontin analysis, the area, $\underline{S}$, is held constant, while the other parameters are manipulated.

a stable age distribution, it would grow from 1000 to 2500 in 10 years if $r = 0.091$, whereas it grows to only 1600 if $r = 0.052$.

Lewontin (1965) further developed the themes of Cole in his paper, "Selection for Colonizing Ability." He was concerned with the problem of the establishment of a population in a new geographic or ecological space. Lewontin was able to examine more subtle changes in the life cycle and illustrated how changes in fecundity, longevity, and length of prereproductive period affect r. Using more sophisticated mathematical techniques, he was able to relax some of Cole's simplifying assumptions.

In his analysis, Lewontin summarized l_x and m_x into a function he called, V_x. $V_x = l_x m_x$. He removed Cole's assumptions concerning infinite life span and constant fecundity with age. He then showed that a common form of V_x is a basic triangle with V_x on the y-axis and age on the x-axis (Figure 6.4). He identified four important parameters, α (age at which reproduction begins), ω (age at which reproduction ends), δ (the age of maximum reproductive output), and S (the total reproductive output, which equals the area of the triangle).

The integral form of the Euler equation was modified as shown in equation 6.11. Since the function V_x has the shape of a triangle, Lewontin was able to solve the integral and devised expressions relating r, to α, ω, δ, and S:

$$1 = \int l_x m_x e^{-rx} d_x$$

$$1 = \int V_x d_x / e^{rx} \tag{6.11}$$

Lewontin manipulated the triangle sizes and shapes by varying the values along the x-axis. His findings can be summarized by the following examples. In each case, he investigated how basic life history changes affect the value of r.

1. Rigid translation. This means moving the entire triangle along the x-axis (varying by age classes), leaving S constant. In this case, α, δ, and T are moved simultaneously to younger (to the left along the axis) or older (to the right) age categories.

2. Change the age of sexual maturity (prereproductive period) only. This means moving α to earlier or later ages without changing δ, ω, or S.

3. Change the age of maximum reproduction only. That is, move δ to earlier or later ages while all else remains constant.

4. Change the age of last reproduction, ω, leaving all else constant.

Lewontin's results can be summarized as follows: Let us assume $r = 0.300$ and we wish to know what life history changes will increase r to 0.330. Lewontin showed that doubling the lifetime fecundity from 780 to 1350 would produce such a change in r, all else being held constant. However, holding fecundity constant at 780, a similar increase in r was achieved through one of the following:

1. Rigid translation of the life history. That is, by simultaneously decreasing α, δ, and ω by 1.55 days, the value of r increased from 0.300 to 0.330. This is a change of approximately 13%, over the total span of the life history.

2. If only α, the prereproductive period, is changed, it takes a decrease of 2.20 days (approximately an 18% change), to produce the desired value of r.

3. If only δ is decreased, it must be shortened by 5.55 (a 24% change) days.

4. If the only change is in ω, it must be decreased by 21.0 (a 38% change) days.

In other words, a simultaneous reduction of 13% in the three measured aspects of the fertility schedule was roughly equivalent to a 100% change in fecundity. Furthermore, reducing the single parameter, α, by 18% resulted in the increase of r from 0.300 to 0.330. As was proposed by Cole, a reduction in the prereproductive period is the single aspect of the life history with the greatest influence on the value of r.

Lewontin pointed out that development time or rate is closely regulated by natural selection and varies little among geographic races, whereas less important factors such as fecundity and longevity show substantial variation.

6.5 The theory of r- and K-selection versus fast and slow life histories

We discussed MacArthur and Wilson's theory of island biogeography extensively in the last chapter. Another application of their findings was the theory of r- and K-selection as applied to life histories. MacArthur and Wilson asserted that for "colonizing species," the ability to grow rapidly and to disperse was the major component of fitness. They theorized that new, unpopulated environments would be colonized first by these "r-selected" species. In contrast, once an environment, either an island or the mainland, was completely colonized, it became densely populated (crowded or saturated) with all populations near their carrying capacities (K). They reasoned that such populations were exposed to "K-selection." The major component of fitness in a crowded environment was not the ability to grow quickly but to survive under these highly competitive conditions.

According to the theory, in an environment with no crowding (an r-selecting environment), those genotypes that harvested the most resources and produced the most offspring in the shortest period, even if they are not efficient at using their resources, would be favored. In the crowded, K-selecting environment, those genotypes that replaced themselves at low resource levels would survive and reproduce. Evolution was said to favor efficiency in these K-selecting environments.

Thus r-selected species would be favored in areas where there has been a disturbance of the established community. K-selected species would be favored in areas, such as mature communities, where competition is high. It is natural to think of r- and K-selection in terms of ecological succession, in which colonization and competition are assumed to be the primary forces determining the appearance of plant communities

Table 6.3 Correlates of *r*- and *K*-selection.

Trait	*r*-selection	*K*-selection
Climate	Variable or unpredictable	Predictable, less variable
Mortality	Density independent	Density dependent
Survivorship	Type 3	Type 1 or 2
Population size	Variable, below *K*	Fairly constant; at or near *K*
	Recolonization common	Recolonization uncommon
Competitive ability	Usually poor	Usually keen
Investment in defense	Little energetic investment	Great energetic investment
Parental care	Minimal	Usually great
Length of life	Short	Long
Stage of succession	Early	Late
Rate of development	Rapid	Slow
r_{max}	High	Low
Prereproductive period	Short	Long
Body size	Small	Large
Number of offspring	Many	Few
Size of offspring	Small	Large
Dispersal ability	Excellent	Fair to poor

Source: Adapted from Pianka (1970).

over time. *r*-selected species appear early in succession, to be eventually replaced by *K*-selected species in the later stages.

Biologists were led to consider what aspects of life histories and of the physical environment would be correlated with *r*- or *K*-selection. In 1970, Pianka published a paper entitled, "On *r*- and *K*-selection," which laid out general expectations of *r*- and *K*-selection for a number of life history traits. Table 6.3 is adapted from that paper with modifications and embellishments.

In evaluating the theory of *r*- and *K*-selection, we should recognize that it is based on the assumption that life histories have evolved as a response to the twin selective pressures of competition and colonizing or dispersal ability. As described later, life histories that appear to be *r*-selected may have evolved in response to predation (Reznick and Endler 1982, also see review in Reznick 2011) or to the uncertainties of the physical environment (Wilbur et al. 1974). Caution must therefore be used in making assumptions about the selective pressures that have shaped a particular life history.

Dobson and Oli (2007) have written than the *r*- and *K*-selection theory should be abandoned since the traits associated with *r*- and *K*-selection are primarily the result of body size differences. However, simply associating life history with body size tells us little about the environmental circumstances that produced these different body sizes. Gaillard et al. (1989) suggested a hierarchy of axes producing life history variation. First was body size, followed by high to low turnover of individuals and third a gradient based on an altricial versus a precocial life history. The second of these ideas has led to the proposal by Read and Harvey (1989) to describe life histories on a "slow-fast" continuum. Thus, a fast life history would be associated with early reproduction, large numbers of offspring, small body size of adults, and short generation times. A slow life history would feature delayed reproduction, small numbers of offspring, large adult body size, and long generation times. Data reviewed by Harvey and Purvis (1999) showed this slow-fast continuum can be discerned even when the data are adjusted for body size. Thus, although body size is a major determinant of life histories, there still exists a slow-fast continuum based on environmentally determined factors.

Dobson (2012) has presented a stepwise series of principles to explain the development of a life history. The first principle, based in part on the metabolic theory of Brown et al. (2004), is that body size, one of the "axes" proposed in the preceding paragraph, is an important determinant of life history. Since it takes longer for larger organisms to develop, they will present the suite of "slow" traits described earlier. The second principle is also about tempo. As described earlier, even if body size is accounted for, the slow-fast continuum is still evident among species. Dobson (2012) explains this through differences in extrinsic mortality rates, especially juvenile mortality rates. As discussed in the next section, given a "zero sum game" hypothesis, higher mortality usually leads to higher reproductive effort.

The third step in Dobson's paradigm is a "lifestyle hypothesis" to replace the slow-fast continuum in life histories and is beyond those differences caused by body size (Dobson 2007). Here the evolutionary, phylogenetic background of species is used to make predictions concerning evolved life history traits. What follows are his hypotheses concerning evolution of life histories among birds. For a similar treatment among mammals, see Dobson and Oli (2007). According to Dobson (2012, p. 10747–10748), the following have strong effects on reproductive effort: (i) altricial species have a faster life history than precocial; (ii) monogamous species have a slower life history than cooperative breeders; (iii) resident species are slower or "less productive" than cooperative breeders; (iv) diet affects the life history with nectar feeders the slowest or least productive; (v) pelagic species have the slowest life histories. In summary, life history traits are produced by a combination of body size, environmentally related factors, and the phylogeny of the species under consideration (Dobson 2012).

However, if mass explains most of the variations in life histories, is it logical to exclude it in developing a life history theory? Is it premature to try to explain life histories with a single theory? To do so ignores the wide variety of environmental influences that determine a life history and the fact that there exists a wide variety of life histories within a species. For example, tiger salamanders (*Ambystroma tigrinum*) have two larval morphs and three adult phenotypes, all with different life histories based on low versus high density conditions (Hoffman and Pfennig 1999). Furthermore, parthenium beetles can develop either a slow or fast life cycle from the same brood raised under the same environmental conditions (Pandey et al. 2013). I thank Kara Jones for pointing out these examples to me.

In a review of the *r*- and *K*-selection theory, Reznick et al. (2002) gave it credit for stimulating a great variety of empirical and theoretical work on life-history evolution. An important aspect of the theory was its focus on density-dependent selection as an important force in the shaping of life histories. Now it is clear that age-specific mortality, resource limitation, predation, environmental variation, metabolic rate, and density-independent factors all must be included in models of life history evolution. Nevertheless, although the *r*- and *K*-selection theory has largely fallen into "disrepute" (Dobson 2012), it is ingrained in the ecological lexicon and has a certain utility as handy shorthand for a particular suite of life history traits. As Fox (2011) points out in his essay on "Zombie ideas in ecology," we cannot kill it (the theory of *r*- and *K*-selection) so we might as well live with it.

6.6 Cost of reproduction and allocation of energy

In the paragraphs that follow, we will be discussing tradeoffs with regard to energy and its allocation to growth, maintenance, and reproduction. However, as pointed out by my colleague, Geoffrey Birchard, the vast majority of ecological studies deal with mass as a surrogate for energy. This likely creates biases in that metabolic rates for a particular body mass can vary by an order of magnitude. For example, there is a significant difference in the density of energy among bird eggs of the same mass.

Whether we use the terms *r*- and *K*-selection or not, they illustrate that as life history traits evolve, tradeoffs have taken place among growth, maintenance, and reproduction. Another way of approaching life history evolution is to examine the allocation of energy devoted to reproduction over a life span. This is known as reproductive effort. Energetic tradeoffs are found in all aspects of a life history. When the parent organism devotes energy to reproduction, it has less energy to devote to growth and maintenance. The energy devoted to reproduction can be partitioned in various ways: large versus small offspring, small versus large litter size, and the amount of energy devoted to parental care. Cole (1954) described the important dichotomy of semelparous versus iteroparous reproduction. In semelparity, reproduction is channeled into one major reproductive effort. Most insects, many invertebrates, fish, and many plants (annuals, biennials and some bamboos) have this life cycle. A semelparous life history in a disturbed habitat offers no mystery. More interesting are organisms that are long-lived, yet have semelparous reproduction. For example, mayflies often spend several years as larvae, and periodical cicadas live 13 or 17 years below ground as juveniles, but the adult phases are only a few hours in mayflies and a few weeks in periodical cicadas. Hawaiian silverswords live 7–30 years before flowering once and dying. Some bamboo species delay flowering for 100–120 years before producing a massive seed crop and dying. Janzen (1976) and others have proposed that this life history allows escape of highly vulnerable juvenile stages through predator satiation.

Iteroparity is common among most vertebrates and perennial plants. Iteroparous species, nevertheless, are extremely diverse in their life histories: (i) short versus long prereproductive periods; (ii) annual versus periodical reproduction; (iii) small versus large amount of reproductive effort; and (iv) many small offspring versus a few large offspring.

Since energy devoted to reproduction is not available for growth and maintenance, reproduction itself has a "cost" in terms of increased mortality or decreased growth of the adult organism. Thus, an individual that reproduces in a given year often has reduced survivorship and/or may reproduce at a lower rate in near future. In addition, there is usually a limit to the number of offspring that an adult or pair of adults can successfully raise without causing harm to themselves, their survivorship, their future reproduction, or to the survivorship of this year's brood.

Therefore, an organism may have greater evolutionary fitness over the long term if it postpones reproduction or limits the allocation of energy to current reproduction. This will be true if the energetic allocation to growth and maintenance (as opposed to reproduction) produces a sufficient gain in future reproduction to compensate for losses in the present. In other words, the value of early reproduction on future population growth does not universally apply.

Field studies documenting the effect of reproduction on growth, survival, and future reproduction are not abundant. Primack and Hall (1990) showed that reproduction in pink lady's slipper orchids (*Cypripedium acaule*) is limited by bee pollination, but seed production could be greatly increased through hand pollination. As seed production increased, the cost of reproduction took effect in the third and fourth years. Hand-pollinated plants had lower growth and flowering rates than controls. For example, an average-sized hand-pollinated plant lost 10–13% of leaf area and had a 5–16% lower flowering rate compared with control plants. In Red Deer (Clutton-Brock 1984; Clutton-Brock et al. 1982, 1989) and in lizards (Tinkle 1969), it has been shown that great reproductive effort leads to declining fecundity and reduced survivorship. In American bison (*Bison bison*), sons suckle longer than daughters (up to 15 months). Cows that have produced sons breed later and are more likely to be barren in the next year compared with cows that had females in the previous year (Wolff 1988).

Bernett et al. (2009) showed that reproduction in bloodroot (*Sanginaria canadensis*) in two or more consecutive years negatively affected leaf size and reproduction compared with plants reproducing in only 1 year. Specifically, plants that produced flowers in two consecutive years showed a reduction in leaf surface area ($p < 0.0001$) compared with plants that produced flowers in only 1 year. Plants that produced fruit in both years had significantly smaller leaf surface areas compared with plants that produced fruit in just 1 year. Thus when a bloodroot plant produces fruit, there is a cost in future flower, fruit, and seed production, as well as in its future photosynthetic surface area.

Experimental field studies are often done with birds, in which the clutch size can be manipulated. Reid (1987) studied the optimal brood size of the glaucous gull (*Larus hyperboreus*) in which the natural range of chicks per nest is 1-3. Reid added and subtracted eggs, producing broods of from one to seven chicks per nest. He found that when more than three chicks were present in the nest, adult survivorship declined significantly.

A study by Beissinger (1990) on snail kites at two field sites is also instructive. Brood size was manipulated with similar results. The normal brood range is 1–3 eggs. With only one egg in a nest, since one parent can raise one young successfully, both females and males often desert the nest and try to start another one. Desertion rates are almost 100% with one egg per nest, with females deserting twice as often as males. With two young per nest, desertion rate is about 50%. With three young in the nest, there was no desertion. All broods with one or two young were 100% successful at fledging. However, the kites had difficulty raising three young and were unable to raise four young. Desertion is an adaptive response that allows the kites to adjust their parental investment to the number of chicks present in a nest.

In the Venezuela field site, Beissinger found that no parents raised four young, and only 40% of nests with three young raised them all. At control nests, no broods of three young occurred naturally (205 nests). In Florida, however, 22% of the control nests fledged three young. The investigators found that total food delivery rates increase up to three young per nest, but there was no further increase with four per nest. Thus, natural selection closely controls the clutch size and the amount of parental effort per nest in snail kites.

6.7 Clutch size

To produce viable offspring, a certain amount of energy must be devoted to each young. Lack (1954, 1966) hypothesized that clutch size in birds evolved through natural selection to correspond to the number of young that could be fed. Although this idea is appealing, there are a number of exceptions and complications. For example, in 35 of 60 studies, an increase in brood size resulted in an increase in the number of young fledged, without affecting parental survival (Lessells 1991). However, very often, the juveniles had lower survivorship and fertility than in normal broods. In other studies, birds with larger clutches bred later the next year.

Although the Lack hypothesis is generally supported over the long run, there are many cases where clutch size is modified through brood reduction after egg laying. Studies of red-winged blackbirds (Forbes et al. 1997), for example, confirm the "insurance hypothesis," which states that hatching asynchrony creates "marginal" offspring that serve as replacements for failed earlier-hatched "core" offspring. Forbes et al. (1997) showed that "marginal" offspring had very high mortality rates in control nests but survived well if broods were experimentally or naturally reduced. In asynchronous hatching, the older siblings beg more vigorously for food, and/or the older siblings may simply kill the weaker sibling. If food shortages arise, the late-hatched young usually die of starvation. In such cases, the parents, who cannot predict food resources, can use brood reduction to adjust reproduction to actual food availability.

6.8 Latitudinal gradients in clutch size

Lack (1966), Cody (1966), and Ashmole (1963) all developed hypotheses to explain observed latitudinal gradients in clutch sizes. Most groups of birds show a gradient in which clutch sizes increase from the tropics to higher latitudes (see Table 6.4). Lack proposed that clutch sizes are adaptations to food supplies, and that increasing day length allows more foraging time in the higher latitudes. Cody (1966) developed an allocation of energy argument. He believed that both competition and predation were more intense in the tropics and tropical birds invested more energy both in competitive interactions and in predator avoidance. Therefore, they had less energy to allocate to reproduction than their temperate counterparts. Ashmole (1963) believed that clutch size varies in proportion to the seasonal variation in resources. In birds, population density in the higher latitudes is regulated by winter mortality, when resources in the temperate zone are scarce. Higher winter mortality produces a high ratio of available food per breeding adult. This results in large clutch sizes. Migratory birds can also be thought of as annual colonizers, whose reproductive styles tend to be r-selected. In addition, plant productivity per day is very high in the short growing season of the higher latitudes. Thus, it is not too surprising that clutch sizes are high. Limited data from plants and insects reinforce the general trend of higher reproductive effort at higher latitudes.

Martin et al. (2000) monitored 1331 nests in Argentina and 7284 nests in Arizona. They found that clutch sizes of passerine birds in Argentina averaged 2.41, similar to averages from the humid tropics. As expected, the Arizona site averaged 4.61 eggs per clutch.

Table 6.4 Clutch size in Passerine birds in relation to habitat.

Habitat	Number of species	Average clutch size
Tropical forest	82	2.3
Tropical grass and shrub lands	260	2.7
Tropical deserts	21	3.9
Middle Europe	88	5.6

Source: Data from Lack (1968) as summarized by Southwood et al. (1974).

However, although predation rates were significant at their sites, Martin et al. concluded that nest predation does not explain the smaller clutch sizes at the Argentina site. Furthermore, they concluded, contrary to Ashmole's theory, that clutch size was not related to food delivery rates, and food limitation is not greater in the South American site. Martin et al. (2000) favor the idea that adult mortality is reduced in the tropics. This leads to a life history involving a reduction in reproductive effort and/or an allocation of a greater amount of energy to a smaller number of offspring.

Another interesting pattern is that many island populations have lower clutch sizes than their mainland counterparts. Litter sizes among mammals also follow this pattern (Fons et al. 1997). Islands often have fewer predators, so both juvenile and adult survivorship are higher. With fewer predators and fewer competing species, densities of the island populations will be higher. Since adults live longer, it is likely to be hard for young of the year to find vacant territories, and selection appears to favor lowering the number of young through a lower clutch size. For example, on Santa Cruz Island scrub jays have a clutch size of 3.71, which is significantly lower than that of the mainland (4.34). Adult survivorship on the island is 0.935 compared with 0.833 per year on the mainland (Atwood et al. 1990). Johnston et al. (1997) have shown that tropical passerine birds have higher adult survivorship than do comparable temperate zone species. This provides additional support for the theory that reproductive effort is based on age-specific mortality patterns.

Wiersma et al. (2007) found that tropical birds have evolved a reduced basal metabolic rate compared with temperate species. This fact helps explain why tropical birds tend to be on slow side of the "slow-fast" continuum. Wiersma et al. (2007) hypothesize that basal metabolic rate is heritable, and this combined with the fact extrinsic mortality is lower in the tropics results in adult birds that allocate more energy to self-maintenance and less to reproductive effort.

6.9 The effects of predation and disease on life history characteristics

Predation also plays a role in the shaping of life histories. Studies by Reznick and Endler (1982) and Reznick et al. (1997) (see summary in Reznick (2011) on guppies have confirmed the effect of predation on life histories. Reznick and Endler (1982) came to the following conclusions:

1. When a predator prefers mature fish, the guppies devote a high percentage of their body weight to reproduction, there are short interbrood intervals, and they mature at a small size.
2. When a predator prefers immature stages or no predators are present, the guppies devote a low percentage of body weight to reproduction, there are long interbrood intervals, and they mature at a large size.

In an experimental test of their predictions, Reznick et al. (1997) compared the life histories of two populations of guppies (*Poecilia reticulata*) living in upstream versus downstream habitats. The upstream habitats lacked several species of fish, many of which prey upon adult guppies, which were present downstream. The upstream, low-predation, population had later sexual maturity, lower fecundity, less frequent litters, and larger offspring compared with the high-predation, downstream population. In other words, lack of predation produced the suite of traits listed under point two in the preceding list. If we knew nothing about predation, we might conclude that such traits were consistent with the K-selection syndrome, and that these life histories were shaped by competitive interactions. In a transplant experiment, Reznick et al. (1997)

moved guppies from the high-predation to the low-predation habitat. After 11 years, the transplanted guppies had evolved life histories similar to the resident populations.

In a similar study, Crowl and Covich (1990) showed that freshwater snails (*Physella virgata virgata*) also shift their life history characteristics when exposed to their major predator, the crayfish, *Oncopeltus virilis*. When exposed to water-borne cues released when crayfish fed on other members of the population, the snails exhibited rapid growth and delayed reproduction until they reached a size of 10 mm after 8 months. In the absence of the cue, snails typically grew to about 4 mm in 3.5 months and then began reproduction. Since crayfish selectively prey on the smallest individuals, juvenile mortality is high. By delaying the onset of reproduction and growing to a larger size, snails decrease mortality from size-specific predation.

In both of these studies, when predators concentrated their efforts on immature or smaller stages, the life history of the prey species shifts to a "K-selected" syndrome. Obviously caution must be used when evaluating the potential causes of a particular life history.

Tasmanian Devil (*Sarcophilus harrisii*) populations in the Australian state of Tasmania populations have been declining since 1996 because of a contagious form of cancer known as facial tumor disease. This disease causes death between the ages of 2 and 3 years (Jones et al. 2008). Because of almost complete mortality of adult individuals, data suggest that devils are now breeding at a younger age. In the past, devils began breeding between the ages of two and four. But now there is a 16-fold increase in individuals breeding at the age of 1 year. Not only that, but since individuals usually die before the age of 2, the cancer has fundamentally changed the life history from one of iteroparity to one of semelparity. Thus, not only predation, but also disease can powerfully select for a particular life history.

6.10 Bet-hedging

When the environment is unpredictable, selection may favor the concentration of reproduction early in life. Alternatively, it may favor spreading reproduction over several seasons or years. If recruitment of offspring is unpredictable from year to year, selection should favor adult survival at the expense of present fecundity (Stearns 1992). Akre (2002), for example, found that the life history of many species of fresh-water turtles appears to be a product of a bet-hedging (or unpredictable) environment. When Sandercock and Jaramillo (2002) examined survivorship rates of wintering sparrows, they also found that the life histories were consistent with a bet-hedging hypothesis.

6.11 The Grime general model for three evolutionary strategies in plants

Grime (1977) developed a life history model specifically for plants. In this model, Grime categorized factors that limit plant biomass or productivity into:

1. *Stress.* Stress is any condition that restricts plant production. Examples would be shortage of light, water, nutrients, or low temperatures.
2. *Disturbance.* A disturbance is partial or total loss of plant biomass arising from the activities of herbivores, pathogens, man, or from environmental factors such as wind, frost, desiccation, soil erosion, or fire.

    Grime then put together a chart (Table 6.5) in which stress and disturbance are placed into categories by intensity (low or high), and the result is a two by two table. Each of the four possible interactions is a possible life history strategy, but he claims that there is no viable strategy for high stress combined with high disturbance. This leaves three ecological life history strategies for plants.

Table 6.5 Grime model for three basic life histories for plants.

Intensity of disturbance	Intensity of stress	
	Low	High
Low	Competitive strategy	Stress tolerant strategy
High	Ruderal strategy	No viable strategy

(a) *The competitive strategy.* Competition is the tendency of neighboring plants to use the same quantum of light, ion of mineral nutrient, molecule of water, or volume of space. A wide range of studies has shown that while competitive ability is important in productive conditions, it declines in importance in unproductive habitats. During periods of high stress, characteristics related to rapid potential growth become disadvantageous.

(b) *Stress tolerant strategy.* This strategy is best represented in unproductive habitats. Examples are as follows: (i) Low-temperature arctic or alpine areas – Here, important life forms are low evergreen shrubs, small perennial herbs, bryophytes, and lichens. (ii) Arid habitats – Plants with small hard leaves, succulents, or other water conserving mechanisms are favored. (iii) Shaded habitats – In this case, shade tolerance is important. (iv) Nutrient-deficient habitats – Plants with slow growth rates and mechanisms for conserving leaves and other plants parts have been selected for.

(c) *Ruderal strategy* (= *r* – selected *strategy of MacArthur and Wilson*). This strategy is common in severely disturbed but potentially productive habitats. In this case, rapid colonization, rapid growth, and high reproductive rates are favored.

During succession, Grime sees the plant community changing over time from ruderal to competitive to stress-tolerant. However, a highly productive habitat may stay in the competitive mode. Grime displays the three strategies on an equilateral triangle (Figure 6.5) with each corner representing maximum importance of one strategy or another. The apex is the competitive strategy, the right corner is ruderal, and the left corner is stress-tolerant. Life forms are organized along these three axes. Annual herbs are basically ruderals, lichens

Fig. 6.5 The Grime model for three evolutionary strategies in plants.

are usually stress-tolerant, trees and shrubs are a combination of stress-tolerant and competitive, and bryophytes are a combination of stress-tolerant and ruderal (Grime 1977).

6.12 Summary

An amazing diversity of life histories can be found in almost any ecological community. Since at least 1954, ecologists have been attempting to understand and explain this diversity. A wide range of organisms in a wide variety of environments have shown us that there exists a basic dichotomy between the reproductive rate and the size or life span of an organism. The MTE emphasizes that mass and temperature heavily influence life history traits, producing an inverse relationship between mass and r. Another dichotomy is between dispersal ability and competitive ability, which is emphasized in the theory of r- and K-selection. More modern theories emphasize the primary role of body size (mass), but even holding mass constant, there is a spectrum of life histories (slow vs fast). Dobson (2012) has added the phylogenetic and environmental backgrounds as important determinants of life history.

Early theorists, such as Cole and Lewontin, emphasized life history characteristics that would be most effective in increasing r. As a result, they focused their attention on generation time and the fertility aspects of the life table. In many environments, however, we have found that the major adaptive syndrome revolves around reduction in adult mortality. Tropical environments, islands, and unpredictable survivorship of juvenile stages, all have selected for long-lived adults with low mortality rates and iteroparous reproduction. The Grime stress-tolerant strategy for plants can be viewed as a set of adaptations for environments in which resources are limited, and recruitment of juvenile is uncertain. The result is mature plants that are long-lived and iteroparous, exactly what we normally find in climax stages of vegetation.

It will always be difficult to distinguish among selective factors that might have produced an observed life history: a stochastic physical environment, predation, competition, mating system, or some other factor we have not thought of yet. The amazing variety of life histories will keep ecologists and evolutionary biologists busy for a long time.

References

Agutter, P.S. and D.N. Wheatley. 2004. Metabolic scaling: consensus or controversy? *Theoretical Biology and Medical Modelling* 1: 1–13. URL: http://www.tbiomed.com/content/1/1/13.

Akre, T.S.B. 2002. Growth maturity and reproduction of the Wood Turtle, Clemmys insculpta (Le Conte, 1830) in Virginia. Ph.D. Dissertation. George Mason University, Fairfax, Virginia.

Ashmole, N.P. 1963. The regulation of numbers of tropical oceanic birds. *Ibis* 103b: 458–473.

Atwood, J.L., M.J. Elpers, and C.T. Collins. 1990. Survival of breeders in Santa Cruz Island and mainland California scrub jay populations. *Condor* 92: 783–788.

Beissinger, S.R. 1990. Experimental brood manipulation and the monoparental threshold in snail kites. *American Naturalist* 136: 20–38.

Bernett, C.C., A. Laemmerzahl, and L.L. Rockwood. 2009. Reduction in reproductive output and leaf size in *Sanguinaria cannadensis* as a cost of reproduction. *Journal of the Torrey Botanical Society* 126: 457–464.

Blueweiss, L., H. Fox, V. Kudzma, D. Nakashima, R. Peters, and S. Sams. 1978. Relationships between body size and some life history parameters. *Oecologia* 37: 257–272.

Brown, J.H. and R.M. Sibly. 2012. The metabolic theory of ecology and its central equation. Pages 21–33, *in* R.M. Sibly, J.H. Brown, and A. Kodric-Brown, Editors. *Metabolic Ecology.* Wiley-Blackwell, Chichester, UK.

Brown, J.H., J.F. Gillooly, A.P. Allen, V.M. Savage, and G.F. West. 2004. Toward a metabolic theory of ecology. *Ecology* 85: 1771–1789.

Brown, J.H., R.M. Sibly, and A. Kodric-Brown. 2012. Introduction: Metabolism as the basis for a theoretical unification of ecology. Pages 1–8, *in* R.M. Sibly, J.H. Brown, and A. Kodric-Brown. *Metabolic Ecology*. Wiley-Blackwell. Chichester, UK.

Calder, W.A., III. 1984. *Size, Function and Life History*. Harvard University Press, Cambridge, MA.

Charnov, E.L. and W.M. Schaffer. 1973. Life history consequences of natural selection: Cole's result revisited. *American Naturalist* 107: 791–793.

Clutton-Brock, T.H. 1984. Reproductive effort and terminal investment in iteroparous animals. *American Naturalist* 123: 212–229.

Clutton-Brock, T.H., F.E. Guiness, and S.D. Albon. 1982. *Red Deer: Behavior and Ecology of Two Sexes*. University of Chicago Press, Chicago, IL.

Clutton-Brock, T.H., S.D. Albon, and F.E. Guiness. 1989. Fitness costs of gestation and lactation in wild mammals. *Nature* 337: 260–262.

Cody, M.L. 1966. A general theory of clutch size. *Evolution* 20: 174–184.

Cole, L.C. 1954. The population consequences of life history phenomena. *Quarterly Review of Biology* 29: 103–137.

Crowl, T.A. and A.P. Covich. 1990. Predator-induced life history shifts in a freshwater snail. *Science* 247: 949–951.

Darveau, C.A., R.K. Suarez, R.D. Andrews, and P.W. Hochachka. 2002. Allometric cascade as a unifying principle of body mass effects on metabolism. *Nature* 417: 166–170.

Dobson, S.F. 2007. A lifestyle view of life-history evolution. *Proceedings of the National Academy of Sciences of the United States of America*. 104: 17565–17566.

Dobson, S.F. 2012. Lifestyles and phylogeny explain bird life histories. *Proceedings of the National Academy of Sciences of the United States of America* 109: 10747–10748.

Dobson, S.F. and M.K. Oli. 2007. Fast and slow life histories of mammals. *Ecoscience* 14: 292–299.

Duncan, R.P., D.M. Forsyth, and J. Hone 2007. Testing the metabolic theory of ecology: allometric scaling exponents in mammals. *Ecology* 88: 324–33.

Enquist, B.J. and L.P. Bentley. 2012. Land plants: new theoretical directions and empirical prospects. Pages 164-187, *in* R.M. Sibly, J.H. Brown, and A. Kodric-Brown, Editors. *Metabolic Ecology*. Wiley-Blackwell. Chichester, UK

Enquist, B.J., J.H. Brown, and G.B. West. 1998. Allometric scaling of plant energetics and population density. *Nature* 395: 163–165.

Enquist, B.J., G.B. West, E.L. Charnov, and J.H. Brown. 1999. Allometric scaling of production and life-history variation in vascular plants. *Nature* 401: 907–911.

Enquist, B.J., G.B. West, and J.H. Grown. 2000. Quarter-power allometric scaling in vascular plants: functional basis and ecological consequences. Pages 167–198, *in* J.H. Brown and G.B. West, Editors. *Scaling in Biology*. Oxford University Press, Oxford, UK.

Fons, R., F. Poitevin, J. Catalan, and H. Croset. 1997. Decrease in litter size in the shrews *Crocidura suaveolens* (Mammalia, Insectivora) from Corsica (France): evolutionary response to insularity? *Canadian Journal of Zoology* 75: 954–958.

Forbes, S., S. Thornton, B. Glassey, M. Forbes, and N.J. Buckley. 1997. Why parent birds play favorites. *Nature* 390: 351–352.

Fox, J. 2011. Zombie ideas in ecology: r and K-selection. http://oikosjournal.wordpress.com/2011/06/29/zombie-ideas-ecology-r-and-k-selection/.

Gaillard, J.–M., D. Pontier, D. Allaine, J.d. Lebreton, J. Trouvilliez, and J. Clobert. 1989. An analysis of demographic tactics in birds and mammals. *Oikos* 56: 59–76.

Grime, J.P. 1977. Evidence for the existence of three primary strategies in plants and its relevance to ecological and evolutionary theory. *American Naturalist* 111: 1169–1194.

Hanski, Ilkka. 1999. *Metapopulation Ecology*. Oxford University Press, Oxford, UK.

Harvey, P.H. and A. Purvis. 1999. Understanding the ecological and evolutionary reasons for life history variations: Mammals as a case study, Pages 232–348, *in Advanced Ecological Theory: Principles and Applications*. Blackwell Science, Malden, Massachusetts.

Hoffman, E.A. and D.W. Pfennig. 1999. Proximate causes of cannibalistic polyphenism in larval tiger salamanders. *Ecology* 80: 1076–1080.

Horn, H.S. 2004. Commentary of Brown et al.'s "toward a metabolic theory of ecology". *Ecology* 85: 1816–1818.

Janzen, D.H. 1976. Why bamboos wait so long to flower. *Annual Review of Ecology, Evolution and Systematics* 7: 347–391.

Janzen, D.H. 1983. *Costa Rican Natural History*. University of Chicago Press, Chicago, IL.

Jenni, D.A. 1983. *Jacana spinosa*. Pages 584–586, *in* D.H. Janzen, Editor. *Costa Rican Natural History*. University of Chicago Press. Chicago, IL.

Johnston, J.P., W.J. Peach, R.D. Gregory, and S.A. White. 1997. Survival rates of tropical and temperate passerine birds: a Trinidadian perspective. *American Naturalist* 150: 771–789.

Jones, M.E., A. Cockburn, R. Hamede, C. Hawkins, H. Hesterman, S. Lachish, D. Mann, H. McCallum, and D. Pemberton. 2008. Life history change in disease-ravaged Tasmanian devil populations. *Proceedings of the National Academy of Science* 105: 10023–10027.

Kleiber, M. 1932. Body size and metabolism. *Hilgardia* 6: 315–322.

Kolokotrones, T., V. Savage, E.J. Deeds, and W. Fontana. 2010. Curvature in metabolic scaling. *Nature* 464: 753–756.

Lack, D. 1954. *The Natural Regulation of Animal Numbers*. Oxford University Press, London, UK.

Lack, D. 1966. *Population Studies of Birds*. Clarendon Press, Oxford, UK.

Lack, D. 1968. *Ecological Adaptations for Breeding in Britain*. Metheum Press. London, UK.

Lessells, C.M. 1991. The evolution of life histories. Pages 32–68, *in* J.R. Krebs and N.B. Davies, Editors. *Behavioral Ecology*, 3rd Edition. Blackwell, Oxford and London, UK.

Lewontin, R.C. 1965. Selection of colonizing ability. Pages 77–94, *in* H.G. Baker and G.L. Stebbins, Editors. *The Genetics of Colonizing Species*. Academic Press, New York, NY.

Lonsdale, W.M. 1990. The self-thinning rule: dead or alive? *Ecology* 71: 1373–1388.

Lotka, A.J. 1925. *Elements of Physical Biology*. Reprinted 1956. Dover, New York, NY.

MacArthur, R.H. and E.O. Wilson 1967. *The Theory of Island Biogeography*. Princeton University Press, Princeton, NJ.

Martin, T.E., P.R. Martin, C.R. Olson, B.J. Heidinger, and J.J. Fontaine. 2000. Parental care and clutch sizes in North and South American birds. *Science* 287: 1482–1485.

Packard, G.C. and G.F. Birchard. 2008. Traditional allometric analysis fails to provide a valid predictive model for mammalian metabolic rates. *Journal of Experimental Biology* 211: 3581–3587.

Pandey, P., G. Mishra, and Omkar. 2013. Slow and fast development in Parthenium beetle and its effect on reproductive attributes. *Journal of Asia-Pacific Entomology* 16: 395–399.

Peters, G.L. and R.P. Larkin. 1989. *Population Geography*, 3rd Edition. Kendall/Hunt, Dubuque, IA.

Pianka, E. 1970. On r and K selection. *American Naturalist* 104: 592–597.

Primack, R.B. and P. Hall. 1990. Cost of reproduction in the pink lady's slipper orchid: a four-year experimental study. *American Naturalist* 136: 638–656.

Read, A.F. and P.H. Harvey. 1989. Life history differences among the eutherian radiations. *Journal of Zoology (London)* 219: 329–353.

Reid, W.V. 1987. The cost of reproduction in the glaucous-winged gull. *Oecologia* 74: 458–467.

Reznick, D.N. 2011. Guppies and the empirical study of adaptation. Pages 205–232, *in* J.B. Losos, Editor. *In the Light of Evolution*. Roberts and Company, Greenwood Village, Colorado.

Reznick, D.N. and J. Endler. 1982. The impact of predation on life history evolution in Trinidadian guppies (*Poecilia reticulate*). *Evolution* 36: 160–177.

Reznick, D.N., F.H. Shaw, H.F. Rodd, and R.G. Shaw. 1997. Evaluation of the rate of evolution in Trinidadian guppies: Genetic basis of observed life history patterns. *Evolution* 36: 1236–1250.

Reznick, D., M.J. Bryant, and F. Bashey. 2002. r- and K-selection revisited: the role of population regulation in life-history evolution. *Ecology* 83: 1509–1520.

Sandercock, B.K. and A. Jaramillo. 2002. Annual survival rates of wintering sparrows: assessing demographic consequences of migration. *The Auk* 119: 149–165.

Sibly, R.M. 2012. Life history. Pages 57–66, *in* R.M. Sibley, J.H. Brown, and A. Kodric-Brown, Editors. *Metabolic Ecology*. Wiley-Blackwell, Chichester, UK.

Sibly, R.M., R.H. Brown, and A. Kodric-Brown. 2012. *Metabolic Ecology*. Wiley-Blackwell, Chichester, UK.

Slobodkin, L.B. 1962. *Growth and Regulation of Animal Populations*. Holt, Rinehart, and Winston, New York, NY.

Smith, F.E. 1954. Quantitative aspects of population growth. Pages 277–294, *in* E. Boell, Editor. *Dynamics of Growth Processes*. Princeton University Press, Princeton, NJ.

Southwood, T.R.E. 1976. Bionomic strategies and population parameters. Pages 26–48, *in* Robert M. May, Editor. *Theoretical Ecology, Principles and Applications*. W. B. Saunders. Blackwell Scientific Publications, Philadelphia, PA.

Southwood, T.R.E., R.M. May, M.P. Hassell, and G.R. Conway. 1974. Ecological strategies and population parameters. *American Naturalist* 108: 791–804.

Stearns, S.C. 1992. *The Evolution of Life Histories*. Oxford University Press, New York, NY.

Tinkle, D.W. 1969. The concept of reproductive effort and its relation to the evolution of life histories of lizards. *American Naturalist* 103: 501–516.

Verhencamp, J.L. 1977. Relative fecundity and parenton effort in communally nests anis, *Crotophaga sulcirostris*. *Science* 197: 403–405.

Verhencamp, J.L. 1978. The adaptive significance of communal nesting in groove-billed anis (*Crotophaga sulcirostris*). *Behavioral Ecology and Sociobiology* 4: 1–33.

West, G.B., J.H. Brown, and B.J. Enquist. 1997. A general model for the origin of allometric scaling laws in biology. *Science* 276: 122–126.

West, G.B., J.H. Brown, and B.J. Enquist. 1999a. A general model for the structure and allometry of plant vascular systems. *Nature* 400: 664–667.

West, G.B., J.H. Brown, and B.J. Enquist. 1999b. The fourth dimension of life: fractal geometry and allometric scaling of organisms. *Science* 284: 1677–1679.

West, G.B., J.H. Brown, and B.L. Enquist. 2000. The origin of scaling universal scaling laws in biology. Pages 87–112, *in* J.H. Brown and G.B. West, Editors. *Scaling in Biology*. Oxford University Press, Oxford, UK.

White, C.R., P. Cassey, and T.M. Blackburn. 2007. Allometric exponents do not support a universal metabolic allometry. *Ecology* 88: 315–323.

Wiersma, P., A. Muñoz-Garcia, A. Walker, and J. B. Williams. 2007. Tropical birds have a slow pace of life. *Proceedings of the National Academy of Sciences of the United States of America* 104: 9340–9345.

Wilbur, H.M., D.W. Tinkle, and J.P. Collins. 1974. Environmental uncertainly, trophic level, and resource availability in life history evolution. *American Naturalist* 108: 805–817.

Wolff, J.O. 1988. Maternal investment and sex ratio adjustment in American bison calves. *Behavioral Ecology and Sociobiology* 23: 127–133.

Yoda, K., T. Kira, H. Ogawa, and K. Hozumi. 1963. Self-thinning in overcrowded pure stands under cultivated and natural conditions. *Journal of Biology, Osaka City University* 14: 107–129.

Part 2

Interspecific interactions among populations

At this point we are ready to consider interactions among populations of different species. These interactions may be at the same trophic level (interspecific competition) or between different trophic levels (predator–prey, parasite–host, parasitoid–host, and plant–herbivore interactions). Some of these interactions are considered "symbiotic." A symbiosis is the intimate biotic association of phylogenetically unrelated species, and is thought to develop as a consequence of co-evolution. While symbiosis is sometimes considered exotic or rare in nature, it is actually a rather common phenomenon. For example, the eukaryotic cell appears to be a coevolved symbiotic complex involving organelles such as mitochondria and chloroplasts that were originally free-living organisms. Lichens consist of a symbiotic complex of algae living inside fungi, and the roots of higher plants have symbiotic associations with fungi (mycorrhizae). Moreover, the roots of leguminous plants have nodules within which bacteria (*Rhizobium* and related genera) live. Corals have a symbiotic relationship with algae known as zooxanthellae, and many species of flowering plants have complex relationships with ants.

Types of interactions

The following are symbolic representations of interactions between two species. In each case the symbol indicates whether the interactions is positive (+), negative (−), or neither for each of the two species involved.

−/− = Competition, which is considered to be a reciprocally negative interaction for both of the species involved.

+/− = A biological association in which one individual benefits while the other is harmed. Such interactions include predator/prey, herbivore/plant, and parasite/host relationships.

+/0 = Commensalism. One partner benefits from the interaction, while the other experiences no particular benefit or harm. Examples include: (i) seeds with barbs that stick to the fur of animals and are thereby dispersed; (ii) ectoparasites (mites, bacteria etc.) that live on the skin of animals but do no harm to the host; (iii) a bird nest in a tree; or (iv) a cattle egret which feeds more efficiently in the company of a cow.

+/+ = Mutualism. Both partners benefit from the interaction. Examples include: (i) "cleaners" and their hosts such as ox-peckers and ungulates, or cleaner shrimp and fish; (ii) pollinator-plant interactions; (iii) frugivore-plant relationships; (iv) a variety of ant-plant interactions; and (v) relationships between higher plants and fungi (mycorrhizae) or bacterial (*Rhizobium*) nitrogen fixers.

Introduction to Population Ecology, Second Edition. Larry L. Rockwood.
© 2015 John Wiley & Sons, Ltd. Published 2015 by John Wiley & Sons, Ltd.
Companion Website: www.wiley.com/go/rockwood/populationecology

We do not usually consider the above associations to be "symbiotic" unless there is a definite "living together" of the associates. Thus the algae and fungi that make up lichens are symbiotic, as are *Rhizobium* bacteria that live in the roots of legumes. But plant-pollinator and bird-seed dispersal interactions are not symbiotic.

Predator–prey, parasite–host, and plant–herbivore interactions

These interactions are of such fundamental importance in that they have effects at the population, the community and the ecosystem level.

1. At the population level, predation, herbivory and parasitism often control or help regulate animal and plant populations. Predators and parasites may also change the demographic structure of the prey/host population and thereby drive population cycles (Dobson and Hudson 1992).

2. At the community level, a predator, parasite or herbivore can decimate a community, eliminating certain species entirely, and fundamentally changing the community (Fritts 1988; Rhodda et al. 1992). Alternatively, they may have a positive effect on community diversity by allowing the coexistence of less competitively dominant species (Paine 1966).

3. At the ecosystem level, predator–prey, plant–herbivore, and parasite–host interactions provide means of energy flow from one trophic level to another.

Terminology

Terms used in discussing interactions among species are explicitly defined below.

Facultative: This is in interaction or association among species that is not required for the survival of the individuals involved. A parasite or an herbivorous insect often has several hosts and a mutualist may be able to survive without its partner.

Obligatory: This is an interaction or association that is required for the survival of the organisms involved. In this case, specialized pollinators, parasites, herbivores or mutualists cannot survive without the other species with which it is associated.

Carnivorous predators kill living prey and usually consume them immediately.

Parasitoids, on the other hand (examples are certain wasp species), immobilize their prey, and lay eggs on them. The host is eventually consumed and killed by the wasp larvae as they mature.

Parasites do not usually kill their hosts, but harvest energy from the hosts over a period to time. During this time the parasite reproduces copiously.

Herbivores are similar to parasites in that they consume plant parts, but usually not entire plants, and they do not usually kill the plants they feed upon.

Seed predation, on the other hand, is an example of true predation if the seed embryo is consumed. On the other hand, *fruit eating* is usually an example of mutualism. The fruit is consumed, but individual seeds are dispersed without the embryo being killed. *Scatter hoarding,* such as a squirrel-acorn interaction, is a fascinating combination of seed predation and seed dispersal.

Types of parasites

Parasites that live outside the host are termed *ectoparasites*, while those that live inside the host are *endoparasites*. Parasites also differ with regard to: (i) the level of harm caused to the host (virulence); (ii) their ability to move from one host to another (transmissibility); and (iii) the degree to which they are restricted to a given type of host species (specialization or host-specificity).

The table below summarizes ecological characteristics of parasite–host, herbivore–plant and predator–prey relationships. In the next few chapters, we will first discuss interspecific competition before moving on to mutualism, followed by parasite–host,

Comparisons of ecological characteristics along a spectrum from Parasite to Predator. Modified from Dobson (1982) and Toft (1986).

Organism type	Micro-parasite	Macro-parasite	Parasitoid	Small herbivore	Large herbivore	Solitary predator	Pack predator
Examples	Viruses/Bacteria	Tick	"Parasitic" wasp	Insect	Howling monkey or ungulate	Lynx or mountain lion	Wolf
Body size	Much smaller than host	Smaller than host	Adult stages are similar in size	Smaller than host	Larger than host when feeding on grasses or herbs; Smaller if host is a shrub or tree	May be larger or smaller than prey	Individuals are usually smaller than prey
Intrinsic rate of increase, r, as compared to host	Much larger than host	Larger than host	Slightly smaller than host	Larger than host	Comparable to smaller	Usually smaller than prey	Comparable to smaller than prey
Effect of interaction on the host individual	No effect to seriously deleterious	Mildly to seriously deleterious in vertebrate host; Can be highly virulent in intermediate host	Eventually fatal	Little effect to seriously deleterious	Little effect to very deleterious	Immediately fatal	Immediately fatal

predator–prey, and herbivore–plant relationships. In the final chapter of the book we will discuss multi-trophic interactions.

References

Dobson, A.P. 1982. Comparisons of some characteristics of the life histories of microparasites, macroparasites, parasitoids, and predators. Page 5 *in* R.M. Anderson and R.M. May, Editors. *Population Biology of Infectious Diseases*. Springer-Verlag, New York, NY.

Dobson, A.P. and P.J. Hudson. 1992. Regulation and stability of a free living host-parasite system: *Trichostrongylus tenuis* in red grouse. II. Population models. *Journal of Animal Ecology* 61: 487–498.

Fritts, T.H. 1988. The brown tree snake, *Boiga irregularis*, a threat to Pacific Islands. U.S. Fish and Wildlife. Service, Biological Report 88(31). 36 pp.

Paine, R.T. 1966. Food web complexity and species diversity. *American Naturalist* 100: 65–75.

Rhodda, G.H., T.H. Fritts, and P.J. Conry. 1992. Origin and population growth of the Brown Tree Snake, *Boiga irregularis*, on Guam. *Pacific Science* 46: 46-57.

Toft, C.A. 1986. Communities of species with parasitic life cycles. Pages 465-463 *in* J. Diamond and T.J. Case, Editors. *Community Ecology*. Harper and Row, New York, NY.

7 Interspecific competition

7.1 Introduction

7.1.1 The ecological niche

Our understanding of interspecific competition and the development of the concept of the niche have shared an interesting history. The term "ecological niche" can be traced to both the American ecologist, Joseph Grinnell, and to the British ecologist, Charles Elton. In a 1927 book, Elton defined niche in the context of trophic position and feeding habits of an animal. Elton viewed the niche as a subdivision within the traditional trophic grouping of herbivore, carnivore, and so on. Grinnell, however, described the niche in terms of the potential distribution of a species over habitat types, and in his paper on the feeding habits of the California thrasher, he connected competitive exclusion to the term "niche." "It is, of course, axiomatic that no two species regularly established in a single fauna have precisely the same niche relationships" (Grinnell 1917, p. 433). However, because various authors used the term "niche" differently, and because there was no formal definition of the niche, many ecologists used the term informally. For example, in his ecology text, Krebs (1994) defined niche as "the role or 'profession' of an organism in the environment; its activities and relationships in the community." In their ecology text, Begon et al. (1986) defined the niche more formally as, "the limits, for all important environmental features, within which individuals of a species can survive, grow and reproduce."

Ricklefs (1997) used both the informal and the formal definition of the niche: "The ecological role of a species in the community" (informal) and "the ranges of many conditions and resource qualities within which the organism or species persists, often conceived as a multidimensional space" (formal).

The second, formal, definition is based on the concept, devised by Hutchinson (1957), of the niche as an N-dimensional hypervolume. In Hutchinson's definition, important environmental features such as temperature, pH, nutrient availability, food types, and/or sizes are depicted as niche dimensions. For each dimension, the limits are identified. For example, what are the minimum and maximum temperatures within which the species can survive? Or, what are the smallest and largest food-particles upon which a species can feed? The range within which the species can survive represents the fundamental niche for the species, for that niche dimension. All of these niche dimensions combined represent the overall *fundamental niche* of the species. If we take three niche dimensions into account, the niche can be represented as a three-dimensional volume. If we consider N dimensions, the resultant niche was described by Hutchinson as an *N-dimensional hypervolume*.

The fundamental niche is the largest ecological niche that an organism or species can occupy. It is based mostly on interactions with the physical environment and is always in the absence of competition. *The realized niche*, in contrast, is that portion of the fundamental niche that is occupied after interactions with other species. That is, the niche after competition. The realized niche must be part of, but smaller than, the fundamental niche.

7.2 Interspecific competition: early experiments and the competitive exclusion principle

Early in the twentieth century, Tansley (1917) experimentally demonstrated the potential power of interspecific competition in shaping ecological communities. Tansley had observed that closely related plant species living in the same region were often found in different habitats or different soil types. For his experiment, he selected two species of an herbaceous perennial, bedstraw, in the genus *Galium* (Rubiaceae). One species, *G. saxatile,* is normally found on peaty, acidic soils, while the second species *G. sylvestre* is an inhabitant of limestone soils. Tansley obtained soils from both areas, planted each species singly in each soil type, and then placed the two species together in each soil. He found that each species, when planted alone, was able to survive in both soils. Therefore, the fundamental niche for both species includes both acidic, peat-rich soil and limestone soil. However, growth and germination were best on the soil where the *Galium* species was normally found. When grown together on limestone soil, *G. sylvestre* overgrew and outcompeted *G. saxatile*. The opposite was true in the acidic peat soil. At this early date, Tansley had established that competitive exclusion could be demonstrated, and that the results differed by environment.

The work by Tansley, however, was not developed further until the publication by Gause (1934) of "The Struggle for Existence." Through a series of experiments with yeast (Gause 1932) and protozoans, Gause found that competitive exclusion is observed most often between two closely related species (e.g., two species in the same genus), when grown in a simple, constant environment. For example, see Figure 7.1a. Gause prepared organic extracts and introduced bacteria as food. When either *Paramecium caudatum* or *P. aurelia* was introduced alone, each flourished and grew logistically, leveling off at a carrying capacity. When placed together, however, *P. caudatum* diminished and eventually went extinct, while *P. auerlia* grew to a steady level (Figure 7.1b). There are two lessons from this experiment. First, two closely related species were unable to coexist in the simple test tube environment. Second, even though we declare *P. aurelia* the "winner," notice that its steady state of approximately 300 per 0.5 ml sample (Figure 7.1b) is less than the carrying capacity of 500 when this species was grown alone (Figure 7.1a). Recall the definition of competition as a reciprocally negative interaction, meaning that competition has a negative effect, even on the winners.

Gause's laboratory work inspired many others who worked with yeast, grain beetles, fruit flies, and other organisms easily grown in the laboratory. Crombie (1945, 1946, 1947), Park (1948), and others did some particularly interesting work with grain beetles. Different species were grown together in vials or other simple environments,

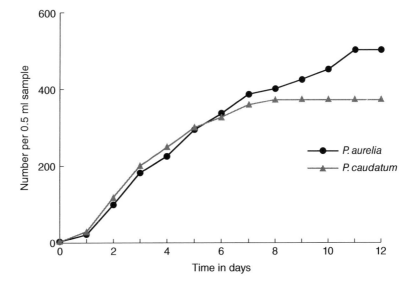

Fig. 7.1a Population dynamics of *P. aurelia* and *P. caudatum* when grown separately. (Source: Adapted from Gause 1934.)

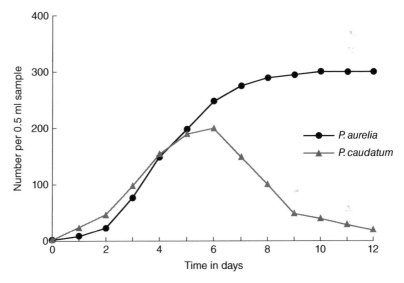

Fig. 7.1b Population dynamics of *P. aurelia* and *P. caudatum* when grown together. (Source: Adapted from Gause 1934.)

usually resulting in competitive exclusion. Crombie, however, showed that the species excluded could change depending on the temperature under which the experiment was run. And when he added glass tubing as a refuge, he found that a grain beetle in the genus *Oryzaephilus* was able to coexist with a related species in the genus *Tribolium*. Without the glass tubing, *Tribolium* drove *Oryzaephilus* extinct. All of this work suggested that closely related species, whose niches are very similar, are unlikely to coexist in a simple environment. From his research with *Paramecium*, Gause (1934) proposed what became known as Gause's theorem or principle.

(A) "Two species cannot coexist unless they are doing things differently."

   This was eventually rephrased such that competition and the niche concept became integrated.

(B) "No two species can occupy the same ecological niche."

Based on such results, Hardin (1960), three years after the publication of Hutchinson's definition of the niche as an *N*-dimensional hypervolume, proposed the *competitive exclusion principle.*

"Species which are complete competitors, that is, whose niches overlap completely, cannot coexist indefinitely."

The concepts of the niche and of competitive exclusion seem to have been known to Theodor Geisel (better known as Dr. Seuss) who obviously enjoyed writing on ecological topics (e.g., *The Lorax,* 1971). In his book *On Beyond Zebra,* Geisel (1955) wrote the following:

> "And NUH is the letter I use to spell Nutches
> Who live in small caves known as Nitches for Hutches.
> These Nutches have troubles, the biggest of which is
> The fact there are many more Nutches than Nitches.
> Each Nutch in a niche knows that some other Nutch
> Would like to move into his Nitch very much.
> So each Nutch in a Nitch has to watch that small Nitch
> Or Nutches who haven't got Nitches will snitch."

So there you have it: the perfect poetic explanation of the competitive exclusion principle evidently published before Hutchinson's 1957 paper! Is there possibly a reference to Hutchinson when he refers to "Hutches" in the second line?

When comparing the fundamental niches of competing species, it becomes obvious that their niches overlap on many, if not all, dimensions. In contrast, if we look at enough niche dimensions, since all species are genetically differentiated from one another, each species will have a unique niche. Therefore, complete niche overlap between two species is virtually impossible. Yet competitive exclusion does occur. A question we might pose is, how closely can niches overlap before competitive exclusion occurs? The other side of that question is, under what conditions do potential competitors coexist?

7.3 The Lotka–Volterra competition equations

The first important attempts to model interspecific competition were devised by Lotka (1925) and Volterra (1926). They based their models on the logistic equation, which is, of course, a model of intraspecific competition.

In modeling interspecific competition, Lotka and Volterra assumed that the growth rate of each species would be decreased as the population of its competitors increased. Thus, the impact of species two on the growth rate of species one is expressed as a modification of the logistic equation. Simultaneously, the number of individuals of species one modifies the growth of species two. To model competition between two species, Lotka and Volterra wrote two simultaneous equations, one for each species. Each equation is based on the logistic but includes a new term, the competition coefficient (α_{ij}), which describes the effect of one species on another. In the equations that follow:

$N_1 =$ the number of individuals of species one;
$N_2 =$ the number of individuals of species two;
$r_1 =$ the intrinsic rate of increase of species one;
$r_2 =$ the intrinsic rate of increase of species two;
$K_1 =$ the carrying capacity of species one;
$K_2 =$ the carrying capacity of species two;
$\alpha_{12} =$ the competition coefficient: effect of species two on species one;
$\alpha_{21} =$ the effect of species one on species two;
$t =$ time.

$$dN_1/dt = r_1 N_1 \left[\frac{K_1 - N_1 - \alpha_{12} N_2}{K_1} \right] \qquad (7.1)$$

$$dN_2/dt = r_2 N_2 \left[\frac{K_2 - N_2 - \alpha_{21} N_1}{K_2} \right] \qquad (7.2)$$

The competition coefficients try to model the impact of adding one individual of species i on species j. The value of the competition coefficient is usually between 0 and 1, for the following reasons. (i) A competition coefficient of zero would mean that there is no competition between the two species. If that were the case, there is no reason to try to model this interaction. (ii) If the competition coefficient were negative, the implication would be that species two actually benefits the growth rate of species one. The interaction between species one and two would then be *mutualistic*. (iii) Notice that the number of individuals of both species one and two decreases the carrying capacity. N_1 (taken directly from the logistic equation) is unmodified and represents intraspecific competition. N_2 is modified by the competition coefficient and represents interspecific competition. The implied α-value for *intraspecific* competition is therefore 1.0. To express this for the numerator on the right side of the equation, we could write:

$$K_1 - \alpha_{11} N_1 - \alpha_{12} N_2$$

Thus, when we say that the usual value of the interspecific competition coefficient is usually less than 1, we are saying that interspecific competition is almost always less intense and diminishes growth less than intraspecific competition, to which we have assigned the coefficient value of 1.0.

If we analyze this assumption in terms of niche overlap, the preceding makes intuitive sense. That is, in intraspecific competition, a species is competing with members of its own species, with which it has total niche overlap. (Actually, since there should be genetic variation within the species, there would not be total niche overlap, and in some species, males and females have different niches, particularly when it comes to food sizes or prey taken.) In interspecific competition, in contrast, the two species should have different niches. Thus, an individual of species two should compete less intensely with an individual of species one than with an individual of its own species. Growth rate should be diminished less by interspecific than by intraspecific competition. Hence the α_{ij} values are normally less than one and greater than zero.

Solving the Lotka–Volterra equations presents a number of problems. In addition to the large number of parameters that would have to be estimated (r, K, and α for each species), we would be assuming that changes with density were linear in both sets of differential equations. However, it is possible to do an "equilibrium analysis." That is, we can analyze the equations after the results of competition are complete, and both species are no longer growing. At equilibrium, the left sides of the equations, which represent growth, can be set to zero. These equations can then be written as follows:

$$dN_1/dt = 0 \quad \text{and} \quad dN_2/dt = 0 \qquad (7.3)$$

$$0 = r_1 N_1 \frac{K_1 - N_1 - \alpha_{12} N_2}{K_1} \quad \text{and} \quad 0 = r_2 N_2 \frac{K_2 - N_2 - \alpha_{21} N_1}{K_2} \qquad (7.4)$$

These equations, again, have no interesting explicit solutions. They can be solved if r_1 or $r_2 = 0$, but that simply means one of the species is not viable in this environment. If N_1 or $N_2 = 0$, then we have competitive exclusion. If K_1 or $K_2 = 0$, we have an undefined situation. What is left is to analyze the following expressions (7.5 and 7.6):

$$0 = K_1 - N_1 - \alpha_{12}N_2 \tag{7.5}$$

$$0 = K_2 - N_2 - \alpha_{21}N_1 \tag{7.6}$$

An ensuing graphical analysis, based on the steps outlined below, produces four possible solutions.

First, we solve equation 7.5 for N_1 and for N_2, producing equations 7.7a and 7.7b:

$$N_1 = K_1 - \alpha_{12}N_2 \tag{7.7a}$$

$$N_2 = \frac{K_1 - N_1}{\alpha_{12}} \tag{7.7b}$$

If species one wins in competition and species two goes extinct, we have equation 7.7c:

$$N_1 = K_1 \tag{7.7c}$$

If species two wins and species one goes extinct, we have 7.7d:

$$N_2 = \frac{K_1}{\alpha_{12}} \tag{7.7d}$$

From equation 7.6, we get the following two equations.

$$N_1 = \frac{K_2 - N_2}{\alpha_{21}} \tag{7.8a}$$

$$N_2 = K_2 - \alpha_{21}N_1 \tag{7.8b}$$

If species two wins and species one goes extinct, we get:

$$N_2 = K_2 \tag{7.8c}$$

If species one wins and species two goes extinct, we get:

$$N_1 = \frac{K_2}{\alpha_{21}} \tag{7.8d}$$

Therefore, if species two is extinct, equations 7.7c and 7.8d tell us that:

$$N_1 = K_1 = K_2/\alpha_{21}$$

and

$$\alpha_{21} = \frac{K_2}{K_1} \tag{7.9}$$

Similarly, if species one is extinct, equations 7.8c and 7.7d tell us that

$$N_2 = K_2 = K_1/\alpha_{12}$$

and

$$\alpha_{12} = \frac{K_1}{K_2} \tag{7.10}$$

According to these results, the competition coefficients are determined by the ratio of the carrying capacities.

To begin the graphical analysis, we start with equation 7.5. As we have seen from equation 7.7c, when species two $= 0$, then $N_1 = K_1$. Similarly, when species one $= 0$, $N_2 = K_1/\alpha_{12}$ (7.7d). If we graph N_1 versus N_2 (Figure 7.2), we can use the two points $(K_1, 0)$ and $(0, K_1/\alpha_{12})$ to produce a line that represents saturation levels for species one. Combinations of populations 1 and 2 below (to the left of) the line are such that population 1 continues to increase. Combinations above the line (to the right) lead to a decrease in the size of population 1 (Figure 7.2). The line itself is known as the zero isocline for population 1.

In a similar fashion, from equation 7.8c, we know that if population $1 = 0$, then $N_2 = K_2$. From equation 7.8d, we have that if population $2 = 0$, then $N_1 = K_2/\alpha_{21}$. From the resultant points $(0, N_2$ and $K_2/\alpha_{21}, 0)$, we can draw a second line or zero isocline, representing saturation levels for species two. Again, combinations of populations 1 and 2 to the left or below the line, result in increases toward the carrying capacity for population 2; combinations above the line lead to a decrease in population 2 (Figure 7.3).

We now place these zero isoclines or saturation levels on the same graph. Four combinations are possible when drawing these lines. (i) $K_1/\alpha_{12} > K_2$ combined with $K_1 > K_2/\alpha_{21}$ (Figure 7.4); (ii) $K_2 > K_1/\alpha_{12}$ combined with $K_2/\alpha_{21} > K_1$ (Figure 7.5);

Fig. 7.2 Zero isocline for species 1.

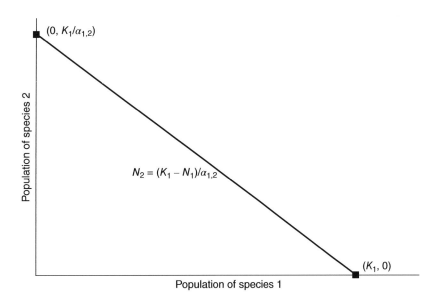

$(0, K_1/\alpha_{1,2})$

Population of species 2

$N_2 = (K_1 - N_1)/\alpha_{1,2}$

$(K_1, 0)$

Population of species 1

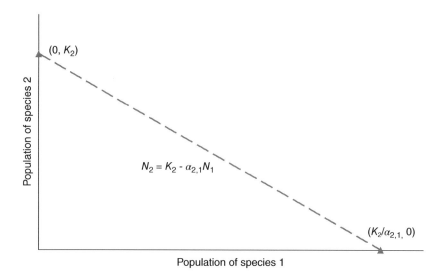

Fig. 7.3 Zero isocline for species 2.

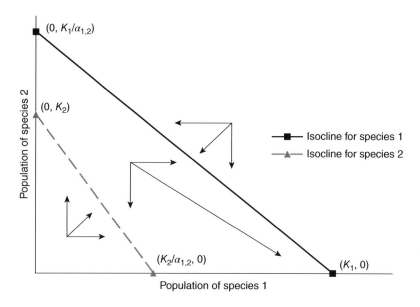

Fig. 7.4 Species 1 wins.

(iii) $K_1 > K_2/\alpha_{21}$ combined with $K_2 > K_1/\alpha_{12}$ (Figure 7.6); and (iv) $K_1/\alpha_{12} > K_2$ combined with $K_2/\alpha_{21} > K_1$ (Figure 7.7).

In Figure 7.4, both populations increase when their numbers are below both saturation lines. However, when combinations of species one and two are above the species-two zero isocline, but below the species-one zero isocline, the resultant vectors move the combined number of individuals toward only one equilibrium point, and that is when $N_1 = K_1$. In case one, then, we have competitive exclusion and species one is the winner.

Figure 7.5 illustrates the opposite situation. When combinations of species one and two are between the two saturation lines, the resultant vector moves the combined

Fig. 7.5 Species 2 wins.

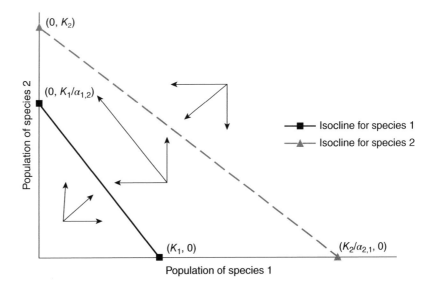

Fig. 7.6 Unstable
equilibrium (S) or
competitive exclusion
with an indefinite
winner.

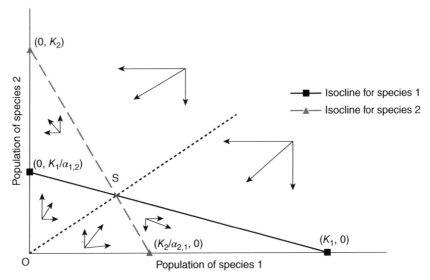

number of individuals toward an equilibrium at $N_2 = K_2$. Again we have competitive
exclusion, but now species two is the winner.

   Figure 7.6 represents a more complex, ambiguous situation. If the combination of
N_1 and N_2 is along the line 0–S, the resultant vector moves along the line to a temporary, unstable equilibrium at S, where both species coexist. However, if anything in the
environment moves the combinations of individuals to the right or the left of the line
0–S, then the resultant vectors move rapidly toward competitive exclusion. However,
in this case, there are two stable equilibrium points. If the combinations of the two
species produce points to the left of the line 0–S, the result is competitive exclusion
with species two as the winner. If the points are to the right of the line 0–S, species one
is the winner. Therefore, this set of conditions will produce an unstable equilibrium
or competitive exclusion with an indefinite winner (a stochastic result).

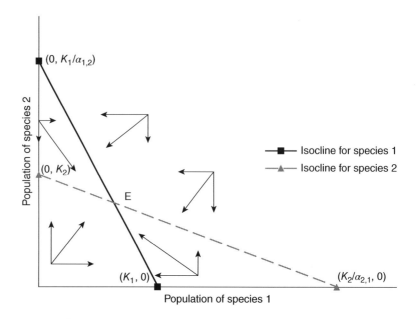

Fig. 7.7 Coexistence of both species at the stable equilibrium, *E*.

Finally, in Figure 7.7, we have a situation in which each species slows its own growth more than that of its competitor. This allows a stable equilibrium with coexistence of the two species at point E.

7.4 Laboratory experiments and competition

The classic works of Thomas Park and his colleagues (Park 1954; Neyman et al. 1956) illustrate cases 1–3, as well as the fact that competitive outcomes are not necessarily deterministic (see Table 7.1). In Park's most often cited study, he raised these two species of flour beetles *Triboium castaneum* and *confusum*, in vials of sifted flour under different temperature and humidity regimes (Table 7.1).

In these experiments, the initial numbers of individuals were equal. That is, $N_1 = N_2$ at $t = 0$. In the top and bottom rows of the chart, we have typical deterministic results illustrating the first two graphical analyses (Figures 7.4 and 7.5) in which competitive exclusion occurs. The other four results are stochastic in that the result of any single experiment is unpredictable. Neyman et al. (1956) conducted further experiments using the cold-moist regime. They found that one species or the other was a deterministic winner when given a large initial numerical edge (Figure 7.6). However,

Table 7.1 Results of competition between two species of flour beetles (*T. confusum* and *T. castaneum*) in six different temperature and humidity combinations.

Temperature in °C	Percent relative humidity	"Climate"	Percent of replicates in which one species or the other wins	
			T. confusum	*T. castaneum*
34	70	Hot-Moist	0	100
34	30	Hot-Dry	90	10
29	70	Warm-Moist	14	86
29	30	Warm-Dry	87	13
24	70	Cold-Moist	71	29
24	30	Cold-Dry	100	0

Source: Data from Park (1954).

there still existed an "indeterminate" zone (in the region of $N_1 \approx N_2$) where either species could still win.

A natural extension of these two species' competitive interactions was the work of Vandermeer (1969). Building on the work of Gause (1934), Vandermeer raised four species of protozoans in monocultures, thereby determining their r_m and carrying capacities. Next, he grew each of the species in pairwise combinations. From these experiments, he estimated the pairwise competition coefficients. The general results were as follows: *Paramecium aurelia* depressed the growth of *P. caudatum*, drove *P. bursaria* extinct, and depressed the growth of *Blepharisma* sp. *P. caudatum* also drove *P. bursaria* extinct and drove *Blepharisma* sp. to a very low population level. *Blepharisma and P. bursaria* had little effect on each other and coexisted.

Based on the pairwise competition coefficients and K values, Vandermeer predicted the outcome of placing all four species together in a community, including predicted growth rates and carrying capacities. His predictions and the actual outcomes were surprisingly similar, leading to the conclusion that higher-order interactions (the combined effects of two species on a third) and nonlinear relationships were not significant in this community.

Other multispecies studies, however, did not confirm Vandermeer's findings. For example, Neil (1974) conducted a series of replicated removal experiments in a laboratory microcosm containing four species of microcrustaceans and associated algae and bacteria. Each species of crustacean was grown in pairs, and estimates of population density were made under each regime. Computation of competition coefficients showed that the α-values depended on the community composition. The joint effects of two species on a third (in a three-species community) were not as predicted from the separate interactions in the two-species systems. Wilbur (1972) examining a natural amphibian community also found that the community composition could not be predicted from simple two-way competition experiments. Furthermore, Brenchley (1979) stated that the assumption by Vandermeer of simple resource (exploitative) competition was "dubious" since interference competition via metabolic by-products may be important. In the end, Brenchley concluded that there were subtle higher-order interactions in Vandermeer's experiment.

Such results helped push ecologists to look for a different theoretical approach to competition.

7.5 Resource-based competition theory

David Tilman (1976, 1987) and others pointed out that the Lotka–Volterra equations were "phenomenological" and not "mechanistic." That is, competition coefficients were merely measures of the effect of one species on the growth rate of another. They are estimated from experiments in which two species are grown together. Therefore, they are not an independently derived value that allows one to predict coexistence or competitive exclusion, or in the latter case, which of two species should win. Furthermore, a competition coefficient does not help determine the mechanism of competition; we have no information on what resource the species might be competing for. If competition is really concerned with a resource in short supply, we need to understand what the resource is and how each species is using it before we can understand the potential competitive interaction.

Tilman, a particularly strong advocate of a mechanistic approach to competition (1976, 1981, 1982), developed what is now known as resource-based or resource-ratio competition theory. In so doing, he brought together ideas from a variety of disciplines, including microbiology, enzyme kinetics, and agricultural chemistry. For example, the idea that population growth is constrained by the depletion of critical resources can be traced to the agricultural chemist Liebig (1855) and his law of the minimum. Liebig asserted that a population increases until the supply of a single critical resource becomes limiting. For example, plant growth may continue until the amount of phosphorus, nitrogen, light, or soil moisture becomes limiting. According to Liebig's law, if plant growth is constrained by phosphorus and a farmer adds phosphorus fertilizer, plant growth will continue until another resource, such as nitrogen, becomes limiting. If the farmer adds nitrogen, then soil moisture may become the limiting factor. Liebig's law is overly simple in that two or more resources may interact to limit a population, but it puts resource supply into the context of population regulation and therefore competition.

In the resource-based approach to competition, we need to couple the availability of resources to population growth. We begin by considering a renewable resource. If the supply rate is not affected by the population of potential consumers, the rate of supply (e.g., phosphorus arriving at a small lake via a local stream) could be considered a constant (equation 7.11), where R_i is the quantity of the resource i and k_{Ri} is the supply rate.

$$dR_i/dt = k_{Ri} \tag{7.11}$$

If this one resource sets the limit of growth for a population (as in Liebig's law of the minimum), the growth rate depends on both the resource level and the density of the population itself. That is, dN/dt is a function of both the resource supply rate, k_{Ri}, and population size, N. Assume that each individual must consume the resource at rate, q, to maintain itself. We ignore the possibility of storing the resource for later use. The population of N individuals will consume the resource at the rate qN. The remaining resources may eventually be lost downstream or in the lake sediments. Alternatively, they may be taken up by the consumers and used for the production of new individuals in the population. The supply rate of the resource available for reproduction is therefore $k_{Ri} - qN$. Suppose each individual converts the resource into new individuals with efficiency b. Population growth can now be written as follows:

$$dN/dt = bN(k_{Ri} - qN) \tag{7.12}$$

We can rearrange this equation to

$$dN/dt = bk_{Ri}N \left(1 - \frac{qN}{k_{Ri}} \right) \tag{7.13}$$

Equation 7.13 is a form of the logistic equation with $r = bk_{Ri}$ and the carrying capacity equal to k_{Ri}/q. We have now coupled resource supply with population growth rate. The growth rate of the population is proportional to the supply of the critical resource, and the carrying capacity is the resource supply divided by the amount needed for maintenance per individual.

Now envision the relationship between resource availability and population growth. As described earlier, in a density-dependent population, the per capita growth rate declines with population size, yielding a negative slope. Logistic-like equations assume that the resources become scarcer as populations grow. However, if we graph per capita growth versus an increasing supply of resources, the predicted slope is positive (Figure 7.8). As the resource becomes more abundant, the population growth

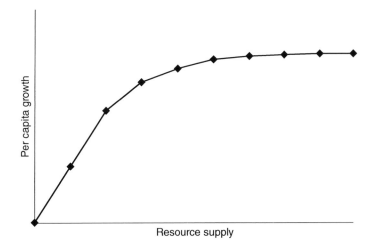

Fig. 7.8 Per capita growth as a function of resource availability.

rate increases. Per capita growth, however, finally levels off and declines to zero when some other resource limits the population (as proposed by Leibig).

The next step is to introduce a mortality rate, m. Instead of assuming that resources are simply turned into births, we assume, that a certain minimum level of the resource is needed to maintain the population. Therefore, as shown in Figure 7.9, we introduce the concept of R^*. R^* is the level of the resource needed to balance mortality. If the resource is provided at the rate R^*, then $dN/dt = 0$, growth just balances mortality, and the population maintains itself. If the resource is provided at a level less than R^*, growth is less than mortality, and the population declines. Conversely, if the resource is provided at a level greater than R^*, we have growth ($dN/dt > 0$).

Examine Figure 7.10. As the population grows, the resource is increasingly depleted, and once the resource declines to the level R^*, the population should stop growing. The population thereafter remains steady at a size determined by the resource quantity R^*.

Tilman (1976) realized that an independently derived R^* could be used to predict population dynamics and ultimately, competitive interactions. He first used the

Fig. 7.9 Per capita growth as a function of resource availability, but with a constant mortality rate. R^* = the amount of resource producing a per capita growth rate of zero.

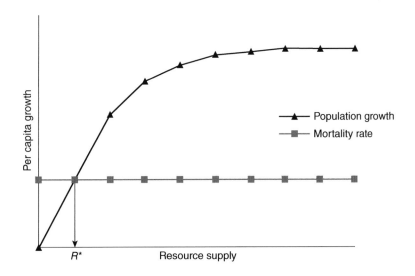

Fig. 7.10 Population and resource dynamics.

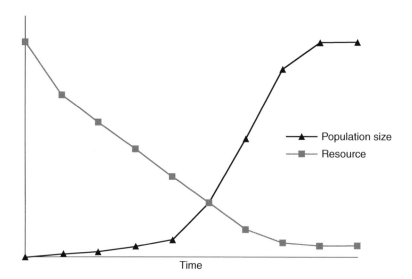

Michaelis–Menten enzyme kinetics equation, which is normally employed to describe the relationship between cellular metabolism and substrate concentrations. This equation had also been used by microbiologists to describe the growth rate of bacteria on organic substrates (Monod 1950).

In the Michaelis–Menten equation, the growth rate μ on a given substrate or resource, R, is set equal to the maximum growth rate modified by the concentration of the resource and by a value known as K_μ, the half-saturation constant for the resource in question. K_μ is the concentration of the resource that produces half the maximum growth rate. If μ_{max} is the maximum growth rate of the population, and R is the resource or substrate, the resultant growth rate (μ) according to the Michaelis–Menten equation is as follows:

$$\mu = \mu_{max} \left(\frac{R}{R + K_\mu} \right) \tag{7.14}$$

In Monod's version of this equation, instead of μ we substitute dN/dt, and we use b instead of μ_{max}. We will also now define the half-saturation constant for any given resource as K_i, rather than K_μ. The result is equation 7.15:

$$dN/dt = \frac{bNR}{K_i + R} \tag{7.15}$$

Per capita growth is shown in equation 7.16:

$$dN/Ndt = \frac{bR}{K_i + R} \tag{7.16}$$

This equation tells us that at very high resource level, the expression $R/(K_i + R)$ is very close to one, and the per capita growth is simply b. This is the maximum growth rate, ignoring the mortality rate, m, and is equivalent to the unmodified intrinsic rate of increase r_{max}. We can add a constant mortality rate to Monod's equation as follows (equation 7.17):

$$dN/Ndt = \frac{bR}{K_i + R} - m \tag{7.17}$$

We can set dN/Ndt (per capita growth) equal to zero and solve for R^*, which is the resource level at which growth stops. The solution, as shown in equation 7.18, allows us to predict R^* if we know three variables: (i) the half-saturation constant, K_i, (ii) the maximum growth rate, b, and (iii) the mortality rate, m. The advantage of this approach, as opposed to that of the Lotka–Volterra equations, is that a resource is identified, and a variable, R^*, can be derived from simple experiments, which can then be used to compare how populations of different species respond to different resource levels.

$$R^* = \frac{mK_i}{b - m} \tag{7.18}$$

For example, Tilman (1976) identified critical resources for which two species of planktonic algae were likely to be competing. Two freshwater diatom species were grown under different levels of the important limiting nutrients, phosphate (PO_4^{-3}) and silicate (SiO_4^{-4}). Using a growth vessel known as a chemostat, he determined the growth rates of each species cultured singly and also determined the limiting concentrations of phosphate and silicate. From these experiments, he determined the values of the half-saturation constants for each species for both phosphate and silicate. By examining the ratios of silicate to phosphate utilizations, he was able to establish boundaries for competition. He found that when both species were limited by phosphate, the diatom *Asterionella* won in competition with *Cyclotella*. When both species were limited by silicate, *Cyclotella* won. However, when both nutrients were simultaneously in short supply, the two species coexisted. Tilman minimized mortality rates and was able to predict which species would win based on growth rates and the half-saturation constants. He found that the species with the lower half-saturation constant (and therefore lower R^* in this case) would win if both species were limited by a single nutrient. However, if each species was limited by a different nutrient, the density of each species was held in check through intraspecific competition and the two species coexisted. The incorporation of the equations originally developed by Michaelis–Menten and Monod and the use of half-saturation constants, allowed Tilman to predict the results of competition involving two species and two resources.

Subsequent papers by Tilman (1981), Tilman et al. (1982), Hansen and Hubbell (1980), and a book by Tilman (1982) have become the foundations for "resource competition" as a distinct theory and an important area of inquiry. Tilman's work expanded from the laboratory work on diatoms and other algae to competition studies of terrestrial plant communities at the Cedar Creek Natural History Area in Minnesota. Hansen and Hubbell (1980) and Tilman (1982) elaborated on the Monod equation, but they still emphasized the critical parameter, R^*. This parameter can be used to predict which one species will survive in a mixed species culture when there is a single limiting nutrient. According to what is now known as the R^*-rule, for any given resource (R), if we determine the R^* value for each species when grown alone, the species with the lowest

R^* should competitively exclude all other species, given enough time and a constant environment.

In deriving their version of the R^*-rule, Hansen and Hubbell (1980) assumed that two competitors are grown in a continuous culture with a continuous input of a nutrient (R) as well as an effluent rate, which is equivalent to a death rate, m. The growth rates for two competing species were defined as follows:

$$dN_1/dt = \frac{b_1 RN_1}{K_1 + R} - mN_1 \tag{7.19}$$

and

$$dN_2/dt = \frac{b_2 RN_2}{K_2 + R} - mN_2 \tag{7.20}$$

where

b_i = maximum cell division rate (= r_{max});
R = the concentration of the one limiting resource in the culture;
K_i = half-saturation constant for the limiting resource;
m = death rate, here due to outflow;
N_i = concentration of cells in the culture (population size).

If we do an equilibrium analysis, and set $dN_i/dt = 0$, the result is $\frac{b_i R}{K_i + R} - m = 0$. If we set, $K_i = R$, then $b_i/2 = m$.

Thus, one solution is that growth stops when the concentration R equals the half-saturation constant. At that time, the cell division rate is at half of its maximum level ($r_a = 0.5 r_m$) and just equals the death rate. Thus as explored earlier in a graphical analysis, the equilibrium resource availability, R^*, occurs when the growth function intersects the line m representing mortality. This is the amount of resource needed to just sustain the population (growth just offsets mortality).

When Hansen and Hubbell (1980) analyzed equations 7.19 and 7.20, they found that they are globally stable when either (i) all competitors die out or (ii) one species survives while the second species dies out. That is, when competitive exclusion occurs. Which species survives depends on the critical parameter, R^*, which we already saw in equation 7.18 as $R^* = \frac{mK_i}{b-m}$.

R^* must be less than R_0, otherwise all species die out because of lack of resources. If all R^* values for all species are less than R_0, then, according to the R^*-rule, the species with the lowest R^* wins. Again, three parameters, the half-saturation constant, the intrinsic rate of increase, and the death rate combine to determine which species wins in competition for any given resource. This is not predicted from classical Lotka–Volterra competition theory. A species with a high affinity for the resource can still lose if it has a low growth rate (r-value) and a high death rate.

Hansen and Hubbell (1980) confirmed the expected results with several species of bacteria auxotrophic for tryptophan. In the example in Table 7.2, based on Hansen and Hubbell (1980), we expect population two to outcompete the other three species for this resource. Species three is expected to outcompete both populations 1 and 4, and population one should win in competition with population 4.

Tilman and others have tested the R^*-rule on algae, other microorganisms, higher plants, and zooplankton. For example, Tilman (1982) grew two species of diatoms,

Table 7.2 Example of R^* calculations based on Hansen and Hubbell (1980).

Population, i	K_i(g/l × 10^{-6})	m per hour	b per hour	$r_a(b-m)$ per hour	R^*(g/l × 10^{-6})
One	4.0	0.05	0.25	0.20	1.00
Two	4.1	0.05	0.50	0.45	0.46
Three	6.5	0.05	0.50	0.45	0.72
Four	20.0	0.05	0.25	0.20	5.00

K = half-saturation constant, m = mortality rate, b = maximal growth rate, r_a = actual growth rate = $b-m$.
$R^* = mK_i/(b-m) = mK_i/r_a$.

Asterionella formosa and *Synedra ulna*, which require SiO_2 for cell wall structure, in a laboratory chemostat. The R^*-values for *A. formosa* and *S. ulna* were 1.0 and 0.4 μM, respectively. When grown together, as the silicate levels eventually were reduced to 0.4 μM, *Synedra* hung on at its equilibrium number, whereas *Asterionella* went extinct.

See Grover (1997) for a relatively recent review covering both the status of the theory and relevant field and laboratory studies.

7.5.1 *Resource competition: critiques and conclusion*

Applying the R^*-rule and resource models to higher animals has proved less useful and, as usual, theories developed for simple laboratory settings run into a number of problems and complications when applied to field situations. In addition, we should recognize that other modes of competition, such as interference competition, do not follow the simple rules of resource-based models.

Miller et al. (2005) reviewed 20 years of publications utilizing or referencing the resource-ratio theory. After reviewing 1333 papers, they found that the vast majority did not actually test the theory and only 26 provided well-designed tests. Most of these were laboratory or microcosm studies. Of 13 papers examined by Miller et al., only 8 confirmed the theory. In contrast, Wilson et al. (2007) found the analyses by Miller et al. were flawed. After a reexamination of the 13 papers cited by Miller et al., Wilson et al. found that 12 of the 13 provided evidence for the resource-ratio theory.

They conclude that the resource-ratio approach is a good guide to competition experiments involving phytoplankton, zooplankton, and bacteria. For a reply to Wilson et al. (2007), see Miller et al. (2007).

Of course, competition on any given site does not necessarily involve all species in the community. The R^*-rule would only apply to the species that have "shown up" in any given patch. Therefore, as elaborated later (also see Tilman 1994, 1999; Hubbell 2001), a community is not limited to the species that are superior competitors. Hubbell, a proponent of the "Neutral Theory" (Hubbell 2001), asserts that the R^* or resource-ratio theory does not work for his *Barro Colorado Island* (*BCI*) tree species community. According to Hubbell (2010), the R^* theory assumes that species are nutrient specialists, and this is not applicable to the vast majority of BCI tree species. A core idea of the neutral theory is that species in a community are mostly ecological equivalents. Rather than finding evidence for resource competition, Hubbell (2010) has found evidence for dispersal and recruitment limitation in the BCI tree community. A footnote of historical interest is that according to Chase and Leibold (2003), the original Island Biogeography theory of MacArthur and Wilson (1967) treated all species as identical in spite of the prevailing niche theories at that time. In the end, Hubbell (2010) calls for a new approach to replace the resource-ratio competition theory.

7.6 Spatial competition and the competition-colonization trade-off

The idea that multiple species can coexist in a community without yielding to the superior competitors can be traced to the competition-colonization trade-off idea first proposed by Levins and Culver (1971) and elaborated by Hastings (1980), Tilman (1994), Yu and Wilson (2001), and others. Recall that in a metapopulation, two species can coexist if one is a superior competitor, and the other is a better colonizer. Remember also that in a metapopulation, the growth in sites occupied by species p was based on the colonization rate, cp times the proportion of sites unoccupied and available $(1 - p)$, minus the local extinction or mortality rate ε. When equation 7.21 is set equal to zero and we solve for $\hat{p}$, we have the proportion of habitat sites occupied at equilibrium (equation 7.22). The colonization rate necessary for equilibrium is displayed as equation 7.23:

$$\frac{dp}{dt} = cp(1 - p) - \varepsilon p \tag{7.21}$$

$$\hat{p} = 1 - \frac{\varepsilon}{c} \tag{7.22}$$

$$c = \frac{\varepsilon}{1 - \hat{p}} \tag{7.23}$$

We can generate equations for two species, one for the superior competitor, p_1 (equation 7.24, based on equation 7.21), and one for the superior colonizer, p_2 (equation 7.25). c_1, c_2, ε_1, and ε_2 represent colonization and mortality rates for species one and two, respectively. Equation 7.24 is the same equation as if the species lived by itself, since it is assumed to be unaffected by the inferior competitor. In contrast, the inferior competitor can only colonize sites occupied by neither species $(1 - p_1 - p_2)$. Thus, its successful colonization of new sites is a product of $c_2 p_2$ times $(1 - p_1 - p_2)$. This success rate is decreased by its natural patch extinction rate $(\varepsilon_2 p_2)$ and its displacement through competition with species $1(c_1 p_1 p_2)$.

$$\frac{dp_1}{dt} = c_1 p_1 (1 - p_1) - \varepsilon_1 p_1 \tag{7.24}$$

$$\frac{dp_2}{dt} = c_2 p_2 (1 - p_1 - p_2) - \varepsilon_2 p_2 - c_1 p_1 p_2 \tag{7.25}$$

As noted in equation 7.23, in order to attain its equilibrium, the colonization rate of the superior competitor must satisfy $c_1 \geq \frac{\varepsilon_1}{1 - \hat{p}_1}$. To find the equivalent value for species two, set equation 7.25 = 0, and divide all terms by p_2. The result is equation 7.26a. The equilibrium value of p_2 is found by the following steps 7.26b–7.26d.

$$0 = c_2 - c_2 p_1 - c_2 p_2 - \varepsilon_2 - c_1 p_1 \tag{7.26a}$$

$$c_2 - c_2 p_2 = c_2 p_1 + \varepsilon_2 + c_1 p_1 \tag{7.26b}$$

$$c_2 (1 - p_2) = c_2 p_1 + \varepsilon_2 + c_1 p_1 \tag{7.26c}$$

$$\hat{p}_2 = 1 - \frac{c_2 p_1 + \varepsilon_2 + c_1 p_1}{c_2} \tag{7.26d}$$

For species two to remain viable $\hat{p}_2$ must be greater than zero. Therefore, we set the right side of equation 7.26 > 0, and set $p_1 = \hat{p}_1$ producing the inequalities in equations 7.27a–7.27d.

$$1 > \frac{c_2 \hat{p}_1 + \varepsilon_2 + c_1 \hat{p}_1}{c_2} \tag{7.27a}$$

$$c_2 > c_2 \hat{p}_1 + \varepsilon_2 + c_1 \hat{p}_1 \tag{7.27b}$$

$$c_2(1 - \hat{p}_1) > \varepsilon_2 + c_1 \hat{p}_1 \tag{7.27c}$$

$$c_2 > \frac{c_1 \hat{p}_1}{1 - \hat{p}_1} + \frac{\varepsilon_2}{1 - \hat{p}_1} \tag{7.27d}$$

By using equation 7.23 and replacing $1 - \hat{p}_1$ with ε_1/c_1 on the far right term, we obtain equation 7.28:

$$c_2 > c_1 \left\{ \frac{\hat{p}_1}{1 - \hat{p}_1} + \frac{\varepsilon_2}{\varepsilon_1} \right\} \tag{7.28}$$

Based on equation (7.23), $c_1 = \frac{\varepsilon_1}{1 - \hat{p}_1}$ we know that for $\hat{p}_1$ to be positive, $1 > \varepsilon_1/c_1$. Therefore, c_1 must be greater than ε_1 (7.29a) for the equilibrium abundance of species one to be stable. Similarly, if we take equation 7.28 and substitute $1 - \frac{\varepsilon_1}{c_1}$ for $\hat{p}_1$, we end up with 7.29b.

$$c_1 > \varepsilon_1 \tag{7.29a}$$

$$c_2 > \frac{c_1(c_1 + \varepsilon_2 - \varepsilon_1)}{\varepsilon_1} \tag{7.29b}$$

As stated by Tilman (1994), these are the "necessary and sufficient conditions for the stable coexistence of a superior competitor and an inferior competitor in a subdivided habitat."

If the mortality rates were equal for the two species, equation 7.29b simplifies to $c_2 > c_1^2/\varepsilon$ found by Hastings (1980). Since we know that $c_1 > \varepsilon_1$, it follows that the ratio $c_1/\varepsilon > 1$. Therefore, $c_2 = a$ number greater than one times c_1 and c_2 must be greater than c_1. If the mortality rate of the inferior competitor (ε_2) is greater than or equal to that of the superior competitor, an examination of equation 7.29b should demonstrate to you that again, species two only persists if $c_2 > c_1$. If the inferior competitor has a lower mortality rate, however, Nee and May (1992) have shown that it may coexist with the superior competitor, but only if at least half of the habitat is left available for colonization. That is, the ratio of $\frac{\hat{p}_o}{1 - \hat{p}_1}$ must be less than one. This becomes clear if we assign $\hat{p}_1 = 0.5$ in equation 7.28, resulting in $c_2 > c_1 \left\{ 1 + \frac{\varepsilon_2}{\varepsilon_1} \right\}$. No matter how small the mortality rate of ε_2 (even if it is zero), we must still have $c_2 > c_1$.

These basic equations have been generalized to multispecies situations by Tilman (1994) and others. Termed the "spatial competition hypothesis," this theory proposes stable coexistence for inferior competitors in a diverse community. Coexistence by spatial competition does assume a two- or even three-way trade-off among competitive ability, colonization ability, and longevity. Since there is actually little evidence for a trade-off between longevity and competitive ability, the major trade-off is assumed to be between competitive ability and colonizing ability. This is, of course, not inconsistent with the MacArthur and Wilson (1963, 1967) proposal for r- and K-selected species, in which *r*-selected species were assumed to sacrifice competitiveness for colonizing capacity.

As pointed out by Yu and Wilson (2001), the displacement competition model (equals the "spatial competition hypothesis" of Tilman 1994) assumes that a propagule of the superior competitor will displace an adult of the inferior competitor. Second, the displacement of the adult occurs rapidly enough to prevent reproduction of the inferior competitor. Yu and Wilson (2001) stressed that in many ecosystems neither of these assumptions is true. Rather, the juveniles compete among themselves while waiting for an adult to die. In a forest, for example, seedlings have little, if any, effect on the death rates of adult trees. Once an opening in the canopy occurs, these juveniles compete with each other for the light gap. Yu and Wilson (2001) designated this replacement or "lottery" competition and asserted that the competition-colonization trade-off is not sufficient to produce coexistence of inferior with superior competing species. We will not elaborate on their models here. However, they propose that in a lottery system, all that is necessary to allow coexistence of inferior with superior competitors is some form of environmental heterogeneity in patch density combined with either the competition-colonization trade-off or a trade-off between dispersal and fecundity. Since many species produce small numbers of large offspring with low dispersal abilities (plants with a small number of large seeds, for example), the basic idea is that spatial variability is essentially a "niche-axis" (Yu and Wilson 2001) that facilitates the high species richness found in many ecosystems. This brings us back to the Hubbell (2001, 2010) idea of diversity generated through dispersal and recruitment limitation for the BCI tree species.

7.7 Evidence for competition from nature

7.7.1 Connell's barnacles

The classic experimental demonstration of competition in the field was done by Joseph Connell (1961a, 1961b) on the barnacle species *Chthamalus stellatus* and *Semibalanus balanoides* (previously *Balanus balanoides*). These two species are found growing in the rocky intertidal zone off the coast of Scotland. Intertidal zones frequently show vertical zonation of species based on their abilities to survive periods of exposure to the air during low tides and wave action followed by submersion during high tides. *Semibalanus* is consistently found on lower rock surfaces, usually near mean tide level or slightly above. *Chthamalus*, however, is found on the upper rocks, between high mean neap tide and high mean spring tide. While the adults of these two barnacle species have nonoverlapping distributions, the larvae of both species settle over a wide variety of rock surfaces, showing a great deal of overlap. The question Connell posed was, is the distribution of adults the result of competition, or is there a difference in the fundamental niches of the two species? Connell performed a variety of experiments in which he moved the barnacles to different levels of the intertidal zone. He also experimentally removed one species or the other where the two were growing together and observed the results of putting the two species together. He found that whenever he removed *Semibalanus*, *Chthamalus* was able to survive in the lower regions of the intertidal zone. However, in the presence of *Semibalanus*, *Chthamalus* was overgrown and eventually displaced. In the upper regions of the intertidal, however, *Semibalanus* was unable to survive the long exposures to air during low tides. Since *Chthamalus* was able to survive this exposure, it survives in the upper intertidal. Thus the two species occupy mutually exclusive microhabitats due to a combination of competition and differences in their fundamental niches.

7.7.2 Direct observations of competition in ants

Because both worker and soldier ants are numerous, easy to observe, and usually diurnal, aggressive interactions among ant species, demonstrating interference competition, can be documented throughout the world (Holldobler and Wilson 1990). Placing a food bait of tuna or sugar water will provoke competitive interactions in a matter of minutes to hours. Once bait is put out in the West Indies, where there are few ant species, there is a kind of predictable sequence, reminiscent of ecological succession (a kind of "ant succession"). As described by Holldobler and Wilson (1990), first to arrive are workers of *Paratrechina longicornis*, known locally as "hormigas locas" (crazy ants). These workers are very adept at locating food and often are the first to arrive at newly placed baits. They fill their crops rapidly and hurry to recruit nestmates with odor trails laid from the rectal sac of the hindgut. However, they are also very timid in the presence of competitors. As soon as more aggressive species begin to arrive in force, the *Paratrechina* withdraw and search for new, unoccupied baits. *Paratrechina* is an example of an "opportunist" species. They are poor competitors but excellent dispersers. Next to arrive are species known as "extirpators." These species recruit other workers by odor trails and fight it out with competitor species. Examples include species in the genera *Pheidole* and *Crematogaster,* the fire ant (*Solenopsis geminata*), and the "little fire ant" (*Wasmannia auropunctata*). Some of these species have well-developed soldier castes, which play a key role in the aggressive interactions. Injury and death are commonplace, and one species eventually dominates the bait. Preemption is usually the deciding factor. The colony whose foragers arrive first typically wins; foragers recruit nestmates who surround the bait. When worker scouts encounter a large number of workers from another colony, they are easily repulsed (Holldobler and Wilson 1990). Species with a third strategy called "insinuators" also arrive at the baits. These are small colonies with small-sized worker ants such as *Tetramorium simillimum* and species of *Cardiocondyla*. A scout, who discovers the bait, will recruit only one nestmate at a time. Small size and stealthy behavior allow these individuals to take some of the bait without provoking a response from the extirpator species, which is reminiscent of small animals sneaking in and removing bits of food at a lion kill.

Holldobler and Wilson also emphasize that territorial fighting and "ant wars" are common, especially among species with large colonies. Numerous cases have been documented in which introduced ant species have eliminated other species over a few years time. For example, on Bermuda *Iridomyrmex humilis* has been replacing *Pheidole megacephala* since the former was introduced in 1953, although the two species may be reaching equilibrium short of extinction of *Pheidole* (Lieberburg et al. 1975). Finally, the red imported fire ant, *Solenopsis invicta*, has virtually eliminated the native fire ant, *S. xyloni* from most of its range (Holldobler and Wilson 1990).

7.7.3 Literature reviews of field studies on competition

In the 1980s, the importance of interspecific competition in the field was questioned by a number of biologists. Strong et al. (1983), among others, challenged much of the evidence usually cited for the prominent role assigned to competitive interactions in structuring natural communities. They asserted that the data were often indistinguishable from random models. Others charged that there were few experimental studies of competition from the field, and that predation was a much more significant ecological interaction. Still others, such as Wiens (1977), asserted that competition is a temporally sporadic, often impotent, interaction. Schoener (1982, 1983) decided to review the literature to determine whether competition had been affirmed as an important interaction in the field. He found, to his surprise, over 150 experimental field studies, many of which had been conducted in the previous 5 years. Schoener carefully defined an interspecific competition experiment as a manipulation of the abundance of one or more hypothetically competing species. All such experiments had to have proper controls. Before these experimental studies, there had been a dependence on "natural experiments," which will be discussed later (Diamond 1983).

The "field" was defined as a study in which some major natural factors extrinsic to the organism remain uncontrolled. Schoener did not allow laboratory or greenhouse setups but did count experiments involving fenced exclosures or caged portions of shorelines to fit the definition of a field study.

Through 1982, Schoener found that 164 published studies fit the criteria. Of those 164, 90% (148) of the studies and 76% of the species involved did show positive evidence of interspecific competition. In a separate analysis and using different criteria, Connell (1983) found evidence of competition in 40% of the experiments and 50% of the species. There were, however, few studies involving herbivorous insects that demonstrated interspecific competition. Schoener suggested, as had Hairston et al. (1960), that herbivores, which occupy an intermediate position in the food web, are controlled by predators, and therefore, competition is a less important interaction for this trophic position. Schoener's literature review found little evidence to support Wiens idea that there is a great deal of temporal variability in competition. Competitive variability is especially rare in marine ecosystems. Variability was mostly found in dry, continental habitats. This is interesting in that Wiens developed his ideas after carrying out research on bird communities in North American arid or semiarid shrub habitats.

A decade after the analyses of Schoener and Connell, Gurevitch et al. (1992) analyzed competition studies carried out from 1980 to 1989, using a statistical approach. They found "medium" effects of competition on primary producers, carnivores, and herbivorous marine mollusks. Larger effects of competition were detected on some herbivores and stream arthropods. As found by Schoener, however, studies on herbivorous terrestrial insects usually failed to show significant effects of competition.

One can conclude from these literature reviews that competition is a common event in nature that contributes to the organization of ecological communities. It is, however, not the only important interspecific interaction.

7.8 Indirect evidence for competition and "natural experiments"

Because competition has been difficult to demonstrate directly in the field, and because the historical or evolutionary result of competition may have resulted in co-evolution between species to minimize competition (sometimes called the ghost of competition past), Diamond (1978, 1983) proposed that indirect evidence, often the result of "natural experiments," should be given the same credibility as experimental studies in the field or laboratory. As important as field experiments are, a single variable is manipulated; nothing else can be controlled once the field sites have been selected. Diamond lists five different drawbacks or weaknesses of field experiments. (i) The outcome of the experiment often varies from year to year and season to season since weather and predators are uncontrolled. (ii) Most field experiments are not run for enough time. This deficiency is, however, being remedied. For example, the National Science Foundation is addressing this problem in its Long-Term Ecological Studies program. (iii) The importance of large temporal and spatial scales cannot be addressed in contemporary time and space. (iv) A manipulation of two species may incorrectly ignore the importance of a third species. (v) The kinds of experiments that might reveal important information, such as the removal or introduction of a species from an ecosystem, are often "technically impossible, morally reprehensible and politically forbidden" (Diamond 1983).

In order to solve these problems, Diamond (1983) extolled the virtues of "natural experiments" and other kinds of data gathered from field observations as opposed to experiments. According to Diamond, natural experiments have three advantages. First, they permit an ecologist to rapidly gather data. As an example, he described the work of Schoener and Toft (1983). They surveyed spider populations on 92 small Bahamian islands, 48 of which lacked lizards and 26 of which were occupied by at least one species of lizard. They found that spiders were ten times more abundant on the islands without lizards. The explanation was that lizards are both competitors with and predators on spiders. Diamond's point, however, was that this natural experiment (lizards present on some islands, absent on others) would have been very difficult and time-consuming to set up, and we would have waited a very long time (up to several years) before the spider populations reached new equilibrium values. Using the natural experiments, Schoener and Toft completed their fieldwork in 20 days!

Second, natural experiments allow ecologists to examine situations they would not be allowed to set up experimentally. It is likely, for example, that the Bahamian government would have objected to having lizards removed from 48 islands. In another example, Brown (1971a, b) has shown that two species of chipmunk (genus *Eutamias*) divide the forest by altitude when they are sympatric on mountains in the Sierra Nevada range. However, on several mountains, probably due to chance colonization or extinction events, only one species is present. When only one species occupies the mountain, without its competitor, it is found at all elevations. A field experiment, in which one species or the other was eliminated from an entire mountain, would never have been approved by the US Fish and Wildlife Service or by any granting agency. Yet this natural experiment is an elegant demonstration of the phenomenon known as ecological release (see below).

Finally, natural experiments allow us to examine the end results of ecological or evolutionary processes operating over a longer period than the usual field experiment.

The weakness of natural experiments is, of course, that the investigator does not know what created the observed situation. There are often multiple explanations for the observed differences, and it may be impossible to distinguish among them. Furthermore, the use of simple field observations, as opposed to rigorous, controlled experiments is unacceptable to many scientists. Diamond, however, asserted that a variety of evidence derived from a variety of methodologies provides the most robust results.

Below we review five different kinds of indirect evidence related to the role of competition, many coming from natural experiments.

7.8.1   *Ecological release*

In ecological release, a species occupies a broader niche or geographical area in the absence of a closely related competitor. The chipmunk distribution in the Sierra Nevada mountain range fits this description. Another example is the distribution of two species of *Planaria* in streams. When found alone in a stream (allopatric distribution), each species occupies a wide range of stream temperatures. When both species are found in the same stream (sympatric distribution), however, the distribution of both species is restricted. *P. montenegrina* is found from 5 to about 13.5 °C, whereas *P. gonocephala* occupies the warmer portions of the stream from 13.5 to approximately 23 °C (Beauchamp and Ullyott 1932).

Ants also show ecological release. For example, *P. longicornis* normally nests under objects on the ground in open environments in southern Florida and in the Florida Keys. However, the Dry Tortugas, the outermost islands of the Florida Keys, have been colonized by very few ant species. On the Dry Tortugas, released from competition, *P. longicornis* is extremely abundant and nests in tree boles and open soil, sites normally occupied by other species in the rest of southern Florida. In contrast, *Pseudomyrmex elongatus*, which also has colonized the Dry Tortugas, has neither expanded its population nor its normal nesting habitat of thin twigs near the top of the tree canopy (Holldobler and Wilson 1990).

7.8.2 *Contiguous allopatry*

In this phenomenon, two species occupy distinctly different geographical areas directly adjacent (contiguous) to one another. Another study on chipmunks on the east slope of the Sierra Nevada Mountains in California provides an example of both contiguous allopatry and ecological release. Heller and Gates (1971) found four species of chipmunks living at different altitudes. The least chipmunk (*Eutamias minimus*) is found at the lowest elevations in the sagebrush range. When all other chipmunks are absent, the least chipmunk can occupy all altitudes up to the alpine. However, the yellow pine chipmunk (*E. amoenus*) through aggressive behavior restricts the least chipmunk to the hot, dry sagebrush areas. If the least chipmunk is absent, the yellow pine chipmunk does not invade the hot, dry habitats. Evidently this dry area is not part of the fundamental niche of the yellow pine chipmunk. The lodgepole pine chipmunk (*E. speciosus*) is the most aggressive of the four species, but it is most vulnerable to heat stress and is restricted to shady, cool forests. It apparently limits both the upper distribution of the yellow pine chipmunk and the lower distribution of the alpine chipmunk (*E. alpinus*). Therefore, these four species, when all are present, show contiguous allopatry by altitude. When one or another is absent, the least chipmunk shows ecological release.

Diamond (1978) has described other examples of contiguous allopatry and ecological release. In one case, he mapped the distribution of two species of warblers in the genus *Crateroscelis* on Mt. Karimui in New Guinea. The first species, *C. murina*, became relatively more abundant as Diamond hiked from sea level to 5400 ft. From that altitude to the summit of the mountain *C. murina* was abruptly replaced by *C. robusta*. This second species had its greatest relative abundance at just over 5400 ft. Diamond found a similar situation when he examined the distribution of three species of nectar-drinking parrots on several New Guinea mountains. When the species *Vini placentis* was present, it occupied forests from sea level to about 2000 ft. The sibling species *V. rubronotata* and *V. rubrigularis* were confined to higher elevations when in competition with *V. placentis*. The distributions were contiguously allopatric. When *V. placentis* was absent from a mountain on a different island, however, both species were able to expand their range to lower elevations. The opposite is also true. In the absence of the two sibling species, *V. placentis* is able to expand its range to higher elevations.

Other examples include (i) the distribution of two species of shrews in the Southern Appalachians. The Southern shrew is found in lower elevations where the forest is dominated by species of *Pinus*, while the masked shrew lives in higher elevations in mixed hardwood and riparian forests (Ford et al. 2001) and (ii) according to Anthony and Smith (1977) in Southeastern Arizona, mule deer are found in lower elevations with associated desert grasslands while white-tailed deer live in higher elevation woodland forest and chaparral.

7.8.3 *Niche partitioning*

In niche partitioning, two or more species coexist while sharing one or more resources in such a way that the niche overlap apparently violates the competitive exclusion principle. Upon closer investigation, the resources, although shared, are used with different frequencies or are used in different ways so as to allow coexistence. For example, the root systems of coexisting annual plants can be shown to partition the soil by depth, thereby avoiding direct resource competition (Wieland and Bazzaz 1975). In his classic study, MacArthur (1958) showed that five species of *Dendroica* warblers coexisted by foraging in different portions of trees in a coniferous forest. Although there was overlap, each species spent the majority of its foraging time in a unique portion of the trees. Other examples come from Diamond (1978). In one case, he describes the coexistence of four species of fruit pigeons on islands of the Bismarck Archipelago. The pigeons range in size from 91 to 722 g. The fruit sizes they consume range from 3 to 50 mm in diameter. Again, while there is a great deal of overlap among the pigeons in terms of fruit sizes consumed, the mean usage rates are clearly different. The smallest pigeon consumes fruits that are on average, 8 mm in size; the 135 g pigeon consumes fruits that average 11 mm; the 470 g pigeon, 18 mm; and the 722 g pigeon, 25 mm. In a similar analysis, Diamond (1978) shows that eight species of fruit pigeons in New Guinea coexist not only by specializing on different sizes of fruit but also by the position of the fruit on the branches.

A particularly interesting example of niche partitioning involves the four species of antbirds of the genus *Myrmotherula* that coexist on BCI in Panama. These birds forage by following swarms of army ants as the army ants move through the forests, insects, lizards, and other small prey flee. Antbirds swoop in and capture unsuspecting prey intent upon escaping from the army ants. This would seem to be a limited niche, yet four bird species of the same genus make their living in this manner. MacArthur (1972), using data supplied by John Terborgh, showed that the mean foraging heights among the four species differed. Not only did these means differ, but they were also separated from each other by at least one standard deviation.

Niche partitioning can also occur within a species. Male red-eyed vireos forage for insects primarily in the upper canopy of the forest, while females concentrate their foraging in the lower canopy and near the ground (Williamson 1971).

Holldobler (1986) found that two species of ants in Australia display niche differentiation based on foraging at different times of day. *Iridomyrmex purpureus* and *Camponotus consobrinus* use the same food sources and are found nesting near each other. However, *Iridomyrmex* forages mostly during the day and *Camponotus* mostly at night. When each species is found alone, its foraging time is increased by 1–2 h per day. Strangely, these shortened foraging periods are the result of direct interference. In the morning, *Iridomyrmex* workers congregate around nest exits of *Camponotus* and close them with pebbles and soil. At dusk, the situation is reversed with *Camponotus* workers harassing *Iridomyrmex* at its nests. In another example of niche partitioning among ants, Levins et al. (1973) showed that *Pheidole megacephala* dominates baits in the shade, whereas *Brachymyrmex heeri* takes over the same baits when they are in direct sun. This situation reverses itself regularly, within 30 min, accompanied by occasional aggressive encounters among workers.

Other examples of niche partitioning include (i) habitat and food partitioning among feral pigs (*Sus scrofa*), the white-lipped peccary (*Tayassu pecari*), and the collard peccary (*Pecari tajacu*) in the Brazilian Pantanal (Desbiez et al. 2009); (ii) microhabitat selection among two species of darters in an Arkansas stream (Weston et al. 2010);

and (iii) coexistence of several species of Anolis lizards in Bimini by ambushing insects from heights within the canopy (Schoener 1968).

7.8.4 Character displacement

Character displacement is defined as a situation in which two species, when living in separate geographical ranges (allopatric distributions), have nearly identical physical characteristics (i.e., beak sizes in birds, overall body sizes in lizards and snails, canine sizes in the cat family). When sympatric, however, these physical or morphological characteristics diverge in one or both species. This divergence minimizes competition for food and allows the two species to coexist. Brown and Wilson (1956) appear to have introduced this idea. When examining the overall size and beak lengths of specimens of the eastern (*Sitta tephronota*) and western rock nuthatches (*S. neumayer*), they found that the allopatric populations were almost identical in both average size and the range of sizes. However, these two species become sympatric in Iran. In sympatry, the eastern rock nuthatch is larger, while the western species has become smaller. In this sympatric zone, their beak and body sizes are completely nonoverlapping. This allows them to feed on different sized prey items and therefore coexist.

Another example (Lack 1947; Schluter et al. 1985; Grant 1999) comes from studies of Galapagos finches (also known as Darwin's finches). The medium ground finch (*Geospiza fortis*) and the small ground finch (*G. fuliginosa*) are allopatric on the islands of Daphne Major and Los Hermanos. On Santa Cruz, they are sympatric. As in the earlier case, when they are allopatric, their beak sizes are very similar (mean sizes approximately 10 mm in *G. fortis* and 8.5 mm in *G. fuliginosa*). On Santa Cruz, however, the sizes of their beaks do not even overlap. The average for *G. fortis* is between 11 and 12 mm, while that of *G. fuliginosa* is reduced to about 7.5 mm. Beak size correlates with diet in the Galapagos, and there is good evidence for food competition (Grant 1999).

Fenchel (1975) has demonstrated character displacement in shell lengths of two species of mud snails (*Hydrobia ulvae* and *H. ventrosa*). The overall size of the shell is indicative of the particle sizes consumed by the snails. As in the earlier cases, in allopatry, the snail sizes were almost identical, but in sympatry, the *H. ventrosa* population is greatly reduced in size. Simultaneously, the *H. ulvae* population includes many very large individuals, not present in the allopatric population. The means and modes of the two species are distinctly different in sympatry. For example, the population mode for *H. ulvae* is 5 mm in sympatry, but only 3 mm in allopatry.

Returning to the Galapogos, the Grants have actually observed character displacement in Darwin's finches in contemporary time (Grant and Grant 2006). The ecological theater for these events is the island of Daphne Major in the Galapogos. The characters in the evolutionary play are the normally 18 g medium ground finch, *G. fortis,* the normally 30 g large ground finch (*G. magnirostris),* and the approximately 12 g small ground finch (*G. fulginosa*), which is a character in this play only because it is missing from Daphne Major.

Before 1977, *G. fortis* was the only seed-eating finch on Daphne Major. As such *G. fortis* had free rein to eat whatever seeds were available, and they actually preferred eating smaller seeds since they were easier to consume. What resulted was a classic case of character release through evolution (Boag and Grant 1984). Released from competition from the small ground finch, *G. fortis* evolved a smaller beak and body size, allowing it to take more efficiently eat the smaller seeds preferred by *G. fulginosa* on other islands.

A drought in 1977 ended these happy times for the small beaked population of *G. fortis*. The drought resulted in a shortage of small seeds, especially in the dry season when *G. fortis* individuals were forced to feed upon the seeds of *Tribulus cistoides*. In this plant, seeds are exposed only after a finch cracks or tears away the woody outer covering of the hard mericarp. Carrying out this maneuver requires a large, strong beak. Naturally, larger finches with larger beaks were selected.

Enter now the large ground finch (*G. magnirostris*), a breeding population of which arrived in 1982. The large ground finch is a serious competitor with *G. fortis*, especially in the dry season when seed supplies are limited. Under such conditions, the large-beaked *G. fortis* are completely outcompeted for seeds from the *Tribulus* fruits, and the small-beaked population of *G. fortis* does not even attempt to crack the fruit open.

At first, there was little competition between the two species since the population of the large ground finch was small. Eventually, however, the large ground finch population grew, and their numbers reached over 350 by 2003. Next came a significant drought in both 2003 and 2004. The populations of both species plunged downward as supplies of seeds were exhausted. In the *G. fortis* population, the death toll was most severe among those individuals with large beaks, only 13% survived. The individuals with medium-sized beaks were more efficient at feeding on smaller seeds when they were present, while the large ground finch monopolized the large seeds. In the end, after 2005, the beak size of *G. fortis* had significantly diverged from that of the large ground finch, showing the development of character displacement in contemporary time (Grant and Grant (2006).

In the earlier examples, only pairs of species were examined. However, size displacements may occur among several coexisting species. Strong et al. (1979) called this "community-wide character displacement." A pattern of regular differences in some size-ranked sequence has been accepted as evidence that past competition helped to mold niche differences. In fact, Hutchinson (1959) proposed that a ratio of only 1.1–1.3 among closely related species in a size-ranked sequence was necessary for coexistence. This is known in the literature as Hutchinson's ratio or rule. There are many who have questioned the meaning or significance of such a rule (Simberloff and Boecklen 1981, Strong and Simberloff 1981). However, an interesting paper by Dayan et al. (1990) has shown that among wild felines in Israel, character displacement of canine diameters is remarkably constant at 1.10–1.14. These size ratios not only include three species of *Felis* but also include male and female sexual dimorphism in canine sizes (Table 7.3).

Table 7.3 Mean canine diameters for male and female coexisting species of *Felis* in Israel.

Species/gender	Mean canine diameter (in mm)	Ratio between adjacent pairs of canine diameters
F. silvestris, female	4.8	–
F. silvestris, male	5.5	1.15
F. chaus, female	6.3	1.15
F. caracal, female	6.9	1.10
F. chaus, male	7.6	1.10
F. caracal, male	8.4	1.11

Source: Data from Dayan et al. (1990).

Dayan et al. (1990) hypothesized that intraspecific and interspecific competition for food has selected for these canine ratios. The assumption is that canine size is highly correlated with the size of prey taken by these cats. In separating males and females, they were treating each species/gender combination as a distinct "morphospecies" in the community. A great deal more information is needed on (i) the size distribution of potential prey; (ii) whether prey populations are limiting to predator populations; (iii) whether larger canines correlate with the capture of larger prey; and (iv) habitat use by the cats. Nevertheless, the data suggest that competition has played an important role in the evolution of this feline community.

Taper and Case (1992) have cautioned that any study purporting to demonstrate character displacement should meet the following criteria:

(i) Morphological differences between a pair of sympatric species must be statistically greater than the differences between allopatric populations; (ii) The observed differences between sympatric and allopatric populations must have a genetic basis; (iii) Differences between sympatric and allopatric populations must have evolved on the site and not be due to different founder populations; (iv) Variation in the character must have a known, ecologically important function; (v) Competition must be known to occur for a resource in short supply and the character must play an important role in that competition; and (6) Differences in the character cannot be explained by differences in resources available to the sympatric and allopatric populations.

Ecologists have confirmed, however, that the Galapagos finch studies by the Grants (Grant and Grant 2006; Grant 1999) satisfy all of these criteria.

7.8.5 *Historical replacement*

Finally, over human history, there are documented cases of one species invading and eliminating another species from its original range. For example, Diamond (1978) has described both historical replacement and niche segregation in two species of tits (*Parus*). The blue tit (*P. caeruleus*) was found in Europe west of the Ural Mountains where it occupied a wide variety of habitats from forest to riparian thickets. The azure tit (*P. cyanus*) was formerly found in Asia east of the Urals. However, in the late nineteenth century, it spread 1000 miles west across Russia to the Baltic Sea. Over the next 10 years, however, the azure tit retreated several hundred miles eastward from the Baltic. At present, the two species overlap on the eastern edge of the geographical range of the blue tit. However, in the overlap zone, the two species segregate by habitat. The azure tit is found in riparian thickets, while the blue tit is in upland forest.

In recent years, the black duck (*Anas rubripes*) has been declining and is being replaced by the mallard in many areas (*Anas platyrhynchos*). It is unclear whether this is being caused by hunting, environmental and habitat changes, captive breeding and release programs, or competition between the two species. A study by Merendino et al. (1993) tentatively concluded that mallards were competitively excluding black ducks from the most productive wetlands.

There are many known examples where introduced species have outcompeted the native flora or fauna. Starlings and house sparrows are infamous for evicting native cavity nesting birds such as bluebirds and flickers from nest sites. The purple loosestrife (*Lythrum salicaria*), a native of Europe, has invaded wetlands in temperate North America and is crowding out native wetland plants. Finally, Australia and Hawaii are living laboratories demonstrating the effects of introduced competitors (as well as predators and herbivores) on native flora and fauna.

7.9 **Summary**

Competitive interactions have had an influential role in ecology since its inception. There has been a long tradition of theory, laboratory, and field studies that have emphasized the potential importance of competition in population regulation and in the shaping population and community relationships among species. Darwin, Liebig, Tansley, Lotka, Volterra, Gause, Park, Connell, MacArthur, Tilman, and many others have contributed to the rich literature on competitive interactions. The competitive exclusion principle established an expectation that competitive interactions shape not only present-day ecological interactions but also have directed past evolutionary events (the so-called ghost of evolution past). Resource-based competition theory has also laid out certain ground rules (the R^*-rule) for coexistence or exclusion in resource limited environments. In contrast, the competition-colonization trade-off concept sets up conditions whereby inferior competitors may persist in communities. For example, ant species known as "opportunists" and "insinuators" are able to coexist with the aggressive dominant competitor species, known as "extirpators," based on their ability to find food sources first or by their stealthy foraging style. And even among the dominant competitors, the species that finds a food resource first and dominates it through recruitment of nest mates is the winner at that resource. Therefore, as suggested by Tilman (1994) and Yu and Wilson (2001), coexistence of inferior competitors should be expected in a spatially diverse environment.

Some ecologists have complained that the fascination with competition has resulted in an underestimation of the importance of parasitic, predatory, or even mutualistic relationships in shaping ecological communities. American politicians have even turned economic competition into a kind of religion. Let us now, therefore, turn to other kinds of interactions.

References

Anthony, R.G. and N. Smith. 1977. Ecological relationships between Mule Deer and White-Tailed Deer in Southeastern Arizona. *Ecological Monographs* 47: 255–277.

Beauchamp, R.S.A. and P. Ullyott. 1932. Competitive relationships between certain species of fresh-water triclads. *Journal of Ecology* 20: 200–208.

Begon, M., J.L. Harper, and C.L. Townsend. 1986. *Ecology: Individuals, Populations and Communities*. Sinauer Associates, Inc., Sunderland, MA.

Boag, P.T. and P.R. Grant. 1984. Darwin's finches (*Geospiza*) on Isla Daphne Major, Galapagos: breeding and feeding ecology in a climatically variable environment. *Ecological Monographs* 54: 463–489.

Brenchley, G.A. 1979. Community matrix models: inconsistent results using Vandermeer's data. *American Naturalist* 113: 456–459.

Brown, J.H. 1971a. Mechanisms of competitive exclusion between two species of chipmunks. *Ecology* 52: 305–311.

Brown, J.H. 1971b. Mammals on mountaintops: non-equilibrium insular biogeography. *American Naturalist* 105: 467–478.

Brown, W.L. and E.O. Wilson. 1956. Character displacement. *Systematic Zoology* 5: 49–94.

Chase, J.M., and M. A. Leibold. 2003. *Ecological Niches: Linking Classical and Contemporary Approaches*. University of Chicago Press, Chicago.

Connell, J.H. 1961a. Effects of competition, predation by *Thais lapillus* and other factors on natural populations of the barnacle, *Balanus balanoides*. *Ecological Monographs* 31: 61–104.

Connell, J.H. 1961b. The influence of interspecific competition and other factors on the distribution of the barnacle, *Chthamalus stellatus*. *Ecology* 42: 710–723.

Connell, J.H. 1983. On the prevalence and relative importance of interspecific competition: Evidence from field experiments. *American Naturalist* 122: 661–696.

Crombie, A.C. 1945. On competition between different species of graminivorous insects. *Proceedings of the Royal Society of London, Series B*, 132: 362–395.

Crombie, A.C. 1946. Further experiments on insect populations. *Proceedings of the Royal Society of London, Series B* 133: 76–109.

Crombie, A.C. 1947. Interspecific competition. *Journal of Animal Ecology* 16: 44–73.

Dayan, T., D. Simberloff, E. Tchernov, and Y. Yom-Tov. 1990. Feline canines: community-wide character displacement among the small cats of Israel. *American Naturalist* 136: 39–60.

Desbiez, A.L.J., S.A. Santos, A. Keuroghlian, and R.E. Bodmer. 2009. Niche partitioning among white-lipped peccaries (*Tayassup ecari*), collard peccaries (*Pecarit ajacu*) and feral pigs (*Sus scrofa*). *Journal of Mammalogy* 90: 119–128.

Diamond, J.M. 1978. Niche shifts and the rediscovery of interspecific competition. *American Scientist* 66: 322–331.

Diamond, J.M. 1983. Laboratory, field and natural experiments. *Nature* 304: 586–587.

Fenchel, T. 1975. Character displacement and coexistence in mud snails (Hydrobiidae). *Oecologia* 20: 19–32.

Ford, W.M., M.A. Menzel, T.S. McCay, and J. Laerm. 2001. Contiguous allopatry of the masked shrew and southeastern shrew in the southern Appalachians: Segregation along an elevational and habitat gradient. *Journal of the Elisha Mitchell Scientific Society* 117: 20–28.

Gause, G.F. 1932. Experimental studies on the struggle for existence. I. Mixed populations of two species of yeast. *Journal of Experimental Biology* 9: 399–402.

Gause, G.F. 1934. *The Struggle for Existence*. Williams and Wilkins, Baltimore. Reprinted in 1964 by Macmillan (Hafner Press), New York, NY.

Geisel, T. Seuss ("Dr. Seuss"). 1955. *On Beyond Zebra*. Random House, New York, N.Y.

Grant, P.R. 1999. *Ecology and Evolution of Darwin's Finches*. Princeton University Press, Princeton, NJ.

Grant, P.R. and B.R. Grant. 2006. Evolution of character displacement in Darwin's Finches. *Science* 313: 224–226.

Grinnell, J. 1917. The niche-relationships of the California thrasher. *The Auk* 34: 427–433.

Grover, J. P. 1997. *Resource Competition*. Chapman & Hall, London.

Gurevitch, J., L.L. Morow, A. Wallace, and J.S. Walsh. 1992. A meta-analysis of competition in field experiments. *American Naturalist* 140: 539–572.

Hairston, N.G., F.E. Smith, and L.B. Slobodkin. 1960. Community structure, population control and competition. *American Naturalist* 94: 421–425.

Hansen, S.R. and S.P. Hubbell. 1980. Single-nutrient microbial competition: qualitative agreement between experimental and theoretically forecast outcomes. *Science* 207: 1491–1493.

Hardin, G. 1960. The competitive exclusion principle. *Science* 131: 1292–1297.

Hastings, A. 1980. Disturbance, coexistence, history, and competition for space. *Theoretical Population Biology* 18: 363–373.

Heller, H.C. and D. Gates. 1971. Altitudinal zonation of chipmunks (*Eutamias*): Energy budgets. *Ecology* 52: 424–433.

Holldobler, B. 1986. Food robbing in ants, a form of interference competition. *Oecologia* 69: 12–15.

Holldobler, B. and E.O. Wilson. 1990. *The Ants*. Belknap Press, Harvard University, Cambridge, MA.

Hubbell, S.P. 2001. *The Unified Theory of Biodiversity and Biogeography*. Princeton University Press, Princeton, NJ.

Hubbell, S.P. 2010. Neutral theory and the theory of island biogeography. Pages 264–292, *in The Theory of Island Biogeography Revisited*, J.B. Losos and R.E. Ricklefs, Editors. Princeton University Press. Princeton, NJ.

Hutchinson, G.E. 1957. Concluding remarks. *Cold Spring Harbor Symposium on Quantitative Biology* 22: 415–427.

Hutchinson, G.E. 1959. Homage to Santa Rosalia or why are there so many kinds of animals? *American Naturalist* 93: 145–159.

Krebs, C.J. 1994. *Ecology*. 4th Edition. Harper Collins College Publishers, New York, NY.

Lack, D. 1947. *Darwin's Finches*. Cambridge University Press, Cambridge, UK.

Levins, R. and D. Culver. 1971. Regional coexistence of species and competition between rare species. *Proceedings of the National Academy of Sciences of the United States of America* 68: 1246–1248.

Levins, R., M.L. Pressick, and H. Heatwole. 1973. Coexistence patterns in insular ants. *American Scientist* 61: 463–472.

Lieberburg, I, P.M. Kranz, and A. Seip. 1975. Bermudian ants revisited: the status and interaction of *Pheidole megacephala* and *Iridomyrmex humilis*. *Ecology* 56: 473–478.

Liebig, J.F. 1855. *Principles of Agricultural Chemistry*. Wiley, New York, NY.

Lotka, A.J. 1925. Elements of physical biology. Dover, New York, NY Reprinted 1956.

MacArthur, R.H. 1958. Population ecology of some warblers of northeastern coniferous forests. *Ecology* 39: 599–619.

MacArthur, R.H. 1972. *Geographical Ecology: Patterns in the Distribution of Species*. Harper and Row, New York, NY.

MacArthur, R.H., and E.O. Wilson. 1963. An equilibrium theory of insular zoogeography. *Evolution* 17: 373–387.

MacArthur, R.H. and E.O. Wilson. 1967. *The Theory of Island Biogeography*. Princeton University Press, Princeton, NJ.

Merendino, M.T., C.O. Ankey, and D.G. Dennis. 1993. Increasing mallards, decreasing black ducks: More evidence for cause and effect. *Journal of Wildlife Management* 57: 199–208.

Miller, T.E., J.H. Burns, P. Munguia, E.L. Walters, J.M. Kneitel, P.M. Richards, N. Mouquet, and H.L. Buckley1. 2005. A critical review of twenty years' use of the Resource-Ratio Theory. *American Naturalist* 165: 439–448.

Miller, T.E., J.H. Burns, P. Munguia, E.L. Walters, J.M. Kneitel, P.M. Richards, N. Mouquet, and H.L. Buckley. 2007. Evaluating Support for the Resource-Ratio Hypothesis: A Reply to Wilson et al. *American Naturalist* 169: 707–708.

Monod, J. 1950. La technique de culture continue, theorie et applications. *Annales d'Institut Pasteur* 79: 390–410.

Nee, S. and R.M. May. 1992. Dynamics of metapopulations: habitat destruction and competitive coexistence. *Journal of Animal Ecology* 61: 37–40.

Neil, W.E. 1974. The community matrix and interdependence of the competition coefficients. *American Naturalist* 108: 399–408.

Neyman, J., T. Park, and E.J. Scott. 1956. Struggle for existence. The *Tribolium* model: biological and statistical aspects. Pages 41–79, *in Proceedings of the Third Berkeley Symposium on Mathematical Statistics and Probability*. Vol. IV. University of California Press, Berkeley, CA.

Park, T. 1948. Experimental studies of interspecific competition. I. Competition between populations of the flour beetles, *Tribolium confusum* Duvall and *Tribolium castaneum* Herbst. *Ecological Monogement* 18: 267–307.

Park, T. 1954. Experimental studies of interspecific competition. II. Temperature, humidity, and competition in two species of *Tribolium*. *Physiological Zoology* 27: 177–238.

Ricklefs, Robert E. 1997. *The Economy of Nature*. 4th Edition. W.H. Freeman, New York, NY.

Schluter, D., T.D. Price, and P.R. Grant. 1985. Ecological character displacement in Darwin's Finches. *Science* 227: 1056–1059.

Schoener, T. 1968. The Anolis lizards of Bimini: resource partitioning in a complex fauna. *Ecology* 49: 704–726.

Schoener, T. 1982. The controversy over interspecific competition. *American Scientist* 70: 586–595.

Schoener, T. 1983. Field experiments on interspecific competition. *American Naturalist* 122: 240–285.

Schoener, T.W. and C.A. Toft. 1983. Spider populations: extraordinarily high densities on islands without top predators. *Science* 219:1353–1355.

Simberloff, D. and W.J. Boecklen. 1981. Santa Rosalia reconsidered: size ratios and competition. *Evolution* 35: 1206–1228.

Strong, D.R., Jr., and D.S. Simberloff. 1981. Straining at gnats and swallowing ratios: character displacement. *Evolution* 35: 810–812.

Strong, D.R., Jr., L.A. Szyska, and D. Simberloff. 1979. Test of community-wide character displacement against null hypotheses. *Evolution* 33: 897–913.

Strong, D.R., Jr., D.S. Simberloff, L.G. Abele, and A.B. Thistle (Editors). 1983. *Ecological Communities, Conceptual Issues and the Evidence*. Princeton University Press, Princeton, NJ.

Tansley, A.G. 1917. On competition between *Galium saxatile* L. (*G. hercynicum* Weig.) and *Galiuim sylvestre* poll. (*G. asperum* Schreb.) on different types of soil. *Journal of Ecology* 5: 173–179.

Taper, M.L. and T.J. Case. 1992. *Coevolution Among Competitors*. Oxford Series in Evolutionary Biology. Oxford University Press, Oxford, UK.

Tilman, D (published as D. Titman). 1976. Ecological competition between algae: experimental confirmation of resource-based competition theory. *Science* 192: 463–466.

Tilman, D. 1981. Tests of resource competition theory using four species of Lake Michigan algae. *Ecology* 62: 802–815.

Tilman, D. 1982. *Resource Competition and Community Structure*. Princeton University Press, Princeton, NY.

Tilman, D. 1987. The importance of the mechanisms of interspecific competition. *American Naturalist* 129: 769–774.

Tilman, D. 1994. Competition and biodiversity in spatially structured habitats. *Ecology* 75: 2–16.

Tilman, D. 1999. Diversity by default. *Science* 283: 495–496.

Tilman, D., S.S. Kilham, and P. Kilham. 1982. Phytoplankton community ecology: the role of limiting nutrients. *Annual Review of Ecology, Evolution and Systematics* 13: 349–372.

Vandermeer, J.H. 1969. The competitive structure of communities: an experimental approach with protozoa. *Ecology* 50: 362–371.

Volterra, V. 1926. Fluctuations in the abundance of a species considered mathematically. *Nature* 118: 558–560.

Weston, M.R., R.L. Johnson, and A.D. Christian. 2010. Niche partitioning of the sympatric yellowcheek darter *Etheostoma moorei* and rainbow darter *Etheostoma caeruleum* in the Little Red River, Arkansas. *American midland naturalist* 164: 187–200.

Wieland, N.K. and F.A. Bazzaz. 1975. Physiological ecology of three co-dominant successional annuals. *Ecology* 56: 681–688.

Wiens, J.A. 1977. On competition and variable environments. *American Scientist* 65: 5990–5997.

Wilbur, H. M. 1972. Competition, predation, and the structure of the *Ambystoma – Rana sylvatica* community. *Ecology* 53: 3–21.

Williamson, P. 1971. Feeding ecology of the red-eyed vireo (*Vireo olivaceous*) and associated foliage-gleaning birds. *Ecological Monographs* 41: 129–152.

Wilson, J.B., E. Spijkerman, and J. Huisman. 2007. Is there really insufficient support for Tilman's R* concept? A comment on Miller et al. *American Naturalist* 169: 700–706.

Yu, D.W. and H.B. Wilson. 2001. The competition-colonization trade-off is dead; long live the competition-colonization trade-off. *American Naturalist* 158: 49–63.

8 Mutualism

8.1 Introduction

8.1.1 Mutualism or parasitism?

As outlined earlier in this book, mutualism is an interaction in which both species benefit. In facultative mutualism, individuals in a population are able to survive and reproduce without their presumptive mutualist, although their fitness is enhanced when they participate in the mutualism. By contrast, in obligatory mutualism, individuals in a species are unable to survive without their mutualistic partner. As pointed out by Vandermeer and Goldberg (2003), mutualistic relationships are complex and do not necessarily fit into the two categories of facultative and obligatory. For example, a mutualism may be obligatory for one partner but not for the other. The mutualism may be very weak (provide few benefits) and, therefore, may only be found in very specific environments. In fact, many types of mutualisms are much more common in tropical environments. For example, Neotropical ants and African termites both raise mutualistic fungi in "gardens" within their nests. Although fungus gardening does occur outside of the tropics, it is much more common and conspicuous in the humid tropics. In ant–plant mutualisms, ants defend plants from herbivores or perform other services in exchange for nest sites and nutrition provided by the plants. Although these mutualisms exist in the temperate zone, almost all of the obligatory ant–plant mutualisms are found within the tropics. The number of plant species providing extra-floral nectar, a low cost method of attracting ants to plants, declines with increasing latitude and altitude and is rare in the North Temperate Zone (Bentley 1977; Keeler 1979, 1980).

Animal pollination and fruit dispersal are also much more common in tropical latitudes. For example, bats, which provide pollination and fruit dispersal for higher plants, are only found south of 33°N latitude. No bees that are obligatory pollinators of orchids are found north of 24° latitude. At Monteverde, a cloud forest in Costa Rica, animals as opposed to wind, pollinate more than 90% of the dicots and 88% of the monocots. A study by Murray et al. (2000) of fruit dispersal at Monteverde showed that more than 81% of tree species are adapted for seed dispersal by vertebrates. This compares well with 89% at Alto Yunda, Colombia; 92% at La Selva, Costa Rica; and 92% at Rio Palenque, Ecuador. At Monteverde, about 80% of animal dispersed trees are specifically adapted for bird dispersal. Adaptations for bat and ant dispersal are less common. Not all life forms, however, are adapted for animal dispersal of their fruit

Introduction to Population Ecology, Second Edition. Larry L. Rockwood.
© 2015 John Wiley & Sons, Ltd. Published 2015 by John Wiley & Sons, Ltd.
Companion Website: www.wiley.com/go/rockwood/populationecology

at Monteverde. A majority of epiphytes (66%) and herbs (73%) are adapted for wind or other abiotic means of dispersal. Most of the wind-dispersed seeds, however, are orchids (among epiphytes) and weedy species of Asteraceae (among herbs). Lianas and shrubs are intermediate; the fruits of the majority of species are bird dispersed while 35%–45% is abiotically dispersed.

Mutualistic relationships are wonders of natural history and appear to be prime examples of co-evolution. Yet, as Bronstein (2000, p. 273) put it, "Mutualism is the most poorly understood form of interspecific interactions." Others such as Law (1988) and Watkinson (1997) have made the same point while Watkinson (1997, p. 395) lamented that "There is not even a sound theoretical framework for the treatment of mutualistic interactions." Herre et al. (1999) have agreed that there is no general theory of mutualism, while Foster and Wenseleers (2006) have even proposed that mutualism actually presents a challenge for evolutionary theory. Mutualism resembles the prisoner's dilemma where two players can benefit from cooperation but gain most from cheating its partner. How does selection favor cooperation among species rather the cheaters?

Recent research emphasizes that the in most mutualisms, the relationship is simultaneously beneficial and harmful to one or both participants. Rather than thinking of the mutualistic species as happily entering into a partnership, Bronstein (2002) has asserted that mutualism is more likely a "reciprocal parasitism" in which each partner obtains what it can at the lowest possible cost to itself.

Foster and Wenseleers (2006) describe the following factors that may favor mutualists remaining true to their partners:
1. Cooperation will be favored when species stay in stable associations.
2. If there is more than potential mutualistic partner, then a species will choose to interact with the more cooperative of the potential partner species.
3. Similarly noncooperative partners are sanctioned and receive less aid.
4. There may be no incentive to cheat if cooperation carries little or no energetic cost.

According to a model developed by Foster and Wenseleers (2006), the following factors should favor the evolution of mutualisms: (1) high benefit to cost ratio, (2) high within-species genetic relatedness, and (3) high between-species fidelity.

In exploring mutualisms, we should be mindful that in most cases, the relationship can be simultaneously beneficial and harmful to one or both participants and may change over time or with environmental conditions. Furthermore, rather than focusing on mutualism as a one species on one species relationship in one slice of time, biologists need to begin examining mutualism in the context of a community of organisms and over longer periods.

8.2 Ant–plant mutualisms

Ant–plant interactions illustrate the complexities of mutualistic interactions and the themes of cheaters versus cooperators. Janzen's 1966 paper is a classic example of a mutualism between species of *Acacia* and species of *Pseudomyrmex* ants in the American tropics. As has been described in numerous other textbooks, *Pseudomyrmex* ants regularly patrol the leaves, particularly the new leaves, of certain species of *Acacia*. In so doing, they protect the plant from almost all herbivores, vertebrate, and invertebrate alike. In addition, by chewing of leaves of other plant species, they prevent *Acacia* leaves from being shaded by competitors. Finally, they even weed areas on the ground under the tree. In return, *Acacia* plants have evolved the following traits to support *Pseudomyrmex* ant colonies.

1. Thorns as places to live for the ants.
2. Extrafloral nectaries: sources of water and carbohydrates.
3. "Beltian bodies" that provide lipids and protein to the ants.
4. Finally, these *Acacia* trees remain evergreen in the tropical dry forest life zones, which feature dry seasons of 6 months.

As Janzen (1966) proved, this is an obligatory mutualism. These ants live nowhere else but on *Acacia* trees, and when an *Acacia* tree was deprived of its ant colony, it died within 6 months. Without an ant colony, the leaves of *Acacia* trees are quickly attacked by a variety of herbivores and suffer severe defoliation. Meanwhile, they are just as quickly shaded out by competing species whose leaves are no longer attacked by the ants. Nearby species of nonant acacias have cyanogenic glycosides in their leaves to deter herbivores while the leaves of acacias housing ant colonies are usually completely edible. The evolution of the traits necessary to support ant colonies has resulted in a progressive evolutionary loss of defensive chemicals in the leaves of these ant–*Acacia* species (Rehr et al. 1973, Seigler and Ebinger 1987).

However, this simple picture is deceiving. First, we must note that there are at least 11 species of *Acacia* in Central America and 10 species of *Pseudomyrmex* ants that participate in these potential mutualisms. At least one of these ant species, *Pseudomyrmex nigropilosus*, has become a parasite. It takes what *Acacia* offers to ant colonies but offers no protection for the plant (Janzen 1975). When approached by a potential herbivore, workers hide. Furthermore, workers are not active at night, and *P. nigropilosus* tries to prevent other *Pseudomyrmex* species from occupying the *Acacia*. *Acacia* occupied by *P. nigropilosus* have low fitness since they are likely to die before maturity.

Founding queens of *P. nigropilosus* invade Acacias that have lost the queens of other, more functional, species of *Pseudomyrmex* and live there until the plant dies from herbivory or competition with other plants. Swollen-thorn acacias may lose their ant occupants via fire, predation by birds, queen deaths, or by an exceptional dry season causing the leaves to drop. Thus, this ant species behaves as an invasive species living on an ephemeral resource. As such, it displays an r-selected or "fast" life cycle. Whereas other *Pseudomyrmex* species take at least 2 years before producing brood, *P. nigropilosus* does so within 2 months. There is also a lower standing crop of workers as compared to other *Acacia*-inhabiting ant species. *P. nigropilosus* is also regularly evicted by six species of obligate acacia-ants. Thus, this species fits a classic view of an invasive species inhabiting a temporary resource. It has low competitive ability, harvests the resource quickly, reproduces quickly, and either moves on or dies out locally.

The next complication in this story is the fact that not all *Acacia* species offer the same rewards. For example, Heil et al. (2009) found that *Acacia collinsii* and *A. cornigera* are "high reward" species, while *A. hindsii* and *A. chiapensis* are "low reward" species. They first looked at the occupancy of *Acacia* hosts by different ant species. Similar to *P. nigropilosus*, *P. gracilis* is an "exploiter" species that does little for the plant it inhabits. Heil et al. (2009) determined that these two ant species are found in higher percentages on the two "low reward" plants. *A. collinsii* and *A. cornigera* produce more thorns, *extrafloral nectar* (*EFN*), and food bodies (Beltian bodies) than the other two species of *Acacia* and are usually occupied by a "good" ant species, *Pseudomyrmex ferrugineus*. Heil et al. (2009) showed that higher amounts of EFN secretions lead to higher ant activity when a leaf is attacked. A model developed by Heil et al. (2009) predicts that exploiter species will prosper if the host plant limits the rewards, but mutualists gain advantage when rewards are high (as also proposed by Foster and Wenseleers 2006). When host rewards are limited, colony growth is reduced. They conclude that high and low reward *Acacia* species can coexist because costs and benefits of defense depend on nutrient availability, herbivore pressure, and other factors. Nevertheless, mutualisms are stabilized when increased investment in one partner leads to increased, reciprocal investment in the other partner.

8.2.1 *Ants and Acacias in Africa*

The relationships among ants and the East African whistling thorn acacias (*Acacia drepanolobium*) are even more complex. These acacias are some of the most common plants of the savannas of central Kenya. Almost all of these trees host thousands of ants. The ants provide defense against herbivores in exchange for food and shelter. A pair of thorns lies at the base of every leaf cluster, and each branch is lined with two types of thorns. The slender, white, needle sharp thorns, which may be three inches in length, are the most abundant. Intermingled with these thorns are pairs of thorns with bulbous, hollow bases. These "swollen thorns" house the ant colonies; each thorn can harbor hundreds of ants. As is the case in Central American acacias (Janzen 1966), when an herbivore disturbs the tree, ants stream out of the thorns and bite the intruder. When Stanton and Young (1999) experimentally removed ants from trees, they found that herbivore damage by both browsing mammals and insects increased significantly.

The acacia provides the ants with nectar from glands along the leaves. These extrafloral nectaries are particularly abundant on new leaves. However, *Acacia drepanolobium* does not provide lipid or protein food sources for these ant colonies. The ants must forage for insects and other protein-rich foods. Therefore, the mutualism is weaker than the *Acacia–Pseudomyrmex* relationship described by Janzen (1966) in Central America, where the acacias provide lipid- and protein-rich structures on the tips of the leaflets, known as Beltian bodies.

There are four different ant species that colonize *A. drepanolobium*. The red-and-black cocktail ant (*Crematogaster mimosae*) and the black-and-white cocktail ant (*C. nigriceps*) are most effective against herbivores. Intruders are immediately attacked by a hoard of biting ants. The ants race around emitting alarm pheromones with their abdomens held high (hence the term "cocktail ant"), that recruit more workers to the scene. Herbivores such as goats, which are bitten while attempting to browse on a defended tree, refuse to approach those trees again. Trees defended by these two ant species are rarely damaged by herbivores.

A third species, the slender black acacia ant (*Tetraponera penzigi*) has a nasty sting but is more passive and only attacks if the swollen thorns themselves (the home of larvae, pupae, adults, and winged reproductive ants) are attacked by monkeys or other animals. On the other hand, these ants patrol leaves day and night, removing pollen and probably fungal spores. Therefore, these workers may provide the acacias protection from disease.

The fourth species (*C. sjostedti*), the black cocktail ant, provides no services to the trees whatsoever. Stanton and Young (1999) found that long-horned beetles could girdle stems and kill entire sections of the tree while this ant was present. The reason for this may be that this ant species does not even live in the swollen thorns. Instead, it nests in hollow spaces within dead and drying branches. The beetles actually provide this ant species with nest space, and the mutualistic relationship has shifted to one in which these two insect species "cooperate" in exploiting the acacias.

Table 8.1 Frequency of occupation of *Acacia drepanolobium* by four species of ants in Kenya.

Scientific name	Common name	Frequency of occupation (%)
Crematogaster mimosae	Red and black cocktail ant	52
Crematogaster sjostedti	Black cocktail ant	16
Crematogaster nigriceps	Black and white cocktail ant	15
Tetraponera penzigi	Blacka acacia ant	17

Source: Data from Palmer et al. 2008

Although all four species of ants may occupy trees on a given hectare of land (Table 8.1), an individual tree is almost never occupied by more than one species of ant. The species are intolerant of each other and engage in aggressive, mortal combat. Experiments have shown that the fights continue until one species has wiped out the second on a given tree. Unfortunately for the acacias, the black cocktail ant, which is functionally a parasite, is the dominant competitor, winning most battles with the other three species.

The red-and-black cocktail ant, as well as the black cocktail ant, tends scale insects that feed on the phloem of the acacias. Thus, both species have found an additional method of draining energy from the trees. Worse still is the habit of the black-and-white cocktail ant workers of removing the tips of most growing shoots. These workers also remove stem tissue containing leaf and flower buds. New branches are only allowed to grow in proximity to swollen thorns. These black-and-white cocktail ants have therefore changed the architecture of these trees from one with large open canopies to one of compact masses of branching stems. In addition, by chewing off its flower buds, it prevents these trees from reproducing.

In summary, as Palmer et al. (2010) have shown (Table 8.2), the ant–acacia symbiosis in Africa, although potentially mutually beneficial, is very complex. In some cases, ants function more like parasites than mutualists. As Table 8.2 shows, there is a definite hierarchy among species and a competition/colonization trade-off. Furthermore, acacias of different size/age classes are preferentially colonized by different ant species. For example, *T. penzigi* occupies more very young trees than the other three ant species, but the proportion occupied by this species declines as trees get older and larger. On the other hand, trees occupied by *C. sjostedti* increase as trees grow until over 60% of trees with a tree height of 5.0 m or more are occupied by this species. Tree properties also change as trees age and grow. For example, more leaves produce nectar as trees age and get larger. Palmer et al. (2010) showed that tree fitness is actually enhanced through sequential partnering with all four different ant mutualists over a lifetime. The point made in Palmer et al. (2010) is that natural selection works on lifetime fitness. If one partner is a long-lived species, one must examine the interactions over a long period. As mentioned earlier, *C. nigriceps* sterilizes the tree and it appears to be an evolutionary parasite. Yet Palmer et al. (2010) found that over a 50-year period, trees with the highest reproductive rates had hosted all four ant species, including *C. nigriceps*. These *Acacia* trees undergo several life stages from seedling to small tree, to mature tree to aged tree. At each stage, a different symbiont is beneficial since in the total absence of ants, *Acacia* survivorship is very low at all sizes. The high colonization rates of certain ant species make them valuable even though the benefits provided are rather low. Another significant factor is the low abandonment rates of *C. nigriceps* and *T. penzigi*. These two species combine high colonization and high reliability and thus become

Table 8.2 Variation in cost and benefits to *Acacia drepanolobium* provided by four ant species in Kenya.

Ant species	Dominance Rank (Palmer et al. 2010)	Colonization Rank (Stanton et al. 2002)	Mean Number of Trees per Colony (± S.E.)	Host Plant Defense (Mean number of workers (± S.E.) recruited in response to disturbance) (Palmer et al. 2007)	Percent of shoots browsed (± S.E.) (Palmer and Brody 2007)	Sterilization of Host Plant?	Extrafloral Use (Mean number of workers tending nectaries) (± S.E.)	Beetle Damage Number of beetle scars (±S.E.) (Palmer et al. 2008)
Crematogaster sjostedti	1	4	22.0 (4.8)	Low 1.3 (0.3)	8.0 (1.1)	No	Low 0.1 (0.1)	6.6 (1.3)
Crematogaster mimosae	2	3	4.4 (0.3)	High 17.6 (1.7)	3.3 (0.4)	No	High 2.0 (0.4)	1.4 (1.0)
Crematogaster nigriceps	3	2	2.5 (0.2)	High 15.0 (1.5)	2.5 (0.7)	Yes	High 3.8 (0.4)	0.4 (1.2)
Tetraponera penzigi	4	1	1.3 (0.3)	Medium 5.4 (0.7)	3.8 (0.6)	No	None (host plant nectaries destroyed)	4.5 (1.1)

Source: Data from Palmer et al. 2010

more valuable to the *Acacia* than one might assume. By contrast, *C. mimosae* colonies, which would seem to be the most beneficial to the plants, have high abandonment rates, leaving their *Acacia* host vulnerable. Suffice it to say that this is an intriguing system that has already produced a stream of scientific papers.

8.2.2 *Geographic variation in coevolution of a mutualism*

In addition to ant–plant mutualisms varying in intensity, they also show geographic variation, as illustrated by the *Cecropia–Azteca* ant mutualism. *Cecropia* trees are common in lowland wet environments in the Neotropics and have a facultative mutualism with species of *Azteca* (Janzen 1966, 1969, 1973). The ant colonies live in the hollow stems of *Cecropia,* which produces "Mullerian" bodies at the base of the leaf petioles. Mullerian bodies contain carbohydrates, lipids, and proteins and are harvested daily by the ants. These bodies are associated with hair-like trichilia, and Mullerian bodies are almost never produced without the associated tricilium. The carbohydrate produced by the Mullerian body is in the form of glycogen rather than starch (Rickson 1977). This glycogen is identical to animal glycogen and is probably necessary for the *Azteca* ants because they are descended from carnivorous species.

In return, the ants protect *Cecropia* by stinging potential herbivores and chewing on vines, which might grow on *Cecropia*. This is not an obligatory mutualism, since individual *Cecropia* exist without ants, particularly at high altitudes and on islands where there is less pressure from insect herbivores and vines (Janzen 1969). Janzen noted that in Puerto Rico few ant-related traits were present on *Cecropia*. Rickson (1977) did a survey of *Cecropia* on the Caribbean island chain from mainland South America near Venezuela to Puerto Rico. *Azteca* ants are present on the mainland and on the islands of Trinidad and Tobago but not on any of the other islands in the Lesser Antilles chain. What Rickson found, however, was that the ant-related traits, in particular, the presence of trichilia and/or Mullerian bodies, decreased the further one moved from the mainland and toward Puerto Rico. With the ants absent, there was a progressive loss of Mullerian bodies and trichilia. With the ants absent, production of Muellerian bodies was selected against since the production of glycogen is an energetic drain. Since insect diversity and herbivore damage was also limited on these Caribbean islands, the maintenance of an ant colony as protection from herbivory was also selected against (Rickson 1977).

8.3 **Modeling mutualism**

In his pessimistic remarks on our understanding of mutualistic interactions, Watkinson (1997) points out that when modifications of the Lotka-Volterra competition equations (equations 8.1 and 8.2) are employed to model mutualism, the result is "unbounded exponential growth" of both populations. Robert May (1981b) called this result an "orgy of mutual benefaction." In equations 8.1 and 8.2, as in the Lotka–Volterra equations, the growth rates of the two mutualistic species are determined by their present populations sizes (N_1 and N_2), their intrinsic rates of growth (r_1 and r_2), and diminished by intraspecific competition (again, N_1 and N_2 compared to the carrying capacities (K_1 and K_2)). These carrying capacities, however, are set for each population when living without its mutualist. Obviously, these equations only apply to facultative mutualism, since in obligatory mutualism, there can be no positive carrying capacity in the absence of the partner species. The main difference between these equations and the competition equations is that the terms $c_1 N_2$ and $c_2 N_1$ have a positive sign. The terms "c_1" and "c_2" are mutualism coefficients and replace the competition coefficients in the Lotka-Volterra competition equations.

The term "c_1" measures the rate at which an individual of N_2 benefits the growth rate of population N_1. Similarly, c_2 measures the rate at which an individual of N_1 benefits the growth rate of population N_2:

$$\frac{dN_1}{dt} = r_1 N_1 \frac{K_1 + c_1 N_2 - N_1}{K_1} \tag{8.1}$$

$$\frac{dN_2}{dt} = r_2 N_2 \frac{K_2 + c_2 N_1 - N_2}{K_2} \tag{8.2}$$

An equilibrium analysis can be done, similar to what we did with the competition equations, by setting dN_1 and dN_2 equal to zero. Ignoring situations where r, N, or K is equal to zero leaves $K_1 + c_1 N_2 - N_1 = 0$ and $K_2 + c_2 N_1 - N_2 = 0$. The basic results are as indicated in equations 8.3 and 8.4. As you can see, the population of each species is increased beyond its carrying capacity with additional individuals of its partner species. The more intense the mutualism (the greater the benefits provided to the partner species), the larger the equilibrium population becomes:

$$N_1 = K_1 + c_1 N_2 \tag{8.3}$$

$$N_2 = K_2 + c_2 N_1 \tag{8.4}$$

Furthermore, the "orgy of mutual benefaction" referred to above should be evident. Greater numbers of N_2 produce greater numbers of N_1, which, in turn, produce larger numbers of N_2 and so on. Strangely enough, the interaction is stable when only one partner benefits from the interaction. If $c_2 = 0$, for example, N_2 cannot exceed K_2 and N_1 stabilizes at $K_1 + c_1 K_2$.

Another approach is to substitute the carrying capacity after mutualism into equations 8.1 and 8.2. The new carrying capacities, K_1^* and K_2^*, are shown in Equations 8.5 and 8.6. The resultant equations (8.7 and 8.8), however, are only stable if we specify either a numerical limit to K_1^* and K_2^* or that the mutualism coefficients decline toward zero as the populations approaches the new K^* carrying capacity. In an obligatory mutualism, we would have to specify that both N_1 and $N_2 > 0$. Otherwise, we have mutual extinction. In fact, it is reasonable to assume, as in the Allele effect, that each population has a minimum viable population size, below which the mutualism falls apart and both species go extinct:

$$K_1^* = K_1 + c_1 N_2 \tag{8.5}$$

$$K_2^* = K_2 + c_2 N_1 \tag{8.6}$$

$$\frac{dN_1}{dt} = r_1 N_1 \frac{K_1^* - N_1}{K_1^*} = r_1 N_1 \frac{K_1 + c_1 N_2 - N_1}{K_1 + c_1 N_2} \tag{8.7}$$

$$\frac{dN_2}{dt} = r_2 N_2 \frac{K_2^* - N_2}{K_2^*} = r_2 N_2 \frac{K_2 + c_2 N_1 - N_2}{K_2 + c_2 N_1} \tag{8.8}$$

When trying to model mutualism by starting with the Lotka–Volterra equations, we encounter the same problems as those described in the previous chapter on competition. The Lotka–Volterra approach is a phenomenological one. What we need is a mechanistic approach based on the rates of exchange of relevant resources between the two mutualists. However, since each mutualistic interaction is both unique and complex, a model based on one mutualism would lack generality. While specific models describing specific mutualisms have been written, none has entered the literature as a standard approach.

8.4 Summary: the costs of mutualism

The study of mutualism has passed through a number of stages but is only recently moving from the descriptive phase (Bronstein 2001). Many early naturalists were skeptical about mutualistic relationships. For example, before Janzen (1966) established the obligatory mutualism between *Pseudomyrmex* ants and *Acacia* trees, the ants were said to be of no more use to the plants than "fleas on a dog." Once mutualisms were described, they were often pronounced to be mostly confined to the tropics. However, Janzen (1985) pointed out that every organism is involved in at least one mutualism in its life. Given that the eukaryotic cell evolved as the result of mutualistic relationships, the phenomenon is ubiquitous. And extratropical locations do not lack mutualisms. Forest trees with mycorrhizal fungi dominate boreal habitats, deserts are populated with legumes and their nitrogen-fixing bacteria, and tundras are dominated by lichens.

While mutualistic interactions have become accepted as comparable in importance to ecosystems as competition and predator–prey relationships, the pendulum may have swung to the view that mutualism represents "cooperation" between species. By contrast, Bronstein (2001) has stressed that mutualisms involve costs for each species, as well as benefits. Costs of mutualism are only now being tabulated, and there is little consistency in how data are gathered. Bronstein (2001) cites the following examples: (1) 20% of the total carbon budget of forest trees may be consumed supporting mycorrhizae (Johnson et al. 1997); (2) 3% of the energy budget of many plants is devoted to providing floral nectar for pollinators (Harder and Barrett 1992); (3) EFN costs about 1% of the energy budget of the plants involved (O'Dowd 1979, 1980). In the obligatory interaction between figs and their wasp pollinators, Bronstein (2001) estimated that 53% of ovaries of *Ficus aurea* are lost to the wasps, while yuccas evidently lose 5%–20% of their seeds to their moth pollinators. Finally, Wolfe (2001a,b) estimates that nitrogen-fixing bacteria consume 20% of the carbohydrates produced by legumes.

Rather than assuming that species have somehow entered into permanent, mutually agreeable contracts, we need to analyze mutualism with the following points in mind: (1) mutualism always involves costs as well as benefits; (2) costs set limits on the evolution of mutualisms; (3) there is a conflict of interest between the mutualistic species; (4) organisms that "cheat" on the mutualism by reaping the benefits without reciprocation will often enjoy an advantage; (5) related organisms (such as nonmutualistic ant species) may take advantage of a mutualism and act as parasites on the relationship; (6) mutualisms may evolve toward a host–parasite relationship from a mutualistic one; and (7) continuous co-evolution is necessary to maintain a mutualism.

References

Bentley, B.L. 1977. Extrafloral nectarines and protection by pugnacious bodyguards. *Annual Review of Ecology and Systematics* 8: 407–427.

Bronstein, J.L. 2000. Fig pollination and seed-dispersal mutualisms. *In* N.M. Nadkarni and N.T. Wheelwright Editors. *Monteverde, Ecology and Conservation of a Tropical Cloud Forest*. Oxford University Press, New York, NY.

Bronstein, J.L. 2001. The costs of mutualism. *American Zoologist* 41: 825-839.

Bronstein, J.L. 2002. Profiles. Page 139 *in* P. Stiling Editor. *Ecology: Theories and Applications.* Prentice Hall, Upper Saddle River, NJ.

Foster, K.R. and T. Wenseleers. 2006. A general model for the evolution of mutualisms. *European Society for Evolutionary Biology* 19: 1283–1293.

Harder, L.D. and S.C.H. Barrett. 1992. The energy cost of bee pollination for *Pontederia cordata* (Pontederiaceae). *Functional Ecology* 6: 226–233.

Heil, M., M. Gonzalez-Teuberb, L.W. Clement, S. Kautz, M. Verhaagh, and J.C. Silva Bueno. 2009. Divergent investment strategies of *Acacia* myrmecophytes and the coexistence of mutualists and exploiters. *Proceedings of the National Academy of Sciences* 106: 18091–18096.

Herre, E.A., N. Knolton, U.G. Mueller, and S.A. Rehner. 1999. The evolution of mutualisms: exploring the paths between conflict and cooperation. *Trends in Ecology and Evolution* 14:49–52.

Janzen, D.H. 1966. Coevolution of mutualism between ants and acacias in Central America. *Evolution* 20: 249–275.

Janzen, D.H. 1969. Allelopathy by myrmecophytes: the ant *Azteca* as an allelopathic agent of *Cecropia*. *Ecology* 50: 147–153.

Janzen, D.H. 1973. Dissolution of mutualism between *Cecropia* and Its *Azteca* Ants. *Biotropica* 5: 15–26.

Janzen, D.H. 1975. *Pseudomyrmex nigropilosa*: a parasite of a mutualism. *Science* 188: 936–937.

Janzen, D.H. 1985. The natural history of mutualisms. Pages 40–99 in D.H. Boucher, Editor. *The Biology of Mutualisms*. Oxford University Press, Oxford, UK.

Johnson, N.C., J.H. Graham, and F.A. Smith. 1997. Functioning of mycorrhizal associations along with the mutualism-parasitism continuum. *New Phytologist* 135: 575–585.

Keeler, K.H. 1979. Distribution of plants with extrafloral nectaries and ants at two elevations in Jamaica. *Biotropica* 11: 152–154.

Keeler, K.H. 1980. Distribution of plants with extrafloral nectaries in temperate communities. *The American Midland Naturalist* 104: 274–280.

Law, R. 1988. Some ecological properties of intimate mutualisms involving plants. Pages 315–342 in A.J. Davy, M.J. Hutchings and A.R. Watkinson, Editors. *Plant Population Ecology.* Blackwell Scientific Publications, Oxford, UK.

May, R.M. 1981. Models for two interacting populations. Pages 78–104, *in* R.M. May, Editor. *Theoretical Ecology. Principles and Applications.* 2nd Edition. Blackwell Scientific, Oxford, UK.

O'Dowd, D.J. 1979. Foliar nectar production and ant activity in a Neotropical tree, *Ochroma pyramidale*. *Oecologia* 43: 233–248.

O'Dowd, D.J. 1980. Pearl bodies of a Neotropical tree, *Ochroma pyramidale*: Ecological implications. *American Journal of Botany* 67: 543–549.

Palmer, T.M., M.L. Stanton, T.P. Young, J.R. Goheen, R.M. Pringle, and R. Karban. 2008. Breakdown of an ant-plant mutualism follows the loss of large herbivores from an African savanna. *Science* 319: 192–195.

Palmer, T.M., D.F. Doak, M.L. Stanton, J.L. Bronstein, E.T. Kiers, T.P. Young, J.R. Goheen, and R.M. Pringle. 2010. Synergy of multiple partners, including freeloaders, increases host fitness in a multispecies mutualism. *Proceedings of the National Academy of Sciences* 107: 17234–17239.

Rehr, S.S., P.P. Feeny, and D.H. Janzen. 1973. Chemical defense in Central American non-ant Acacias. *Journal of Animal Ecology* 42: 405–416.

Rickson, F.R. 1977. Progressive loss of ant-related traits of *Cecropia peltata* on selected caribbean islands. *American Journal of Botany* 64(5): 585–592.

Seigler, D.S. and J.E. Ebinger. 1987. Cyanogenic glycosides in ant-Acacias of Mexico and Central America. *Southwestern Naturalist* 32: 499–503.

Stanton, M. and T. Young. 1999. Thorny relationships. *Natural History* 108: 28–31.

Vandermeer, J.H. and D.E. Goldberg. 2003. *Population Ecology: First Principles*. Princeton University Press, Princeton, NJ.

Watkinson, A.R. 1997. Plant population dynamics. Pages 395–400 *in* M.J. Crawley, Editor. *Plant Ecology*. Blackwell Science, Ltd., Oxford, UK.

Wolfe, D.W. 2001a. Out of thin air. *Natural History* 110: 44–53.

Wolfe, D.W. 2001b. *Tales from the Underground: A Natural History of Subterranean Life*. Perseus Publishing, New York, NY.

9 Host–parasite interactions

9.1 Introduction

Given the prominent role of modern medicine today, it is amazing to realize that the theory that microbes (germs) were the cause of many diseases was not really established until the 1870s. Attempts to create models of epidemiology did not begin until the 20th century. Hamer (1906) formulated a discrete time model in an attempt to understand epidemics of measles. In 1927, Kermack and McKendrick had published models showing that the density of individuals susceptible to a disease must exceed a critical number before an epidemic was possible (Hethcote 2000). A rich literature of mathematical epidemiology developed during the middle of the 20th century. However, most ecologists were only dimly aware of this literature until the late 1970s when Anderson and May published a series of articles with titles such as "Population dynamics of infectious diseases" (Anderson 1982); Anderson and May 1978, 1979, 1981, 1982). A review by Hethcote (2000) is recommended for students interested in the history of infectious disease models.

In this chapter, we will examine models for interactions between hosts and microparasites such as viruses and bacteria. Parasitoid–host interactions (the Nicholson–Bailey models) will be covered in the chapter on predator–prey relationships. Here we are primarily concerned with models for microparasitic diseases such as measles in which the parasite reproduces quickly and reaches tremendous populations within the host. The duration of the acute infection is limited by either host defenses or host mortality and is short relative to the host life span. Recovered hosts may have lifelong immunity. The dynamics of this relationship are driven by transmission between hosts and the essentials can be modeled by classifying host individuals as *susceptible, infected, or recovered (immune)*. This is known as the SIR model, although there are many modifications and elaborations on the basic SIR model (Hethcote 2000). For example, in the MSEIR model, individuals are born with passive immunity (M); over time, they lose this immunity and transfer to the susceptible class (S). Individuals exposed (E) to the disease then become part of the infected (I) class. Those who survive become

Introduction to Population Ecology, Second Edition. Larry L. Rockwood.
© 2015 John Wiley & Sons, Ltd. Published 2015 by John Wiley & Sons, Ltd.
Companion Website: www.wiley.com/go/rockwood/populationecology

part of the recovered (R) class. The actual abundance of the parasite within the host is ignored. Recall that many metapopulation models also ignore population size and dynamics within habitat patches.

In contrast, macroparasites (such as intestinal worms) typically cause chronic, persistent infections. Disease severity depends on the number of parasites present. Not only do infected vertebrate and invertebrate hosts accumulate parasites throughout their adult lives, but both the number and diversity of parasitic species can also increase with host age (Dobson et al. 1992). For example, in Brown Pelicans (*Pelecanus occidentalis*), the number of helminth parasite species increases by age class, reaching 13 for birds aged over 3 years. The same study showed that the number of individual parasites peaked at 8000 for 1-year-old pelicans and then declined to about 4000 in birds older than 3 years (Humphrey et al. 1978). In one study, zebras in Southern Africa hosted 75 species of parasites, including 41 species of nematodes and 42 species of arthropods. The numbers of individual parasites were estimated as 1–4 million (Krecek et al. 1987; Malan et al. 1997). A survey by Dobson et al. (1992) of North American mammals showed that the average individual carried 369 parasites of three different species. The survey examined four orders: carnivores, lagomorphs, rodents, and artiodactyls with carnivores carrying the most diverse parasite fauna.

The observed patterns of infection rate with age are characterized by three hypothetical patterns. In the first pattern, referred to as type I by Hudson and Dobson (1995), the host acquires parasites from the environment at a steady rate over time, and there is a steady increase with age. This assumes no vertical transmission and no reproduction within the host. Thus, a graph of parasite load versus age for a type I curve is linear and resembles the first half (without the leveling off) of the Holling Type I curve for predation rate versus prey density described in the next chapter (Figure 10.9). In contrast, if one assumes that the rates of parasite acquisition and of parasite mortality are constant, the number of parasites per host will increase to an asymptote where there is a balance between the two rates. This is referred to as a type II relationship between parasite load and age. This graph resembles the Type II graph of Holling (Figure 10.10).

Finally, there is a type III relationship between parasite load and host age. In this case, as in the example of Humphrey et al. (1978) cited earlier, parasite loads decline with age after an initial increase. Possible mechanisms include acquired immunity, age-related changes in predisposition to infection, and age-dependent changes in exposure to infections (Wilson et al. 2009). Another factor to consider is the fact that males tend to exhibit higher infection rates of parasites than females (Wilson et al. 2009).

Models of macroparasite dynamics must account for parasite abundance within hosts as well as host-to-host variation in parasite abundance (May and Anderson 1979). A general trend is that parasites show an aggregated (or "clumped") distribution, with the vast majority of parasite individuals concentrated in a minority of the host population. For example, in humans less than 20% of individuals are host to 80% of the individuals of a helminth parasite (Wilson et al. 2009), and although 600 of 690 individuals of *Rana nigromaculata* were free of the nematode, *Spiroxys japonica,* the remaining 296 hosted 896 parasites ranging from one to as many as 109 parasites per individual (Shaw et al. 1998). The vast majority of these aggregated distributions of parasites/host are best described by a negative binomial distribution (Wilson et al. 2009).

Another complication is that many macroparasites such as *Schistosome mansoni*, which causes the disease known as schistosomiasis, have an asexually reproducing stage in an intermediate host. Schistosomiasis is one of the most important human diseases in tropical regions. It is estimated that 100 million people carry at least one worm. The adult schistosomes live in pockets of the intestinal blood vessels or veins of the bladder in the vertebrate host where males and females carry out sexual reproduction. The eggs are passed out of the host through the feces and urine into a body of water where the eggs hatch into free-swimming larvae known as miracidia. A successful miracidium penetrates a snail. Once inside the snail, it undergoes asexual reproduction. After 4–7 weeks, a new free-swimming stage, known as cercariae, is shed from the snail into the water. The cercariae must find the vertebrae host within 48 hours, and each cercaria is capable of penetrating the skin of a vertebrate host. Once inside the host, they travel through the circulatory system in a journey that can take 6–12 weeks. Once they locate the proper tissue, they mature into adult worms, completing the life cycle. An important aspect of this kind of life history is the asexual reproduction stage in the snail. The large numbers of cercariae shed from the snail make it much more likely that the vertebrate host will be located and the life cycle completed. Finally, perhaps because of the complexity of the life cycle of the parasite, vertebrate hosts generally have ineffective immune responses. In fact, the pathology of schistosomiasis is a consequence of the immune response of the host where shed eggs are attacked and calcareous deposits laid down around them. The accumulation of these deposits blocks the spleen and excretory organs of the infected hosts. For models of macroparasites with complex life cycles, such as the parasitic helminthes, see Dobson et al. (1992).

9.2 Factors affecting microparasite population biology

As outlined by Nokes (1992), three main factors govern the interaction between a host and a microparasite: (1) the course of the infection in the host, (2) the mode of transmission between host individuals, and (3) the behavior and demography of the host population.

The course of an infection includes a latent period after exposure to the source of the infection. During this period, the virus, for example, will increase exponentially. The next stage includes the infectious period, during which time the host develops the symptoms of the disease. Meanwhile, as the parasite population is building its numbers, the host immune system begins developing specific antibodies. As the antibody numbers increase, the parasite population plummets and the symptoms of the illness subside. The host ceases to be infectious at some point during the illness and the previously susceptible individual passes from *S* (susceptible) to *I* (infected) to *R* (recovered and immune). Of course, some infected individuals may die during the course of the disease, and some recovered individuals may eventually lose their immunity. Other important factors affecting the natural history of the infection include the length of the infectious period, the time-lag derived from the latent period, the development of immunity by the host (thereby removing susceptible individuals from the host population), and the ability of some parasites to remain latent and undetected in host individuals (e.g., herpes), only to reappear at some later date.

The two basic modes of transmission are as follows: (1) vertical, in which the disease is passed from mother to offspring (cytomegalovirus and hepatitis B virus) and (2) horizontal, in which diseases are passed from one individual to another in the environment. Most diseases are passed through the horizontal method. These include direct and indirect transmission. Direct transmission includes (1) close-contact diseases (common cold, influenza, and measles); (2) sexual contact diseases (Hepatitis B virus, HIV, and syphilis); and (3) contaminative contact diseases (cholera, tetanus, and typhoid). Indirect contact diseases include those that involve transmission from one animal host to another (malaria, rabies, Lyme disease or the plague) or via needles (HIV or Hepatitis B). Animal-borne or transmitted diseases have been the hardest to control, just as macroparasitic diseases with intermediate hosts, such as schistosomiasis, have yet to be successfully controlled in many tropical countries.

Finally, the frequency and severity of disease outbreaks are also highly dependent on the behavior of the host population. As discussed later, many diseases such as measles and influenza are dependent on the size and density of the host population. Large and dense host populations lead to a high disease transmission rate. Public health measures such as isolation of infected individuals, the treatment of water and sewage wastes, and pasteurization of milk have been instrumental in curbing the spread of many diseases. Behavioral changes related to sexual activity are also essential in limiting the spread of other classes of disease.

9.3 Modeling host–microparasite interactions

In the SIR model, we assume the following:

N = total host population density
S = susceptible host density
I = infected host density
R = recovered (immune) host density
b = host birth rate
m = natural host mortality rate unrelated to disease mortality
α = disease-induced mortality rate
β = transmission rate of disease from one host to another
v = recovery rate, or the per capita rate of passage from the infected (I) to the recovered (R) classes. This is usually the inverse of the average infectious period
γ = rate at which recovered individuals lose their immunity. That is, the rates at which individuals return to the susceptible class (S) from the R class.

We assume that the rate of transmission of the disease, which is the rate by which susceptible individuals become infected, equals the product, βSI. This means that the rate of infection depends on the rate of population mixing, which is a simple function of the density of both types of individuals. β is the parasite-specific transmission rate, which will differ depending on the life history and type of the infectious agent and the host. The expression, βSI, is often referred to as the "law of mass action" because of its similarity to the laws governing the mixing of gases (Nokes 1992).

The total population consists of susceptible, infected, and recovered individuals:

$$N = S + I + R \tag{9.1}$$

The growth rate of each segment of the population is written as a differential equation. The increase in the number of susceptible individuals (equation 9.2) is based on the birth rate (bN) and the rate at which recovered individuals lose immunity (γR). Losses are due to the host mortality rate unrelated to the disease (mS) and to the conversion of individuals from the susceptible to infected classes (βSI):

$$\frac{dS}{dt} = bN - mS\text{-}\beta SI + \gamma R \tag{9.2}$$

The growth rate of infected individuals (equation 9.3) equals the product, βSI, minus losses due to the combined effects of natural and disease caused mortality ($m + \alpha$), as well as recovery rate (from I to R), v.

$$\frac{dI}{dt} = \beta SI - (m + \alpha + v)I \tag{9.3}$$

The growth rate of the recovered class (equation 9.4) equals vI minus the death rate unrelated to disease and the loss of immunity γR.

$$\frac{dR}{dt} = vI - mR\text{-}\gamma R \tag{9.4}$$

If we add up all of these equations, we have the following:

$$\frac{dN}{dt} = \frac{dS}{dt} + \frac{dI}{dt} + \frac{dR}{dt}$$

$$\frac{dN}{dt} = b(N) - mS\text{-}\beta SI + \gamma R + \beta SI\text{--}(m + \alpha + v)I + vI\text{--}mR\text{-}\gamma R$$

Let $N = S + I + R$. Expanding, we have:

$$bS + bI + bR - mS\text{-}\beta SI + \gamma R + \beta SI\text{--}mI\text{--}\alpha I + vI\text{--}mR\text{-}\gamma R$$

After canceling terms, we have the following:

$$\frac{dN}{dt} = bS + bI + bR\text{--}mS\text{--}mI\text{--}\alpha I\text{--}mR$$

Rearranging, letting $N = S + I + R$, and substituting the intrinsic rate of increase r, for $b\text{--}m$, we have the following:

$$\frac{dN}{dt} = bS\text{--}mS + bI\text{--}mI\text{--}bR\text{--}mR - \alpha I = (b - m)(S + I + R) - \alpha I$$

The result is equation 9.5:

$$\frac{dN}{dt} = bN - mN - \alpha I = rN - \alpha I \tag{9.5}$$

Therefore, we see that the growth of the population in diminished by the "natural" death rate, m, and by the mortality rate due to infections, αI. As explained by Tompkins et al. (2009), the population of the host is "regulated" when the per capita impact of the pathogen is greater than the intrinsic rate of growth of the host. Starting from:

$$\frac{dN}{dt} = bN - mN - \alpha I = 0, \quad \text{we have} \quad \frac{dN}{dt} = bN - mN < \alpha I$$

If we then convert both the host and the parasite equations to per capita growth, we have the following:

$$b - m < \alpha \tag{9.6}$$

This is sufficient for the parasite to regulate the growth of the host population assuming that the contacts between the susceptible and infective individuals follow the law of mass action (see later; Tompkins et al. 2009).

Another point made by Tompkins et al. (2009) is that if the parasite is the only density-dependent factor capable of regulating the host population, then the mortality rate of the host will be directly proportional to the growth rate of the population. Thus, a parasite can regulate a host population with a low growth rate even if the rate of infection is very low (see Anderson (1995), for an example).

9.4 Dynamics of the disease

Spreading of a disease, that is, an epidemic requires that the number of infected individuals remain steady or increase. This means that $dI/dt \geq 0$. Since $dI/dt = \beta SI - (m + \alpha + v)I$, we have (equation 9.7)

$$\beta SI - (m + \alpha + v)I \geq 0 \tag{9.7}$$

This simplifies to the following:

$$\beta S - (m + \alpha + v) \geq 0 \tag{9.8}$$

The rates m, α, and v are all time dependent. They represent rates at which a susceptible individual either dies or moves from the infected to the recovered class. The inverse of $(m + \alpha + v)$ can be thought of as the length of the infectious period, D (Nokes 1992). If we substitute D for $m + \alpha + v$, we have the following:

$$\beta S - \frac{1}{D} \geq 0$$

Through rearrangement, we then have the following:

$$\beta S > \frac{1}{D} \quad \text{and then,}$$

$$\beta SD > 1 \tag{9.9}$$

We define the product, βSD, as equal to R_0, the basic reproductive number (BRN) or parameter for this disease. The result is equation 9.10.

$$R_0 = \beta SD > 1 \tag{9.10}$$

R_0, *the mean number of new infections caused by a single infective individual*, is an important parameter. If this value is >1, then $dI > 1$, and the disease incidence will increase. If $R_0 < 1$, the epidemic fails. R_0 is directly proportional to the rate at which an infection spreads, βS. It also depends on the mean amount of time an infection is active, D. For an infection to spread, it must have the right combination of susceptible hosts (S), a reasonably high transmission rate (β) and a sufficiently long period of transmission. Infectious periods range from only 2–3 days for influenza up to 19–22 days for Hepatitis B (Anderson and May 1991). One conclusion we may draw from this analysis is that for a disease to succeed, it needs a dense population of hosts. For mosquito-borne diseases such as malaria, BRN depends on (1) vector (mosquito) abundance; (2) focused feeding (the tendency to bite specific hosts and nothing else; and (3) vector longevity (the equivalent of D; Spielman and D'Antonio 2001).

Since the maximum number of susceptible individuals is N, to have an epidemic, we must have the following:

$$\beta ND > 1 \tag{9.11}$$

By rearranging, we find that the *minimum size for an epidemic* is such that:

$$N > \frac{1}{\beta D} \tag{9.12}$$

This allows us to conclude that high infectious rates and long duration diseases can have a low minimum number of infected individuals, but the value of N must be fairly high for directly transmitted microparasitic diseases such as common viral infections, which have short durations of infection and produce long-lasting immunity.

Another way of expressing R_0 is to allow $N = 1/\beta D$, and *redefine the minimum size for an epidemic as S_T*. If $S_T = 1/\beta D$, then $1/S_T = \beta D$. Substituting $1/S_T$ for βD in equation 9.10, we have the following:

$$R_0 = \frac{S}{S_T} \tag{9.13}$$

This means that when the population density of susceptible hosts is above S_T, then $R_0 > 1$ and the infection spreads. There exists a fair amount of data on human epidemics that confirm the theoretical prediction that epidemics occur mainly in areas with populations greater than a particular threshold. For example, measles, mumps, influenza, and polio can be totally absent from small, isolated communities (Nokes 1992). Table 9.1, for example, shows data reported by Anderson (1982) on the number of cases of measles in North American cities of various sizes from 1921 to 1940. Observe that with one exception, Cleveland, no months without reported cases of measles were reported until the city size was below 300 000.

Island populations show the same trend. During the same period, 1921–1940, when the population of an island was less than 100 000, there were no reported cases of measles in about 50% of the months. The proportions increase until, in Hawaii, with a population of 550 000 at that time, cases of measles were reported in 100% of the months (Cliff et al. 1986).

Host population size also has an effect on the average age of infected individuals. In larger populations, the rate of transmission (βSI) and the Basic Reproductive Number, R_0 will be larger since S is larger. Therefore, the mean age at which individuals are infected will be lower. This idea is illustrated later. Table 9.2 is based on data from Smith (1982) and published in Nokes (1992). The mean age of the infection decreases, as communities get larger, although the last two population categories do not show the trend consistently.

Table 9.1 Reported cases of Measles by month in North American cities, sorted by size, in the period 1921–1940.

City	Population size $\times 10^5$	Number of years with at least one month in which no cases of measles are reported
New York	75	0
Chicago	34	0
Philadelphia	19	0
Detroit	16	0
Los Angeles	15	0
Montreal	10	0
Cleveland	9	1
Baltimore	9	0
Boston	8	0
Toronto	7	0
Washington, D.C.	7	0
Pittsburgh	7	0
Milwaukee	6	0
Buffalo	6	0
Minneapolis	5	0
Vancouver	3	20
Rochester	3	3
Dallas	3	18
Akron	2	8
Winnipeg	2	7

Source: Data from Anderson (1982)

Table 9.2 The effect of population density on the mean age (in years) at the time of infection (A) of various childhood diseases in New York State in the period 1918–1919.

Population size	Measles	Whooping cough	Scarlet fever	Diphtheria
Less than 2500	12.9	8.2	12.3	14.2
2500–10 000	10.7	6.9	11.2	12.5
10 000–50 000	9.0	5.7	10.2	11.5
50 000–200 000	9.0	6.3	10.5	10.6

Table 9.3 Mean age of infection and R_0 for measles in England and Wales, 1950–1955.

Population density	Mean age at infection in years, A	Basic reproductive number, R_0
"Rural"	5.0	15.0
<50 000	4.5	16.5
50 000–100 000	4.2	17.5
>100 000	3.9	18.7

Reproductive rate based on an average life span of 69.5 years
Source: Based on Smith (1982) and adapted from Nokes (1992)

Table 9.3 illustrates the drop in mean age of infection for measles in England and Wales during the period, 1950–1955. Notice the increase in the value of R_0 with population density. Note also the contrast between the New York and British data. Children in Britain contracted measles between ages of 4 and 5 on average, compared with ages 9–13 in New York State. The infection rate depends on "mass action" or mixing. The New York populations in the early 20th century were much less mobile than the British populations in the 1950s. In addition, a larger percentage of children were likely attending school regularly in the 1950s. Increased rates of infection accompany school terms when children are aggregated together. In contrast to the data from the United States and England, the average age at which children contracted measles during the 1960s and 1970s in countries such as Ghana, India, and Thailand was 1–3 years (Anderson and May 1985).

Hethcote (2000) asserts that human contact rates, and, therefore, the potential for the spread of diseases, are now less dependent on population size. He believes that the patterns of daily encounters are largely independent of the size of the community within a given area. This may be especially true for school children who are routinely bused long distances to population centers, at least in the United States.

Finally, before massive immunizations, microparasitic diseases were well known for their regular oscillatory patterns. Data from New York, Britain, and elsewhere have shown that in the 20th century, measles had a regular 2-year epidemic cycle, mumps a 3-year cycle, and rubella a 4- to 7-year cycle (Nokes 1992; Hethcote 2000). Infections, such as certain influenzas, are seasonal and are, therefore, on a yearly cycle. These yearly cycles are related to seasonal climactic variations and/or from regular changes in population mixing. The greatest factor in the longer-term oscillations is the reduction in the susceptible population during epidemics as they either die or become immune. As new births increase the density of susceptible individuals, the epidemic returns when the value of S exceeds S_T. The tendency for host–parasite, as well as predator–prey interactions to undergo oscillations is a confirmation of Turchin's (2001) third principle of population ecology (Chapter 1).

9.5   Immunization

One goal of immunization is to reduce the number of susceptible individuals in a host population, thereby lowering the net reproductive number of the disease to less than one. Recall that S_T is inversely proportional to D, the average infectious period, and to the rate of infection (β). Therefore, a high rate of infection and/or a long course of the disease make effective immunization more difficult.

Suppose we immunize a fraction, p, of the population. The new reproductive parameter, R_0^* becomes

$$R_0^* = (1 - p)R_0 \tag{9.14}$$

We need to drive S to a level below S_T to eradicate the disease. If we set

$$R_0^* = \frac{S}{S_T} = 1 = (1 - p)R_0 \tag{9.15}$$

and we solve for p, we have the following:

$$p = 1 - \frac{1}{R_0} \tag{9.16}$$

Therefore, for the disease to be extinguished, p (the proportion immunized) must be greater than $1 - 1/R_0$.

For example, let us reexamine Table 9.2 for New York State in 1918–1919. As you can see from Table 9.4, the proportion of the population that would need to be immunized for successful eradication is between 84% and 90%. In comparison, May (1983,

Table 9.4 The estimated fraction needed for successful immunization against several diseases, in a dense (>50 000) New York State population in 1918–1919.

Disease	A	R_0	p	p from Anderson and May (1991) (%)
Measles	9.0	7.1	0.86	90–95
Whooping cough	6.3	9.7	0.90	90–95
Scarlet fever	10.5	6.3	0.84	82–87
Diphtheria	10.6	6.2	0.84	82–87

A equals average age of infection (from Table 9.2). R_0 equals the reproductive parameter for the disease, and p equals the proportion that would need to be immunized for eradication of the disease. Average life span estimated at 55 years. The last column is based on Table 5.1 in Anderson and May (1991).

as summarized in Alstad 2001) estimated the proportion immunized needed for successful eradication of measles and whooping cough in England and Wales as 92% and 94%, respectively. Successful eradication of rubella requires 86% immunization (Hethcote 2000). These percentages are not atypical for highly contagious diseases. Because about 5% of those vaccinated do not become immune, Hethcote (2000) estimated that eradication of measles requires 99% vaccination and 91% is needed for rubella. In 1998, measles was no longer an indigenous disease in the United States (Hethcote 2000), and in March of 2005, the Centers for Disease Control pronounced rubella eradicated as an indigenous disease in the United States.

More complex models would include equations for a host that has a very long latent period during which time it is not yet infectious, or for hosts that never reach a state of total immunity and continue to be infectious throughout the remainder of their lives (e.g., various venereal diseases or typhoid).

9.6 Endangered metapopulations and disease

As discussed earlier, metapopulation ecology has come to the fore as a theoretical framework for conservation planning. High dispersal rates (high movement between patches) have been predicted to increase the proportion of patches occupied at equilibrium, the time of metapopulation extinction, and the effective population size (Hanski and Gilpin 1997). Conservationists have favored measures, such as habitat corridors, which increase connectivity among patches. Hess (1996), however, has suggested that easy migration can have the negative effect of spreading disease among patches, causing extensive local extinctions. Using a metapopulation analysis, Hess (1996) found that high migration rates, by facilitating the movement of disease organisms, could reduce patch occupancy and increase the probability of metapopulation extinction.

Gog et al. (2002), however, disagreed. They believe that most of the infections that threaten wildlife are not caused by migration of diseased organisms but by "spillover" from other, more abundant hosts, already present in the habitat patches. The reservoir for these "spillover" diseases is often domestic animals. For example, domestic dogs are the probable source of diseases that have threatened African wild dogs (*Lycaon pictus*), African lions (*Panthera leo*), Baikal seals (*Phoca sibirica*), grey wolves (*Canis lupus*), and arctic foxes (*Alopex lagopus semenovi*; Gog et al. 2002).

In the deterministic model of Gog et al. (2002), S is the proportion of susceptible host patches (host population present, no disease), and I is the proportion of infected patches (host population and disease present). The extinction rates of susceptible and infected populations are x_S and x_I, respectively. The migration rate between susceptible and infected populations is ψ. When an infected disperser arrives at a susceptible patch, it infects the resident population with the probability of δ. Infection spreads at the rate of $\psi \delta IS$. The preceding is identical to the Hess (1996) model. What Gog et al. (2002) added is an extension of the Hess model in which they simulated various parameters of an infection rate from an "outside source," g. In the Hess model, they set g at zero, Gog et al. (2002) then ran a number of simulations showing the important effects when g is a nonzero parameter.

The equations for mean proportion of patches occupied in the S and I states are as follows:

$$\hat{S} = \psi S(1 - I - S) - x_S S - \psi \delta IS - gS \tag{9.17}$$

$$\hat{I} = \psi I(1 - I - S) - x_I I + \psi \delta IS + gS \tag{9.18}$$

Equations for stable equilibrium values of S and I for different values of g and m are found in Appendix A of Gog et al. (2002). Figures 9.1 and 9.2 have been produced based on their equations for $g = 0$ (representing no outside sources of disease) and $g = 0.4$ (representing a moderately large background infection rate). Other parameters are the same as in Hess (1996): $x_S = 1.4$, $x_I = 2.4$, $\delta = 0.5$.

As we see from Figures 9.1 and 9.2, Gog et al. (2002) found that when g is zero or very small (as in Hess), occupancy rates first increase but then decrease with increased migration as more and more patches experience extinction due to disease. Eventually, patch occupancy increases again with more migration, as all patches become infected. This result (Figure 9.1) led Hess (1996) and others to suggest that increased migration between patches can have a negative effect on patch occupancy and can increase the probability of metapopulation extinction. However, for larger values of g, increasing the migration rate results in little, if any depression in the rate of patch occupancy. In other words, the decrease in patch occupancy at intermediate levels of migration is minimized (Figure 9.2). Gog et al. (2002) concluded that the net effect of migration is almost always positive, and that at high rates of infection from external sources, the benefits of migration will always outweigh the costs. The major application of these models is that in wild populations suffering from a high rate of infection from alternative host species, patch occupancy should increase, rather the decrease, with migration rate.

Fig. 9.1 Proportion of suitable patches occupied as a function of movement rates with the parameter $g = 0$. (Source: Adapted from Gog et al (2002) and Hess (1996).)

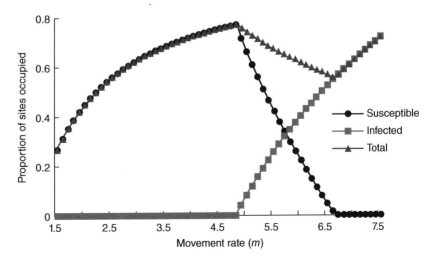

Fig. 9.2 Proportion of available patches occupied as a function of movement rate with the parameter $g = 0.4$. When $g = 0.4$ there is a reasonably large chance of infection from an "outside source." (Source: Adapted from Gog et al. (2002) and Hess (1996).)

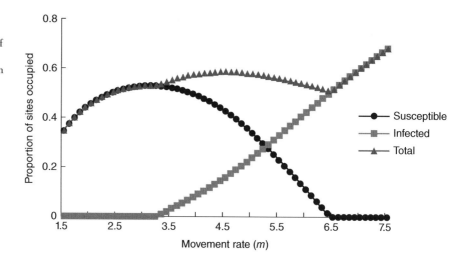

Gog et al. (2002) concluded that corridors between suitable habitats are likely to benefit metapopulation persistence, a conclusion also reached by Laurance and Laurance (2003; Chapter 5). However, they pointed out that Hess' model might apply well to captive populations where transfer of animals from one facility to another is often a cause of disease epidemics. They stressed the importance of veterinary screening and quarantining procedures before transferring animals from one captive population to another.

9.7   Social parasites

By definition, parasites reduce the fitness of the host on which they live. This is usually the result of the consumption of host tissues. There are, however, parasites that reduce fitness by means of behavior. Social parasites include brood or nest parasites among birds and slave making among the Hymenoptera, but especially among ants. Birds such as cowbirds and cuckoos lay their eggs in the nests of other species (Davies 2000). When the eggs hatch, the host parents feed the parasitic chicks, even in preference to their own offspring. The chicks are large and aggressive, and since the parents often make no distinction among the chicks, the cowbird or cuckoo chicks get a majority of the food provided by the parents. In most cases, the host raises few young of their own species when a parasitic chick is present in the nest. Moreover, cuckoo and cow bird females often remove host eggs before laying their own eggs, and in a nest still containing host eggs, a cuckoo chick will eject host eggs from the nest (Davies 2000). Social parasitism among birds has evolved on every continent except Antarctica, and there are about 100 species of obligate brood parasite worldwide (Davies 2000). Brood parasitism has probably evolved six times since it is found in six different bird families. Brood parasites include Old World cuckoos (Europe, Asia, Africa, Australia, and New Zealand), New World cuckoos (North and South America), cowbirds (North and South America), a duck (South America), honeyguides (Africa and Asia), and finches (Africa; Davies 2000).

In some instances, the host makes an attempt to distinguish among the eggs and dumps the parasite eggs from the nest. Social parasites, including several species of cuckoos, cowbirds, African honeyguides, and finches, have responded by laying "mimetic" eggs. That is, eggs that look like those of the host species. Furthermore, according to the "Mafia hypothesis" (Zahavi 1979, Zahavi and Zahavi 1997), some nest parasites retaliate against "dumpers" by destroying all of the eggs in the nest and perhaps even the nest itself. This parasite makes the host birds an offer they "can't refuse." That is, "raise one of my chicks or lose all of your children and your house!" Evidence, however, suggests that common cuckoos do not adopt Mafia-like tactics (Davies 2000), since female cuckoos do not usually revisit nests they have parasitized. Because cuckoos usually kill all host eggs and young before laying their own eggs, they have already exerted control and punished the host. However, Davies suggests this hypothesis could work in parasitic species that leave one or more host eggs in the nest.

Soler et al. (1995a, 1995b) experimentally tested the Mafia hypothesis between great spotted cuckoos (*Clemator glandarius*) and their magpie (*Pica pica*) hosts in Spain. They found that when a magpie accepted a cuckoo egg, the nest was successful, but when the host rejected the cuckoo egg, the nest was often destroyed. Soler and his team experimentally removed cuckoo eggs while simultaneously visiting nests without destroying the cuckoo egg (controls). They found that more than 50% of the experimental nests were preyed upon, either at the egg or at the young chick stage, compared with only 10% for the control nests. The predation was most likely by cuckoos. Furthermore, in a follow-up experiment, Soler et al. (1999) showed that the magpies learned from the experience. If their eggs were destroyed after they removed a cuckoo egg, when they renested they were 50% less likely to remove the cuckoo egg the second time around. Curiously, however, the cuckoo lays eggs that mimic those of the magpie. Davies (2000) suggests that the Mafia tactics do not work that well and on average magpies still do better to reject cuckoo eggs. However, the behavior of individual hosts varies, and some are more susceptible to Mafia tactics than others. Therefore, efforts to intimidate the host are successful often enough to have become part of the relationship between cuckoos and magpies in Spain.

Similar to any host–parasite interaction, social parasitism can result in counter-adaptations by the host species. Such counter-adaptations are likely to provoke a response by the parasite resulting in what has been termed a co-evolutionary arms race between parasites and their host (Krüger 2007, Langmore et al. 2003).

As mentioned earlier, social parasitism also occurs among social insects. In the so-called slave making ants, a queen of the species *Lasius reginae*, enters the nest of another species (*L. alienus*), kills the resident queen, and forces the workers to care for her own offspring. In such a case, the workers are unable to distinguish the foreign ant queen from their own (Faber 1967). This parasitism is temporary, however, during colony foundation. *L. reginae* workers eventually take over foraging and management of the nest. In many other cases, however, the parasitic species produces no workers, a condition termed inquilinism. This socially parasitic species spends its entire life in the nest of its host species. Workers are lacking or are degenerate in normal foraging behavior. Holldobler and Wilson (1990) have described the "ultimate social parasite," *Teleutomyrmex schneideri*, which is a social parasite of *Tetramorium caespitum* and *T. impurum*. This parasite is only found in the nests of its hosts. It lacks a worker class, and the queens contribute nothing to the host colonies. Finally, the parasitic queen spends its life as an ectoparasite on the back of the host queen!

In more dramatic cases of interspecific slave making, species such as *Formica sanguinea* raid other species of the same genus (*F. fusca*). As described by Wheeler in Holldobler and Wilson (1990), *F. sanguinea* heads to an *F. fusca* colony 50–100 m away, following an amazingly direct route. Once they have arrived, they surround the nest and wait for reinforcements to arrive. The raiders snatch larvae and pupae from the nest and from the jaws of the host workers attempting to flee with their young. They kill adults of *F. fusca* only if they offer resistance. Otherwise, the host colony is left to rebuild itself as long as the queen has not been killed. The brood of *F. fusca* is brought back to the *F. sanguinea* colony and enslaved. There are also numerous examples of intraspecific slave making, in which large colonies raid small ones, kill or drive off the queen, and carry or drag larvae, pupae and young workers to the home nest, where they are put to work (Holldobler and Wilson 1990).

Foitzik et al. (2003) studied geographic variation in relationships between an obligatory slave-maker (*Harpagoxenus sublaevis*) and three species of *Leptothorax*). In this remarkable parasite, the life cycle starts with a new queen leaving its natal nest in summer and mating after recruiting males through vocalizations. The newly mated queen then searches for a host colony and enters its nest. She kills or expels all host workers and queens and takes over the nest. When host larvae pupae and emerge as new workers, the work as slaves and raise the parasite's brood. New slaves are recruited through continuous raids and take over all routine chores in the nest.

This continuous recruitment of new slaves is based on two adaptations of the host. First, the slave-marking workers have large mandibles, good for fighting. Second, during the raids, *H. sublaevis* emits a "propaganda substance" from the Dufour gland, the result of which is to entice host workers to attack and kill each other (Regnier and Wilson 1971).

Nevertheless, Foitzik et al. (2003) found geographic differences among host populations. They examined the behavior of host population of *Leptothorax acervorum* from Germany, Russia, and England. Even though the slave-maker ant species is not present in England, these colonies recognized it as an enemy and attacked it more successfully than colonies that have co-evolved with the parasite. The German colonies appear to have evolved resistance characteristics more successfully than the Russian colonies against *L. acervorum*. For example, they position more guards at the nest entrances and were effective at immobilizing the slave-maker by biting. Furthermore, the "propaganda" substance delivered by the slave-maker was less effective in manipulating the host workers to attack each other. In contrast, the host species, *L. muscorum,* showed the opposite trend: the German host populations are more severely parasitized than the Russian population.

What is illustrated earlier, just as outlined in the previous chapter when discussing the *Cecropia-Azteca* ant mutualism, is that co-evolutionary relationships, be they competitive (Benkman et al. 2001), mutualistic (Rickson 1977), parasite–host, or predator–prey (Brodie 2011), may vary geographically. Since Thompson (1999) outlined his geographic mosaic theory of co-evolution, many examples have been found; the work of Foitzik et al. is another excellent illustration.

Finally, let us end this chapter with the news that all is not lost for host species in the co-evolutionary battle between slave-making ants and their hosts. A study by Achenback and Foitzik (2009) has shown "evidence for slave rebellion." In this case, they describe the relationship between the slave-making ant, *Protomognathus americanus,* and its host species of the genus *Temnothorax*. In this typical scenario, colonies of *P. americanus* raid 2–10 host colonies per year. Slave workers raised as larvae imprint on the slave-making queen and take over colony maintenance and care of the parasite brood. What is unique in this interaction is that the enslaved workers of *Temnothorax* have evolved the trait of killing healthy pupae of the slave-making species. Essentially the pupae of females, both workers and potential new queens, are removed from the nest chamber and allowed to die. Yet the male pupae of the host species were cared for and allowed to pupate to adulthood. Achenbach and Foitzik (2009) point out that whereas there would be strong selection to kill female pupae who become slave-making raiders, males, after the mating flight, simply die. Thus, there are few fitness benefits from killing parasitic males. In the end, whereas slaves were thought to be caught in and evolutionary trap with no viable mechanism to increase their fitness, this study shows there is, at least for some species, a way out.

9.8 Summary

The complexities of parasite–host interactions rival those of mutualisms. Parasites range from endoparasitic viruses to ectoparasitic ticks, and we lack an understanding of the life cycle of most parasites in organisms other than humans. What we do know is often based on models of human diseases and their modes of transmission. Parasitism can also involve complex behavioral interactions such as brood parasites or slave-making among ants. As we will discuss in the next chapter, the activities of parasites can have major impacts on populations, communities, and ecosystems. Some ecologists consider parasite–host interactions the newest and least explored ecological frontier. Parasite–host interactions are much harder to study and less obvious than predator–prey relationships, and ecologists may well have been guilty of underestimating their importance.

References

Achenbach, A. and S. Foitzik. 2009. First evidence for slave rebellion: enslaved ant workers systematically kill the brood of their social parasite, *Protomognathus americanus*. *Evolution* 63(4): 1068–1075.

Alstad, D. 2001. *Basic Populus Models of Ecology*. Prentice Hall, Uper Saddle River, NJ.

Anderson, R.M. 1982. Transmission dynamics and control of infectious disease agents. *In* R.M. Anderson and R.M. May, Editors. *Population Biology of Infectious Diseases (Dahlem Conference Report)*. Springer-Verlag, New York, NY.

Anderson, R.M. 1995. Evolutionary pressures in the spread and persistence of infectious agents in vertebrate populations. *Parasitology* 111: S15–S31.

Anderson, R.M. and R.M. May. 1978. Regulation and stability of host-parasite population interactions. 1. Regulatory processes. *Journal of Animal Ecology* 47: 219–247.

Anderson, R.M. and R.M. May. 1979. Population biology of infectious diseases. Part I. *Nature* 280: 361–367.

Anderson, R.M. and R.M. May. 1981. The population dynamics of microparasites and their invertebrate hosts. *Philosophical Transactions of the Royal Society of London B* 291: 451–524.

Anderson, R.M. and R.M. May. 1982. Directly transmitted infectious diseases: control by vaccination. *Science* 215: 1053–1060.

Anderson, R.M. and R.M. May 1985. Vaccination and herd immunity to infectious disease. *Nature* 318: 323–329.

Anderson, R.M. and R.M. May 1991. *Infectious Diseases of Humans. Dynamics and Control*. Oxford University Press, Oxford, UK.

Benkman, C.W., W.C. Holimon, J.W. Smith. 2001. The influence of a competitor on the geographic mosaic of coevolution between crossbills and lodgepole pine. *Evolution* 55: 282–294.

Brodie, E.D., III. 2011. Patterns, process, and the parable of the coffeepot incident: arms races between newts and snakes from Landscapes to Molecules. *in*, J.B. Losos, Editor, *In the Light of Evolution, Essays from the Laboratory and the Field*. Roberts and Company, Greenwood Village, Colorado.

Cliff, A.D., P. Haggett, and J.K. Ord. 1986. *Spatial Aspects of the Influenza Epidemics*. Pion Ltd., London, UK.

Davies, N.B. 2000. *Cuckoos, Cowbirds and Other Cheats*. Academic Press, New York, NY.

Dobson, A.J., P. J. Hudson, and A.M. Lyles. 1992. Macroparasites: worms and others. Pages 329–348 in M.J. Crawley, Editor. *Natural Enemies, the Population Biology of Predators, Parasites and Diseases*. Blackwell Scientific, Oxford, UK.

Faber, W. 1967. Beiträge zur Kenntnis sozialparasitischer Ameisen, 1: *Lasius* (*Austrolasius* n.sg.) *reginae* n.sp., eine neue temporär sozialparasitische Erdameise aus Österreich (Hym. Formicidae). *Pflanzenschutz-Berichte*, 36(5–7): 73–107.

Foitzik, S., B. Fischer, and J. Heinze. 2003. Arms races between social parasites and their hosts: geographic patterns of manipulation and resistance. *Behavioral Ecology* 14: 80–88.

Gog, J., R. Woodroffe, and J. Swinton. 2002. Disease in endangered metapopulations: the importance of alternative hosts. *Proceedings of the Royal Society of London B* 269: 671–676.

Hamer, W.H. 1906. Epidemic disease in England. *Lancet* 1: 733–739.

Hanski, I.A. and M.E. Gilpin. 1997. *Metapopulatio Biology: Ecology, Genetics and Evolution*. Academic Press, London, UK.

Hess, G. 1996. Disease in metapopulations: implications for conservation. *Ecology* 77: 1617–1632.

Hethcote, H.W. 2000. The mathematics of infectious diseases. *SIAM Review* 42: 599–653.

Holldobler, B. and E.O. Wilson. 1990. *The Ants*. Belknap Press, Harvard University, Cambridge, MA.

Hudson, P.J. and A.P. Dobson. 1995. Macroparasites: observed patterns. Pages 144–176, in, *Ecology of Infectious Diseases in Natural Populations*. B.T. Grenfell and A.P. Dobson, Editors. Cambridge University Press, Cambridge, UK.

Humphrey, S.R., C.H. Courtney, and D.J. Forrester. 1978. Community ecology of helminth parasites of the brown pelican. *The Wilson Bulletin* 90: 587–598.

Kermack, W.O. and A.G. McKendrick. 1927. Contributions to the mathematical theory of epidemics, part I. *Proceedings of the Royal Society of London A* 115:700–712.

Krecek, R.C., F.S. Malan, C.E. Rupprecht and J.E. Childs. 1987. Nematode parasites from Burchell's zebras in South Africa. *Journal of Wildlife Diseases* 23: 404–411.

Krüger, O. 2007. Cuckoos, cowbirds and hosts: adaptations, trade-offs and constraints. *Proceedings of the Royal Society of London B* 362: 1873–1886.

Langmore, N.E., S. Hunt, and R.M. Kilner. 2003. Escalation of a coevolutionary arms race through host rejection of brood parasitic young. *Nature* 422: 157–160.

Laurance, S.G.W. and W.F. Laurance. 2003. Bandages for wounded landscapes: faunal corridors and their role in wildlife conservation in the Americas. Pages 313–325 in G.A. Bradshaw and P.A. Marquet, Editors. *How Landscapes Change*. Ecological Studies 162. Springer-Verlag, Berlin.

Malan, F.S., I.G. Horak, V. de Vos and J.A. Var Wyk. 1997. Wildlife parasites: lesson for parasite control in livestock. *Onderstepoort Journal of Veterinary Research* 71: 137–153.

May, R.M. 1983. Parasitic infections as regulators of animal populations. *American Scientist* 71: 36–45.

May, R.M. and R.M. Anderson. 1979. Population biology of infectious diseases: Part II. *Nature* 280: 455–461.

Nokes, D.J. 1992. Microparasites: viruses and bacteria. Pages 349–376 in M.J. Crawley, Editor. *Natural Enemies, the Population Biology of Predators, Parasites and Diseases*. Blackwell Scientific, Oxford, UK.

Regnier, F.E. and E.O. Wilson. 1971. Chemical communication and "propaganda" in slave-making ants. *Science* 172: 267–269.

Rickson, F.R. 1977. Progressive loss of ant-related traits of *Cecropia peltata* on selected caribbean islands. *American Journal of Botany* 64(5): 585–592.

Shaw, D.J., B.T. Grenfell, A.P. Dobson. 1998. Patterns of macroparasite aggregation in wildlife host populations. *Parasitology* 117: 597–610.

Smith, A.D.M. 1982. *Epidemiological Patterns in Directly Transmitted Human Infections*. ICCET (Imperial College), Series E1, London, UK.

Soler, M., J.G. Martinez, J.G. Soler, J.J. Moller, and A.P. Moller. 1995a. Preferential allocation of food by magpies *Pica pica* to great spotted cuckoo *Clemator glandarius* chicks. *Behavioural Ecology and Sociobiology* 37: 243–248.

Soler, M., J.J. Soler, J.G. Martinez, and A.P. Moller. 1995b. Magpie host manipulation by great spotted cuckoos: evidence for an avian Mafia? *Evolution* 49: 770–775.

Soler, J.J., G. Sorci, M. Soler, and A.P. Moller.1999. Change in host rejection behavior mediated by the predatory behavior of its brood parasite. *Behavioural Ecology*10: 275–280.

Spielman, A. and M. D'Antonio. 2001. *Mosquito. The Story of Man's Deadliest Foe*. Hyperion, New York, NY.

Thompson, J.N. 1999. Specific hypotheses on the geographic mosaic of coevolution. *American Naturalist* 153: 1–14.

Tompkins, D.M., A.P. Dobson, P. Arneberg, M. E. Begon, I.M. Cattadori, J.V. Greenman, J.A.P. Heesterbeek, P.J. Hudson, D. Newborn, A. Pugliese, A.P. Rizzoli, R. Rosà, F. Rosso, and K. Wilson. 2009. Parasites and host population dynamics. Pages 45–62. *in*, P.J. Hudson, A. Rizzoli, B.T. Grenfell, H. Heesterbeek and A.P. Dobdon, Editors. *The Ecology of Wildlife Diseases*. Oxford University Press, Oxford, UK.

Turchin, P. 2001. Does population ecology have general laws? *Oikos* 94: 17–26.

Wilson, K., O.N. Bjørnstad, A.P. Dobson, S. Merler, G. Poglayen, S.E. Randolph, A.F. Read and A. Skorping. 2009. Heterogeneities in macroparasite infections: patterns and processes. Pages 6–44. *in* P.J. Hudson, A. Rizzoli, B.T. Genfell, H. Heesterbeck and A.P. Dobson, Editors. *The Ecology of Wildlife Diseases*. Oxford University Press, Oxford UK.

Zahavi, A.1979. Parasitism and nest predation in parasitic cuckoos. *American Naturalist* 13: 157–159.

Zahavi, A. and A. Zahavi. 1997. *The Handicap Principle*. Oxford University Press, Oxford, UK.

10 Predator/prey interactions

10.1 Introduction

The relationships between predators and their prey have provided lively topics of conversation for groups of humans ever since they began gathering together for meals, around the fire or the seminar table. Historically, people have seen large predators not only as dangerous to themselves and their families, but also as competitors for the prey they were seeking. The large number of fables and fairy tales involving wolves ("Little Red Riding Hood," "Peter and the Wolf," "The Three Little Pigs"), lions and tigers attest to their prominent role in human culture. However, in the 20th Century the perception of large predators for many people, particularly in developed, affluent countries, has shifted from "vermin" to "charismatic megafauna."

In population ecology the basic question remains, what determines distribution and abundance? For trophic levels above that of producer, what is the role of predation in controlling herbivore populations? Are prey populations limited primarily by available habitat and food supply, or by their predators? This is, are prey limited by what they eat, or by what eats them? In terms of species diversity, do predators allow more species to exist in a community, or, by limiting prey populations to low levels, do they often drive them locally extinct and thereby limit diversity? Do predators primarily kill very young, very old and/or sick individuals, such that their effect on population growth is insignificant? Do they actually benefit the prey population by eliminating the spread of disease and eliminating genetically inferior individuals? Do predators take prey individuals when the prey population has exceeded the carrying capacity, thereby

Introduction to Population Ecology, Second Edition. Larry L. Rockwood.
© 2015 John Wiley & Sons, Ltd. Published 2015 by John Wiley & Sons, Ltd.
Companion Website: www.wiley.com/go/rockwood/populationecology

helping stabilize the prey population? Or, alternatively, are predators the cause of the periodical population cycles seen in many prey species?

There are no simple answers to these questions and the answers likely change from one ecosystem to the next. Throughout human history, although we have domesticated members of the cat and dog family, we have also tried to extirpate canine or feline predators perceived to be dangerous and/or which interfere with the management of prey species. The Alaska Department of Fish and Game engages in wolf and bear control projects in order to ensure an abundance of moose and caribou for human hunters (NRC 1997). In the Commonwealth of Virginia, during FY 2013, $120,000 was spent on coyote (*Canis latrans*) control, which includes poisoning, snares and foot-hold traps. According to Fox (2013), since 2000, the average annual predation rate of coyotes on domestic sheep was 279 in the western counties of Virginia. In response, the USDA Virginia Cooperative Coyote Damage Control Program killed an average of 347 coyotes per year during the same period of time. Yet, in spite of control efforts throughout the United States, coyotes have expanded their range to include Virginia and other mid-Atlantic states of the Eastern United States (Bozarth et al. 2011).

In Europe, brown bears, leopards, wolves and other large predatory species have been eliminated from most of their prehistoric ranges. On the other hand, predators such as wolves and grizzly bears are major tourist attractions in National Parks in the western United States and Canada, and the US Fish and Wildlife Service reintroduced gray wolves (*Canis lupus*) in Yellowstone National Park in 1995, after they had been deliberately extirpated in the 1920s. One of the rationales for reintroducing wolves was the assertion that, without wolves, the elk (*Cervis elaphus*) population had grown too large. Over-browsing by the elk supposedly has led to a decline in willow (*Salix sp.*) and quaking aspen (*Populus tremuloides*), an increase in stream bank erosion and a decline in cutthroat trout (*Salmo clarki*) (Huff and Varley 1999). According to this scenario, the reintroduction of wolves will have a salubrious effect on the Yellowstone ecosystem. Local cattle and sheep ranchers have begged to differ, and Pyne (1997) might argue that fire suppression is much more likely the cause of a decline in willow and quaking aspen rather than the elimination of wolves. In addition, Meagher and Houston (1998) have noted that willow and aspen are minor components (1–2%) of the vegetation in the northern range of the Greater Yellowstone ecosystem.

Ecosystems are complex and the specific roles played by predators are difficult to disentangle from the complexities of the food web. This point was emphasized by Middleton (2014) in a New York Times editorial entitled, "Is the Wolf a Real American Hero." Referencing work by Kauffman et al. (2010), Middleton explained that aspen are not actually recovering in Yellowstone, even in areas with established wolf populations and a 60% drop in elk numbers. Middleton (2014) wrote that based on his own research, elk rarely change their feeding behavior in the presence of wolves and, in fact, there is no clear link among wolves, aspen and willows, songbirds and beavers. However, (Ripple et al. 2014, Figure 4), proclaim the reality of the same trophic cascade that Middleton (2014) says does not exist.

Opinions about predator-prey interactions have been formed based on the supposed Hare–Lynx cycle first proposed by Elton (1924), by the Isle Royal wolf-moose interaction and, as described in the next paragraph, the history of the Kaibab deer herd. As we will explore in this chapter, none of these interactions really lives up to scientific scrutiny, in spite of their popularity in biology and ecology textbooks. As indicated in the preceding paragraph, the proposed tropic cascade involving Yellowstone wolves may also have to be discarded.

In spite of the much more sophisticated methods of gathering and analyzing data that we have today, when trying to understand predator–prey relationships we are still bedeviled by the simplistic theory, and inadequate analysis of data, published in the first half of the 20th century. For example, the fable of the Kaibab deer herd, first published in a Wisconsin Wildlife Bulletin by Leopold (1943), influenced the opinion of at least one generation of ecologists on the role of predators in ecosystems. The story goes like this. Prior to 1906, a population of mule deer shared the Kaibab plateau in northern Arizona with cattle, sheep, coyotes, wolves, mountain lions, bear, and bobcats. When this area was declared a game refuge as part of the new Grand Canyon National Park by President T. Roosevelt, federal agents not only removed the cattle and sheep, but also did their best to eliminate all the predatory species. In contrast to our views today, in which we advocate the maintenance of an ecosystem, including the restoration of predators, Leopold and other wildlife biologists had a very negative view of predators. Although Leopold changed his view later in life (Leopold 1949; Botkin 1990), he and other government biologists initially agreed with a policy of predator removal. Hunters, in the period 1906–1931, killed approximately 781 mountain lions, 30 wolves, 4338 coyotes, and 554 bobcats on the Kaibab. These are impressive numbers, but no one actually knows what effect this had on the predator populations. The deer population, approximately 4000 strong in 1906, was released from both competition and predation, and rapidly increased. In the fable, as published in many ecology and general biology textbooks as late as 1972, the deer population increased to 100 000 by 1924, crashed due to overgrazing and ended up at a population lower than that of 1906. This was thought to be due to permanent damage to the soil and plant life in the ecosystem. The problem is that no one actually counted the deer herd in a systematic manner. The estimate of 100 000 was by one individual who was visiting the area. Other individuals estimated peak abundance at 50, 60, and 70 thousand. Forest supervisors on site thought the peak population was 30 000. In any event, the data are totally unreliable (Caughey 1970). Unfortunately, Leopold based his graphical analysis on that of Rasmussen (1941), who used 100 000 as the deer population estimate for 1924. Commenting on this story, Botkin writes (1990, pp. 78–80), "an examination of the facts about counts (of mule deer) leaves us up in the air. The famous 'irruption' of mule deer ... may or may not have occurred, and if it did occur the cause may have been completely unrelated to the presence or absence of predators." The Kaibab story was reprinted repeatedly because it fit into the "new" view that predators were part of the "balance of nature" (Botkin 1990).

This dichotomous view of predation lives on. To some people predators are pests, undesirable and unworthy of existing except in someone else's ecosystem. To others, predators are part of the balance of nature and we eliminate them at our peril. Since human modification of ecosystems is so pervasive, it is almost impossible for large predators to survive throughout much of their previous range. Yet prey populations, such as white-tailed deer, have prospered. For example, cervid densities are nearly six times larger in areas without wolves than in areas without wolves (Ripple et al. 2014; Ripple and Beschta 2008, 2012). Hunters who view themselves as filling the necessary role of the wild predator often employ the balance of nature argument. Alaskan hunters see themselves in competition for big game with wolves.

The prevailing views of predation in the first half of the 20th century were colored by the fact that theory and most early studies isolated two species, one prey and one predator, from their community and ecosystem. As we will see, simple, and even many complex models, predict that predator–prey relationships are "inherently oscillatory." An intensively studied phenomenon in population ecology is the regular population cycles displayed by rodents, hares and other small animals living in boreal and arctic communities. After 80 years, there is still disagreement as to whether lemming (*Dicrostonyx groenlandicus*) cycles, for example, are caused by the interaction of lemmings with their food plant (Turchin et al. 2000) or are the result of an interaction with their predators: stoats (*Mustela erminea*), arctic fox (*Alopex lagopus*), snowy owl (*Nyctea scandiaca*) and long-tailed skua (*Stercorarius longicaudus*) (Gilg et al. 2003). What separates these modern studies from the simple models that are discussed below are the assumptions that: (1) Prey populations are at least partially determined by their food supplies, not just their predators; (2) Prey populations respond to the entire community of predators, not just a single species; and (3) Predator populations are affected by factors other than just the prey population density. Although most models and theoretical treatments we will discuss have been based on the single predator-single prey, you should keep in mind that more realistic treatments should be community and ecosystem based.

The overall predation rate, that is the number of prey killed per unit area during a specified time period, is dependent on both the *numerical response* of the predator population, and the *functional response* of individual predators in the population. As prey density increases, a numerical response is an increase in the numbers of predators per unit area based on both immigration and reproduction. A functional response is an increase in the number of prey consumed per unit time by each individual predator as a function of prey density. The *total response*, which is the combined effect of both the numerical and functional responses, is density dependent at low prey densities, but

Fig. 10.1 Stable limit cycle. Prey and predator populations versus time.

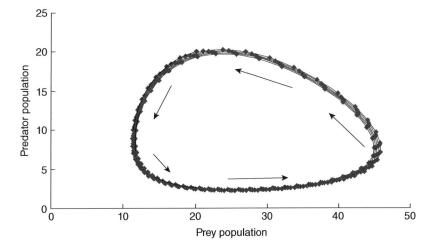

Fig. 10.2 Stable limit cycle. Prey population versus predator population.

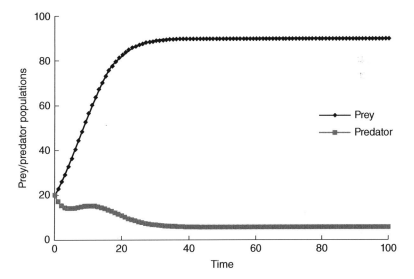

Fig. 10.3 Stable point. Prey and predator populations versus time.

can become inversely density dependent at high prey densities (Holling 1959, Messier 1994) as explained in more detail below.

When Robert May (1976) reviewed existing predator-prey models and theories, he found that they all led to one of the outcomes enumerated below. In a stable limit cycle, both the prey and the predator go though regular, predictable cycles (Figures 10.1 and 10.2). In a stable point, both the prey and the predator populations settle at a fixed number (Figures 10.3 and 10.4). In parentheses are early studies that have illustrated these outcomes:

1. Extinction of the predator, survival of the prey (Gause 1934);
2. Extinction of the prey followed by extinction of predator (Gause 1934; Huffaker 1958);
3. The prey and the predator populations go through oscillations, which dampen to a stable limit cycle or a stable point (Figures 10.5 and 10.6) (Dodd 1940);

Fig. 10.4 Stable point. Prey population versus predator population.

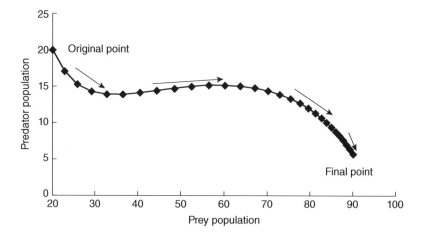

Fig. 10.5 Prey and predator populations versus time, showing dampened oscillations leading to a stable point.

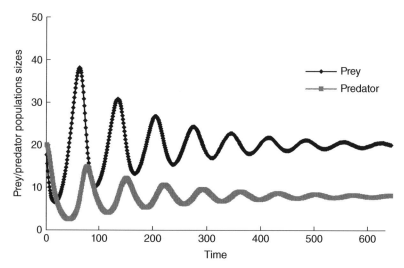

Fig. 10.6 Prey/predator populations showing dampened oscillations leading to a stable point.

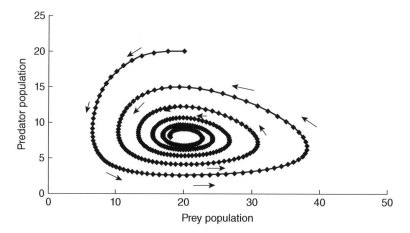

Fig. 10.7 Prey and predator populations versus time showing increasing oscillations leading to extinction of both the prey and the predator population.

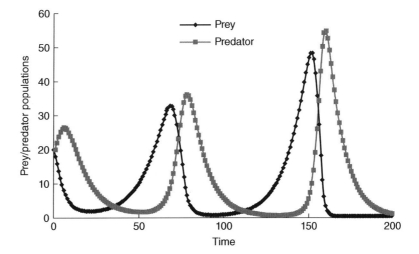

Fig. 10.8 Increasing oscillations leading to extinction of both the prey and the predator populations.

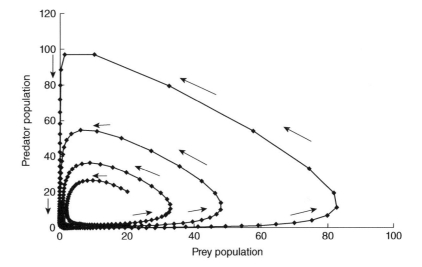

4. The prey and the predator go through increasing oscillations, leading to extinction of the prey and/or the predator (Figures 10.7 and 10.8) (Hassell 1978);
5. Immediate stable limit cycle (Hudson et al. 1998);
6. Immediate stable point.

Metapopulation theory also predicts that a predator-prey interaction may consist of coexistence in a complex heterogeneous environment (Huffaker 1958, Dodd 1940) combined with extinctions of prey and/or predators at any given site. Prey populations have also been shown to fluctuate while the predator population remains stable. For example, a stable population of predators may switch from one prey species to another, depending upon prey availability (Southern 1970).

As we review various attempts to model predator/prey relationships, it will be obvious that a number of simplifying assumptions has been made. Below is a short list of factors that should be considered when evaluating the reality of a predator/prey model.

1. Errington (1946), based on his field work with muskrats, asserted that prey usually have a refuge; only when prey numbers are sufficiently large that individuals must leave the refuge are they subject to predation.

2. Some models assume that predation is random. In fact, predation is almost always non-random. In addition, predators often specialize on certain age-classes of the prey, or on weakened or diseased individuals.

3. The generation times of the predator and the prey populations are often very different. If the growth rate of the prey is much higher than that of the predator, and/or the generation time is much shorter in the prey species, the predator may rarely have an effect on the prey population. Alternatively, there may be a time lag between growth of the prey population and the numerical increase in the predator population. As we saw in Chapter 2, time lags tend to produce population cycles.

4. Predators may be generalists and switch from one prey species to another, depending on prey population size. Such behavior tends to stabilize prey populations. In fact, recent studies discuss the idea of a "predator pit" for the prey (Hudson and Bjørnstad 2003) when there is a rich community of generalist predators. When there are many predator species present, prey populations are controlled at a stable number and do not go through cycles typical of the same prey species in less predator rich communities.

5. Alternatively, the predator population may remain relatively constant in spite of large-scale fluctuations in the prey population. For example, when Southern (1970) studied the relationship between the tawny owl (*Strix aluco*) population and its rodent prey near Oxford, UK, he found a relatively constant breeding population of owls, in spite of the fact that the rodent prey base oscillated. Over a 13-year period the number of rodents per 12 acres varied from fewer than 10 to more than 250. During the same time the number of breeding pairs of the owl changed from a low of 17 to a high of 30. In years of low rodent density, however, no owls attempted to breed.

6. The predator may have a carrying capacity unrelated to the number of prey individuals. Thus the predator may never increase beyond a certain number and the prey population "escapes" from predation once it reaches a certain population threshold.

7. Density independent mortality of the prey and/or the predator population must be evaluated.

8. Predator–prey interactions may have multiple equilibriums. As pointed out by Messier (1994) a predator–prey interaction may have a low-density equilibrium and high-density equilibrium. At low prey densities the predation rate is density dependent and the prey population is maintained at the level K_1. At densities just above K_1 is the "predator pit" and the prey population is pushed back down to K_1 by the functional and numerical responses of its predator(s). However, if the prey population escapes to density K_2 is has escaped the pit. At densities above K_2 the overall predation rate is inversely density dependent due to limitations in the functional and/or numerical responses of the predators. The prey population is able to increase to level K_3, the high-density equilibrium, where it is limited by its food supply or some other environmental variable.

The introduction of predators into "naïve" communities has provided us with evidence that predators can have a powerful destabilizing effect on ecosystems. One well-known example is the inadvertent introduction of the sea lamprey (*Petromyzon marinus*) into the Great Lakes (Baldwin 1964, Smith and Tibbles 1980). Marine sea lampreys are found off the Atlantic Coast of North America, but migrate into fresh water to spawn. Adult lampreys feed by rasping a hole in the host and sucking out fluids. Lampreys were able to move up the St. Lawrence River but were not found in the Great Lakes due to Niagara Falls. But beginning in 1829, various canals were built to allow ships to pass from the St. Lawrence into the Great Lakes. The Welland Canal was finished in 1829, allowing ships and lamprey passage to Lake Erie. However, the systematic movement of lampreys into the Great Lakes did not begin until 1921. Once established in the Great Lakes, the lamprey began preying upon Lake Trout, an economically important fishery. Lampreys were found in Lakes Huron and Michigan in 1936-37 and in Lake Superior in 1938. Catches of Lake Trout declined from a high of 6 million pounds per year in Lake Huron in 1936 to virtual extinction by the late 1950s. The story was repeated in Lakes Michigan and Superior, but took more time. The peak of 7 million pounds of Lake Trout was harvested in the middle 1940s in Lake Michigan and dropped to virtually zero by 1950-51. In Lake Superior the peak harvest was around 5 million pounds in 1950 and dropped to less than 1 million pounds by 1960. Chemicals, traps and electric barriers failed to bring the sea lampreys under control. However, the sterile-male-release technique was developed in the 1980s and sterilized males were introduced into the tributaries of the Great lakes beginning in 1991 (Bergstedt and Twohey 2007). This technique, first used to control the screwworm fly (Baumhover et al. 1955), requires the development of a sterilization technique that does not reduce either the survival or mating competitiveness of sterilized males. If sterilized males successfully compete with normal males for fertilization the result will be a nest of infertile eggs. Results are very positive and lake trout are recovering but the "battle" continues.

Like most Pacific Islands, Guam lacked large predatory snakes. The only native snake was a blind, wormlike snake, which fed on termites and ants (Frits 1988). The brown tree snake (*Boiga irregularis*), a member of the family Colubridae and native to Australia, Papua New Guinea and other islands in north-western Melanesia, arrived in Guam shortly after World War II, probably as a stowaway in military cargo from Papua New Guinea (Frits 1988). These snakes did not become conspicuous until the 1960s, but by 1968 had colonized the entire island. Brown tree snakes feed on a wide variety of bird, mammal and lizard species. The population of snakes increased for over 35 years, reaching a density at one point of 100 per ha in some areas (Rhodda et al. 1992). By 1963, several formerly abundant native bird species had disappeared from areas where the snakes were most populous. By 1986, nine native forest bird species were extinct due to snake predation and several other bird species were endangered. Small mammals are also very rare in most areas inhabited by the snake. Finally, having depleted birds and mammals, the brown tree snake now feeds to a large extent on lizards. Between 3 and 6 lizard species have been or are being extirpated on Guam (Savidge 1988; Rhodda and Fritts 1992). Meanwhile because these snakes love to take up residence in electric substations, they cause an average of 80 power failures per year, costing $4 million in annual repair costs based on an Interior Department estimate from 2005 (NBC News, http://www.nbcnews.com/news/other/two-thousand-mice-dropped-guam-parachute -kill-snakes-f2D11685572). Everything from traps to snake-sniffing dogs have been tried to control the snakes, but nothing really worked and Guam hosts an estimated 2 million of them. The newest attempt to control the snake population is dropping thousands of mice, injected with acetaminophen, with tiny parachutes over the island.

Why acetaminophen? Brown tree snakes are uniquely sensitive to this over the counter painkiller and ingesting as little as 80 mg will kill the snake. If a snake eats one of these mice, which are dead prior to being injected and dropped from helicopters, it will die. The image of dead mice being dropped from helicopters over Guam may be humorous, but scientists and the citizens of Guam are hoping for a positive outcome (NBC News, http://www.nbcnews.com/news/other/two-thousand-mice-dropped-guam-parachute-kill-snakes-f2D11685572).

Over evolutionary time, of course, predators and their prey reach an accommodation through co-evolution or else they cannot coexist. Extinction and re-colonization in metapopulations occurs continuously over shorter periods of time, as does extinction and speciation over evolutionary time. Therefore we should not necessarily expect based on as a "quasi-religious idea of the balance of nature" (McCullough 1997) that predator and prey populations should be "stable." Nevertheless, we can all agree that the rapid extinction of entire communities of native organisms by introduced predators is highly undesirable.

One last word, before we proceed. The term "stable" as applied to a population has a variety of definitions. Here we will use the word to simply mean a population, or a pair of populations when applied to predator/prey relationships, which have a "return tendency," to a particular density. Recall that a density dependent population is one in which there is some mean level of density around which the regulated population fluctuates. In addition, over time the population does not wander increasingly away from this level. If we apply this to several populations simultaneously, each will fluctuate in a cloud of points around some point to which it tends to return. We will distinguish this from a cyclic population in which there is a regular, identifiable pattern of population sizes over time (Figure 10.1). We will call this, as described above, a "stable cycle" or "stable limit cycle." These two patterns are to be distinguished from those of increasing oscillations leading to extinction, or other patterns of extinction of the predator or mutual extinction of the predator and the prey. These interactions are usually called "destabilizing" or "unstable." The cases described above of the introduction of the sea lamprey or the brown tree snake, qualify as destabilizing. Some authors characterize certain predators as destabilizing and other as stabilizing. For example, Gilg et al. 2003 describe the stoat, a specialist predator, as destabilizing, while the generalist predators stabilize the interaction with lemmings.

10.2 The Lotka-Volterra equations

The first well-known models of predator–prey interactions can be traced to Lotka (1925) and Volterra (1926). Although the Lotka–Volterra model has been critiqued on many grounds (May 1975, 1976), it is still well respected, and forms the basis for models still in use today. As Hudson and Bjørnstad (2003, p.797) put it, "the fundamental theory of predator–prey interactions-encapsulated in the worthy Lotka–Volterra model predicts cycles in prey and predator abundance."

The Lotka–Volterra model, like most predator–prey models, consists of two parts. The prey population grows according to a simple exponential or logistic model. Subtracted from this are losses due to predation. These losses are due to the overall predation rate, which itself consists of two parts. The numerical response of the predator is a function of an increased rate of reproduction, an increase in immigration, or both. The second factor, which increases the overall predation rate, is the rate of consumption of prey per individual predator, the functional response. In the prey equation, the rate of growth is decreased by the overall predation rate, which is a function of both the numerical and functional responses of the predator.

The predator equation also consists of two parts. The growth of the predator population is a function of the overall predation rate, and is similar to the negative part of the prey equation. The growth rate of the predator is then decreased by a mortality factor, which can be either density independent or density dependent.

The relatively simple model Lotka–Volterra model was based on the following assumptions.

1. In the absence of predators, the prey population grows either exponentially or logistically.
2. The population growth of the predator is limited only by the availability of the prey.
3. Both predator and prey reproduce continuously, have no age structure, and all individuals are identical.
4. The predation rate is proportional to the rate of encounter between predators and prey. Encounter rate is a random function of population density. That is, both prey and predator individuals move at random.
5. The predator has a density independent, constant mortality rate.

Most of the seven points raised in section 10.1 are ignored in the Lotka–Volterra model. Predation is random; there is no refuge for the prey; and there is no carrying capacity for the predator independent of that set by the prey population.

In the version of the Lotka–Volterra model in which the prey population grows exponentially in the absence of predators, the prey growth rate is:

$$\frac{dN}{dt} = r_n N \tag{10.1a}$$

Where r_n is the prey intrinsic growth rate and N is the prey population size.

If the prey population grows according to the logistic, the prey equation is:

$$\frac{dN}{dt} = r_n N \frac{K_n - N}{K_n} \tag{10.1b}$$

where K_n is a carrying capacity for the prey population.

Without prey, the predator population (P) dies off based on the instantaneous density independent mortality rate, m_p, and the population declines according to formula 10.2.

$$\frac{dP}{dt} = -m_p P \tag{10.2}$$

The chance of an encounter between the predator and prey is the product:

$$ENP \tag{10.3}$$

E is a number less than one, which measures the searching (and capturing) efficiency of the predator. Equation 10.3 assumes that the number of prey taken varies linearly with prey abundance. The coefficient E is a functional response term based on the rate of predation per individual predator per unit time. The increase in the predator population is the encounter term (equation 10.3) times a constant χ_p, which measures the efficiency by which the food (prey) is turned into new predator individuals. In essence this is the assimilation efficiency of the predator. The population growth term for the predator is: $(\chi_p)(E)(N)(P)$. This product, therefore, includes the numerical response to an increase in the prey population, and a linear functional response. The Lotka–Volterra model assumes that an encounter leads to the death of a prey individual. The prey population is thus decreased by the term, ENP. Starting with equation 10.1a, the equations for prey and predator are as shown below.

$$\frac{dN}{dt} = r_n N - ENP \tag{10.4a}$$

$$\frac{dP}{dt} = \chi_p ENP - m_p P \tag{10.5}$$

The behavior of this model at equilibrium can be analyzed by setting both dN/dt and $dP/dt = 0$, leading to equations 10.6 and 10.7.

$$P^* = \frac{r_n}{E} \tag{10.6}$$

$$N^* = \frac{m_p}{\chi_p E} \tag{10.7}$$

Equation 10.6 expresses the predator equilibrium in terms of the growth rate of the prey and searching efficiency of the predator. Equation 10.7 describes the prey equilibrium in terms of the mortality rate, the searching efficiency and the assimilation efficiency of the predator. These equations seem to indicate that the equilibrium values are independent of the numbers of the other population. However, we can also see that the per capita growth rate of the prey becomes zero or negative when the predator population exceeds a fixed number. Similarly, the per capita rate of predator increase becomes zero or negative when the prey population drops below a specific density. Finally, the functional response for the predator is unrealistic in that it assumes that at high prey densities the predator has an unlimited appetite.

As pointed out by May (1975, 1976) the Lotka–Volterra model has a peculiar "neutral stability," that can be compared to that of a frictionless pendulum. Populations are predicted to oscillate forever based on the initial conditions, with no mechanism for increasing or decreasing the amplitude of the oscillations. Furthermore, both the functional and the numerical responses of the predator are fixed. This is a significant flaw in either a predictive or a descriptive set of models.

Long ago, Volterra (1931) recognized that adding a density dependent component to the prey equation would add realism to the model. Therefore, the prey equation (10.4a) was modified to 10.4b. The Volterra model is characterized by a stable equilibrium point (Figures 10.5 and 10.6).

$$\frac{dN}{dt} = r_n N \frac{K_n - N}{K_n} - ENP \tag{10.4b}$$

10.3 Early tests of the Lotka–Volterra models

According to the Lotka–Volterra equations, the response of a predator population to an increase in a prey population is to increase its own numbers. This increase may be through an increase in the birth rate of the predator (perhaps combined with a decrease in death rate) or through immigration. Again, this is termed a numerical response.

According to the Lotka–Volterra equations, the result of this increase in predation is a coupled numerical response in the prey population, which declines. Once the prey population has declined sufficiently, the negative consequences for the predator population results in its decline. Once predator numbers have decreased sufficiently, the prey population begins to recover, leading eventually to an increase in the predator population and so on. A graph of the Lotka–Volterra results versus time look like a stable limit cycle, though it is not (Figure 10.1). In a true limit cycle, if the populations are pushed out of the cycle by density independent factors, the populations return to the original limit cycle. Due to the neutral stability of the Lotka–Volterra equations, they would have no return tendency.

The Lotka–Volterra equations led some ecologists to adopt the view that predator–prey interactions were "inherently oscillatory," and research was directed to test this proposition. As discussed previously, some parasite-host and various small mammal and bird populations in boreal and arctic regions display regular oscillations in number. Do predators cause these cycles? Already in 1924 Charles Elton had published, "Periodic Fluctuations in the Numbers of Animals: Their Causes and Effects." In this and later publications (Elton and Nicholson 1942) he presented the now famous and infamous data on the populations of snowshoe hare (*Lepus americanus*) and the lynx (*Lynx canadensis*). As in the case of the Kaibab deer, these data were not based on systematic population surveys, but rather on the numbers of pelts brought into the Hudson's Bay Company by trappers. The data do show regular fluctuations of great magnitude, but we must ask, how reliable are these data? Are the fluctuations due to predator/prey interactions envisioned by Lotka and Volterra? We will not try to answer these questions now, but note here that after 70 years of field experiments and time series analysis, Krebs et al. (2001) concluded that the hare cycle can only be understood as an interaction involving the hare population, its food supply, and a community of predators (not just the lynx).

As discussed earlier, the competition equations of Lotka and Volterra were tested in the laboratory by Gause (1934). Gause also tried to test the predictions of the Lotka–Volterra predator/prey equations using microorganisms. He attempted to produce the predicted oscillations using as his prey, populations of *Paramecium caudatum* grown in test tubes. To these tubes he introduced another ciliated protozoan, *Didinium nasutum*. *Didinium* is a voracious predator on *Parmecium* and it reproduces by binary fission, just as does its prey. In the simple test tube environment *Didinium* was able to hunt down all of the *Paramecium*. Once its food supply was gone, *Didinium* starved. Thus in any one tube, mutual extinction was assured. Gause next tried adding sediment to the bottom of the tubes as a refuge for *Paramecium*. This ensured the survival of the *Paramecium*, but the *Didinium* population eventually went extinct. With its predator eliminated, the *Paramecium* population rapidly grew to the expected carrying capacity. But Gause had more tricks. He now added one *Paramecium* and one *Didinium* every third day to each test tube. This finally resulted in coexistence of the prey and its predator for more than two weeks. Both the prey and the population went through two oscillations during this period.

Did Gause see his work as confirming the equations developed by Lotka and Volterra? Just the opposite. Gause stated that predator/prey interactions are not inherently oscillatory, and that coexistence was possible only through adding heterogeneity to the simple test tube environment, or through constant interference of the system through the addition of immigrants.

Another early laboratory experiment illustrates the weaknesses in the simple Lotka–Volterra model. In the earlier chapter on metapopulations we described the work of Huffaker (1958), who was a California entomologist interested in biological control of pests in orange orchards. The prey species was the six-spotted mite (*Eotetranychus sexmaculatus*), which feeds on oranges. The predator was a carnivorous mite (*Typhlodromus occidentalis*), which preys on the six-spotted mite. Both species reproduce rapidly through parthenogenesis. In each experiment, Huffaker began with 20 prey females and introduced two predator females eleven days later.

In one experiment, Huffaker concentrated the food (oranges) in one area. The results mirrored those of Gause. The prey population rapidly increased; then was located by the predators, which also rapidly increased. Within a short time (25–30 days) both populations were extinct. Huffaker then began creating a heterogeneous environment. He set up a complex laboratory environment consisting of three 40-cell trays with a total of 120 feeding positions. Although each position contained one orange, he controlled the feeding surfaces by dipping the oranges in wax, leaving only 5% of the orange available for feeding. This forced the herbivorous mite to constantly seek out new feeding surfaces. He added small wooden pegs as launching pads for the six-spotted mites to speed their dispersal from one orange to another. And he added a maze of Vaseline™ barriers across the trays to slow the dispersal of the predatory mites, which could travel only by foot. Once a predator arrived on an orange already colonized by the prey species, it quickly killed and consumed all of the herbivorous mites on that particular orange. But the rapid immigration and emigration of the herbivorous mite, along with the complex, heterogeneous environment created by Huffaker, allowed the two species to coexist in this laboratory environment for over 200 days.

In both of these laboratory studies, the coexistence of the prey with the predator depended upon environmental heterogeneity. Secondly, both systems required regular immigration of the prey and/or the predator to avoid extinction. Neither of these requirements was anticipated by the Lotka–Volterra approach.

Another instructive laboratory experiment was that of Utida (1957). In this case stable oscillations between the Azuki bean weevil (*Callosobruchus chinensis*) and a parasitic wasp (*Heterospilus prosopidis*) were maintained for more than 25 generations in a laboratory Petri dish (1.8 cm high by 8.5 cm diameter). The wasp is actually a specialized predator known as a parasitoid, which paralyzes its prey without killing it. The wasp lays an egg on the paralyzed host, usually an insect larva. The egg hatches and the wasp larva slowly consumes the host, leading to its eventual death. In this case, the wasp only lays eggs on certain instar larvae. The wasp does not parasitize adult or pupal beetles. Therefore, although these two populations undergo the regular oscillations of a limit cycle, both populations persist due to the non-random predation by the wasp. Again, the violation of the Lotka–Volterra assumption of random hunting is what allows the coexistence of prey and predator.

10.4 Functional responses

The Canadian ecologist C.S. Holling concluded that our understanding of predation needed a more realistic, empirical approach. He believed that to understand predator–prey relationships it was first necessary to understand the act of predation. That is, he asserted that the first response of an individual predator to an increase in a prey population is not to increase its growth rate, but rather to increase its per capita predation rate, often by selective behavior. Using terminology from Solomon (1949), Holling (1959, 1961, 1966, 1973) termed this per capita increase in predation the functional response of a predator. He analyzed the act of predation and broke it down into behavioral units termed the components of predation. In his papers Holling described many other potential factors that might influence predation, but we will limit our discussion to the time and/or energy devoted to these four major components: (1) Search, (2) Capture, (3) Handling, and (4) Digestion. Theory and laboratory testing suggest that the functional response of an animal may take one of three forms called the type I, type II and type III functional responses. In all of these functional response graphs, the *x*-axis consists of number of prey per unit area (prey density), while the *y*-axis consists of the number of prey eaten per individual predator per unit time. Thus the functional response is the result of changes in consumption rate per individual predator. In all three types of functional responses the predation rate will rise with prey density, but will eventually level off. At low prey densities, predation rate is influenced mostly by the amount of time and/or energy devoted to search and capture, while at very high prey densities predation rate is bounded by the last two components of predation, time or energy devoted to handling and digestion of prey.

The simplest functional response is one in which prey consumption rises linearly with prey density. This is sometimes called the filter feeder's functional response, and is often pictured as a simple straight line. However, this picture is misleading. It omits the fact that increases in prey density beyond a threshold do not result in an increased per capita predation rate. This is due to the handling and digestion components of predation. For example, Porter et al. (1982, 1983) maintained concentrations of the green alga *Chlamydomonas* in the laboratory at concentrations up to 10^6 cells per cubic cm. They then introduced the arthropod *Daphnia magna* (the "water flea"), a filter feeder. As illustrated by Figure 10.9, *Daphnia* needed almost 1000 (10^3) cells per cm³ to maintain itself. As the concentration of *Chlamydomonas* increased, so did the consumption rate of *Daphnia*. Once about 20 000 cells per cm³ were reached, however, the increase in consumption decreased rapidly and the water flea processed a maximum of about 7 cells per second or 25 000 per hour. There is evidently a limit to what *Daphnia* can process, and the type I functional response has a plateau, as do all functional responses, due to the handling and digestive components of predation. Picture yourself eating popcorn. If individual pieces of popcorn were spread about the house and you had to

Fig. 10.9 Type I functional response. Ingestion of the alga *Chlamydomonas* by *Daphnia magna*. (Source: Data from Porter et al. (1982).)

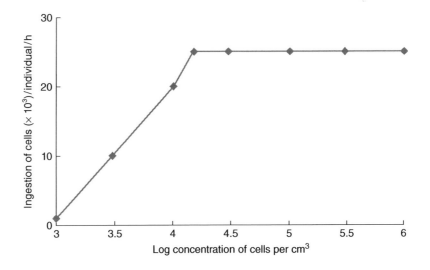

search for them, your consumption rate would be limited by the search component of predation (capture time for popcorn is presumably zero). But if you were surrounded by an unlimited number of tubs of popcorn, your consumption rate would be limited by the time it would take to handle and chew. And eventually, your stomach would tell you, no more! Until you had time to digest what you had stuffed into your mouth, your consumption rate would plummet. Next time you go to a movie theater, take notes on the rates of consumption of your fellow predators!

Porter et al. (1982, 1983) also found that when the algal concentration went up, the amount of effort by *Daphnia*, as measured by filtering rate, went down. As food concentration went up the movement of filtering appendages declined from 6 to 3.5 per second, resulting in a decrease in filtering rate from four to one cm^3 per hour. This means that the food resource was so concentrated that maximum ingestion was possible with much less effort. Therefore, since less energy was expended, the *Daphnia* population was able to turn more assimilated energy into reproduction. As we can see from Table 10.1, an increase in food consumption led directly to a numerical response in *Daphnia magna*, at least up to a point. As the density of *Chlamydomonas* increased, several reproductive parameters also increased. However, at the highest prey density (10^6 cells) reproductive parameters decreased and the intrinsic rate of increase declined from 0.28 to 0.20. These data illustrate the relationship between the functional and the numerical response of a predator, as well as the fact that at very high prey densities other factors may limit the numbers of a predator.

In the Type II functional response, as prey density increases, prey consumption increases more rapidly than in a type I response, and is nonlinear, but also reaches a plateau. This is called the "general invertebrate curve," since Holling found this curve applied to a number of invertebrate predators that he tested (Figure 10.10).

In the Type III response, prey consumption remains low until a threshold density is reached. The predation rate then increases exponentially until it levels out. The shape of this curve is described as "sigmoid" and it looks something like the logistic population growth curve. Sometimes called the "vertebrate" curve or the "learning predator's" curve, the predator ignores the prey when densities are very low in order to make hunting energetically profitable. Another interpretation of the low predation rate at low prey densities is that the predator does not see the potential prey often enough to recognize it as prey. Once the prey population reaches the threshold density, the predator develops a "search image" for the prey and begins actively seeking it. At this point, the predation rate increases exponentially until reaching the maximum rate, which occurs at the plateau due to handling and digestion time.

Table 10.1 Reproductive parameters of *Daphnia magna* as a function of the concentration of its prey, the green alga *Chlamydomonas*.

	Concentration of *Chlamydomonas* in cells per cm^3			
	10^3	10^4	10^5	10^6
Percent reproducing	50	87	97	50
Eggs per brood	2.8	2.6	15.5	21.1
Broods per female	1.7	7.5	8.2	3.4
Days between broods	5.4	3.6	3.1	3.3
Age of first brood (days)	23.4	16.9	9.8	9.1
Net reproductive rate (R_0)	2.25	16.23	99.33	34.80
Intrinsic rate of increase, r	0.03	0.10	0.28	0.20

Experiment conducted at 20 °C.
Source: Based on Porter et al. (1983), and adapted from Ricklefs (1990).

Fig. 10.10 Type II functional response. The larvae of the damsel fly, *Ischnura elegans*, feeding on *Daphnia magna*. (Source: Data from Hassell (1976).)

Fig. 10.11 Type III functional response curves for three predators of lemmings in Greenland. (Source: Data from Glig et al. (2003).)

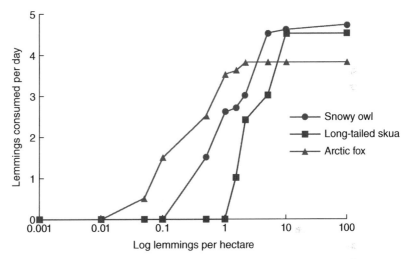

Working in the high-Arctic tundra of Greenland, Gilg et al. (2003) have shown that the snowy owl (*Nyctera scandiaca*), the arctic fox (*Alopex lagopus*), and the long-tailed skua (*Stercorarius longicaudus*) all have type III functional responses to increasing densities of collard lemmings (*Dicrostonyx groenlandicus*) (Figure 10.11)

The next step is to integrate the components of predation into the Lotka–Volterra equations. The functional response, f, is set equal to EN for a type I response as was actually incorporated into the original Lotka-Volterra equations (equations 10.4a 10.4b, 10.5).

The type II and III functional responses involve variables for search and capture rates (E_2N and E_3N^2) and handling and digestion rates (h_2 and h_3). As described above, the plateau or asymptote found in each functional response curve is the result of the handling and digestion rates, h. When prey densities get very large, little time or energy must be spent searching or capturing, and the predation rate depends on handling time and digestive pause. Thus, we have the Holling equation for the type II response (10.8a) (See Case 2000 for a derivation).

$$f_n = \frac{E_2N}{1 + h_2E_2N} \tag{10.8a}$$

If we take its inverse, we have:

$$\frac{1}{f_n} = \frac{1 + h_2 E_2 N}{E_2 N} \tag{10.9}$$

We can rewrite this as:

$$\frac{1}{f_n} = \frac{1}{E_2 N} + \frac{h_2 E_2 N}{E_2 N} = \frac{1}{E_2 N} + h_2 \tag{10.10}$$

The second term simplifies to h_2; when the prey population becomes very large, the first term becomes very small and we can set it to 0. Thus $1/f = h_2$ and:

$$f_n = \frac{1}{h_2} \tag{10.11}$$

Equation 10.11 simply states that the functional response is the inverse of the handling time component. A similar analysis applies to the type III functional response. A useful interpretation here is that the inverse of handling time equals the maximum killing rate (c), when the search and capture components are essentially zero. An alternative form for equation 10.8a sets the parameter $c = 1/h_2$ and substitutes the half saturation constant, d (the prey density at which killing rate is half of the maximum) for $(E_2 h_2)^{-1}$. Substituting $1/c$ for h_2 and c/d for E_2 in equation 10.8a, we have 10.8b, the alternative equation for the functional response. The advantage of this version of the functional response is that we can relatively easily estimate these parameters from field data.

$$f_n = \frac{cN}{d + N} \tag{10.8b}$$

When prey densities are very low, the search and capture rates become very large compared to the handling time. If we set h equal to 0 in equation 10.8a, the denominator becomes equal to one and the functional response equations become:

$$f_n = E_2 N \text{ (Type 2 Functional Response)} \tag{10.12a}$$

$$f_n = E_3 N^2 \text{ (Type 3 Functional Response)} \tag{10.12b}$$

Therefore, at a low prey density, the functional response is completely determined by the search and capture components, and it is essentially linear in the type II response.

10.5 Adding prey density dependence and the type II and III functional responses to the Lotka-Volterra equations

Now let's return to the Lotka–Volterra equations as modified by a type II functional response. For the prey equation, we will not only add the functional response, but we will also add the carrying capacity term as suggested by Volterra (1931). The prey equation again consists of two parts: (1) recruitment and (2) death due to predators. The recruitment section is the traditional logistic, and the losses due to predation are based on the numerical response term (ENP) modified by the type 2 functional response (Equation 10.8a). The resultant equation is 10.13a:

$$\frac{dN}{dt} = r_n N \left(1 - \frac{N}{K_n} \right) - \left(\frac{E_2 N P}{1 + E_2 h_2 N} \right) \tag{10.13a}$$

In the predator equation the left term is again modified by the type II functional response, but is otherwise unchanged (10.14a).

$$\frac{dP}{dt} = \frac{\chi_p E_2 NP}{1 + E_2 h_2 N} - m_p P \tag{10.14a}$$

Recall that χ_p measures the efficiency at which the food (prey) is turned into new predator individuals (=assimilation efficiency of the predator), and m_p is the mortality rate of the predator.

In the type III functional response, the term N is replaced by N^2 in the phrase describing death due to predation in Equation 10.13a. That is, $E_3 N^2 P / 1 + E_3 h_3 N^2$. Similarly, in equation 10.14a N is replaced by N^2 on the right side of the equation.

How does the functional response affect the behavior of the predator/prey interaction? Figures 10.12, 10.13a and 10.13b are based on equations 10.13a and 10.14a. In each case $r_n = 0.2$, $K_n = 100$, $N_0 = P_0 = 20$. E, the coefficient measuring the efficiency of search and capture $= 0.02$. χ_p (assimilation efficiency) is set at 1.0. The mortality rate $m_p = 0.25$.

In Figure 10.12 we have removed the functional response terms from both equations. This means the populations are growing according to the type I functional response, which is equivalent to the basic Lotka–Volterra equations. What we see is mutual extinction such as Gause and Huffaker found. In Figures 10.13a,b by contrast, we have introduced the type II and III functional response terms with $h = 3.0$ in Figure 10.13a and $h = 4.0$ in Figure 10.13b. The populations move toward a stable point in each case. Handling time puts a cap on the number of prey that each predator can consume, thereby limiting both its growth rate and the losses to the prey population. The effect is to help stabilize the interaction.

Fig. 10.12 Predator–prey interaction following the Lotka–Volterra equations with a type I functional response.

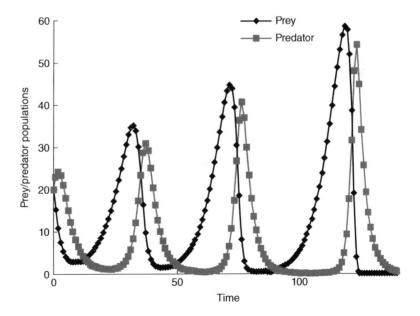

Fig. 10.13 (a) Predator–
prey interaction with a
Type II functional
response. (b)
Predator–prey
interaction with a Type
III functional response.

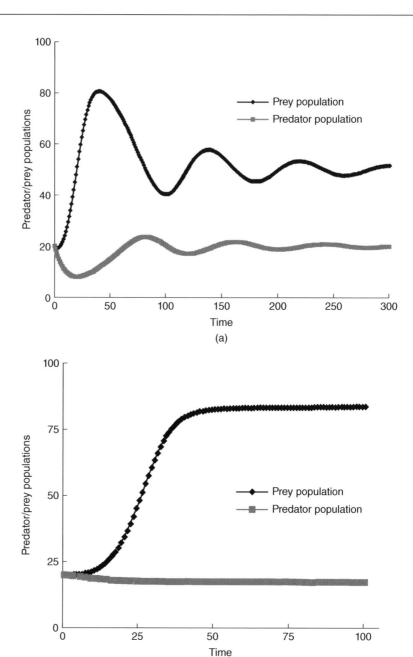

(a)

(b)

10.6 The graphical analyses of Rosenzweig and MacArthur

Rosenzweig and MacArthur (1963) founded a general approach for studying predator–prey interactions known as the isocline or graphical analysis. The graphical analysis consists of a plot whose x-axis is the prey population and whose y-axis is the predator population. Two isoclines are drawn on the graph. The first line, known as the predator isocline, may be a vertical line, or may be modified to any linear representation with a positive slope. This line represents the prey density that results in zero population growth of the predator. Areas to the right of the line allow the predator population to grow; to the left predator populations decline. The second line, the prey isocline, represents the predator and prey combinations that result in zero population growth of the prey. The prey isocline (Figure 10.14) is based on the per capita growth rate of the prey. At the far right is the point $(K, 0)$. There are zero predators, but the prey growth rate is limited to zero by its carrying capacity. The growth rate has a positive slope at low prey densities and reaches its maximum value at $K/2$. Regions above the prey isocline represent combinations of predator and prey resulting in a negative prey growth rate due to predation. The prey isocline is often shown with the Allee effect included (Figure 10.15). As discussed in Chapter 2, very low prey populations may result in negative growth. Thus the prey isocline is shifted away from the point $(0, 0)$ to a point representing a minimum viable population size.

This graphical approach allows us to envision a variety of situations in which extinctions, stable cycles, and stable points are produced. In Figure 10.16 the predator isocline is far to the right, intersecting the prey isocline in an area where it has a negative slope. This represents an "inefficient" predator, which needs large numbers of prey.

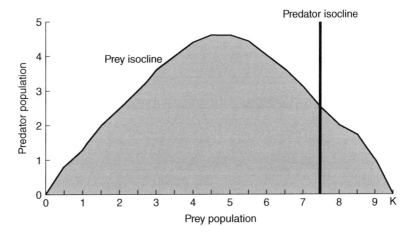

Fig. 10.14 Prey and predator isoclines. The prey population has positive growth in the shaded area. The predator population has positive growth to the right of the vertical line. (Source: Data from Rosenzweig and MacArthur (1963).)

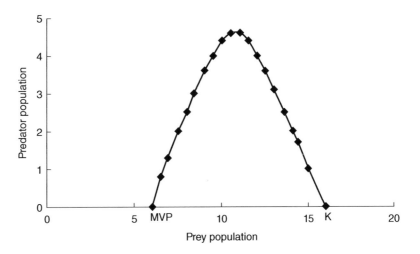

Fig. 10.15 The prey isocline with the Allee effect. Minimum viable population, MVP

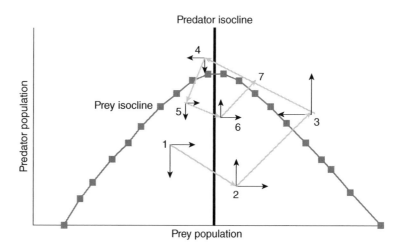

Fig. 10.16 Graphical analysis of predator/prey interactions of Rosenzweig and MacArthur (1963). An inefficient predator isocline produces decreasing oscillations to a stable point, *S*, where the two isoclines meet.

Fig. 10.17 Predator/prey isoclines for a moderately efficient predator resulting in a stable limit cycle.

The result is a series of decreasing oscillations resulting in a stable point at the intersection of the two isoclines. By pushing the predator isocline to the left, however, such that the predator isocline intersections the prey isocline at a right angle at $K/2$, the result is a stable cycle (Figure 10.17), rotating around the intersection point, S. Finally, if we push the predator isocline further left, so that it intersects the prey isocline where it has a positive slope, simulating a "highly efficient predator," the result is mutual extinction of the predator and its prey (Figure 10.18). If the predator is limited by a resource other than the prey population, the result is a stable point as illustrated in Figure 10.19.

This analysis also allows us to examine what is known as the "Paradox of Enrichment." Laboratory experiments and theoretical simulations show us that predator/prey interactions are often stabilized by lower prey growth rates. In his experiments with oranges and orange mites described earlier, Huffaker (1958) found that to stabilize the predator–prey interaction he had to diminish the prey growth rate. He reduced the amount of feeding surface available for the prey on the oranges by dipping them in wax. Similarly, when Luckinbill (1973a, 1973b) repeated Gause's experiments (substituting

Fig. 10.18 A highly efficient predator resulting in increasing oscillations leading to extinction of both the predator and the prey (point E).

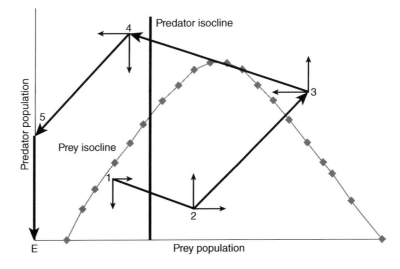

Fig. 10.19 Predator/prey interaction when predator growth is restricted by a factor other than the prey population. The result is stable point, S.

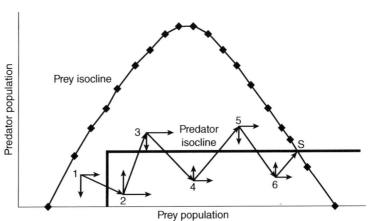

P. aurelia for *P. caudatum*, but still using *Didinium nasutum* as the predator) both prey and predator went extinct within a few hours. However, by adding methyl cellulose to the medium, which slowed down the movement of both predator and prey, the two species coexisted for several days. Then he then cut the food supply for the prey in half, and the two species coexisted for 33 days before the experiment was terminated. Rosenzweig (1969) had predicted this result on theoretical grounds. In Figure 10.20 we see that, although the predator isocline is the same as in Figure 10.17, where the result was a stable point, by moving the prey isocline to the right, simulating a higher growth rate for the prey due to an increase in resources, the result is mutual extinction (Figure 10.20).

Finally, we can simulate an invulnerable prey refuge (Figure 10.21). In this case the result, depending on the efficiency of the predator, is a stable point or a cycle (limit cycle), although the cycle may be rather complex. We can easily add the idea of prey refuge in equations 10.13a and 10.14a by adding a constant M^*. This number represents a minimum prey population which is never exposed to predation. The number of prey available for predation will then be only $N - M^*$ and we replace N by $N - M^*$

Fig. 10.20 The effect of the "paradox of enrichment" on a predator/prey interaction. The predator isocline is identical to that of Figure 10.16. The prey isocline has been shifted to the right, simulating resource enrichment. The final point is E (0,0).

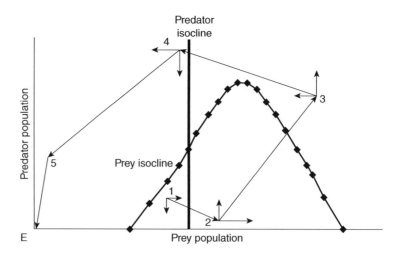

Fig. 10.21 Predator/prey interaction when the prey has a refuge. The result is a stable cycle.

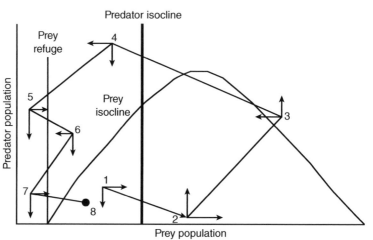

in the appropriate places (equations 10.13b and 10.14b).

$$\frac{dN}{dt} = r_n N \left(1 - \frac{N}{K_n} \right) - \left(\frac{PE\,[N - M^*]}{1 + Eh[N - M^*]} \right)$$ (10.13b)

$$\frac{dP}{dt} = \frac{\chi_p E[N - M^*]P}{1 + Eh[N - M^*]} - m_p P$$ (10.14b)

As discussed by Case (2000) and Crawley (1992), these types of analyses underscore the point that in a predator-prey relationship there can be more than one stable point, and predator-prey relationships that produce cycles in one ecosystem, may result in a stable point in another.

10.7 Use of a half saturation constant in predator/prey interactions

Turchin and Ellner (2000) have asserted that the predator prey equations used by Rosenzweig and MacArthur (1963) should be adopted as "the standard" for predator prey interactions, since they eliminate the assumption of the linear functional response and are "perhaps the simplest model that can actually be applied to real life systems" (Turchin 2003, p. 95). The only differences between these equations and equations 10.13a and 10.14a are that the handling term in the functional response has been replaced by $d + N$, where d is the half saturation (half maximum killing rate) parameter, and E has been replaced by c, the maximum killing rate when the search and capture components have been minimized. In other words, we are using the alternative functional response shown as equation 10.8b.

$$\frac{dN}{dt} = r_n N \left(1 - \frac{N}{K_n} \right) - \frac{cNP}{d + N}$$

(10.15)

$$\frac{dP}{dt} = \frac{\chi_p cNP}{d + N} - m_p P$$

(10.16a)

An alternative form of the predator equation (10.16a) is to add what is known as the ZPG component consumption rate (μ_p). This parameter represents the minimum rate of prey consumption needed for a predator to survive and just replace itself. This is usually easier to estimate that the predator death rate in the absence of prey, m_p. The result is equation 10.16b, a version of which we will use in the chapter on herbivore–plant interactions.

$$\frac{dP}{dt} = \chi_p P \left(\frac{cN}{d + N} - \mu_p \right)$$

(10.16b)

Now compare the predator equation (10.16a) to equations 7.22 and 7.23, which describe the effects of resource depletion on the growth rate of a consumer in the context of competition.

In a revised version of equation 7.23, instead of a competing species, N_2, we are substituting P, the predator population. The death rate, m, is now m_p. We have replaced $\chi_p c$, which measures the maximum rate of conversion of the resource (prey) into predators, by b, which was the maximum growth rate of the competing species.

$$\frac{dP}{dt} = \frac{bRP}{K_R + R} - m_p P$$

(7.23 revised)

The prey population (N in equation 10.16a) is considered a resource, so is replaced by R, the concentration of the resource. The half saturation constant d is replaced by K_R. The two equations are now identical. Equation 7.23 measures the growth rate of a consumer in terms of its maximum growth rate, the concentration of the resource, the half saturation constant for that resource, and the death rate of the consumer. The predator equation (10.16a) measures the growth rate of the predator in terms of maximum efficiency in turning prey into predator individuals ($\chi_p c$), the concentration of the resource (N), the half saturation constant and the death rate of the predator. We now have a general mechanistic equation, useful in both competitive interactions and predator–prey interactions for the growth rate of a consuming population.

Because the half saturation parameter is more easily estimated in the field than is a specific handling time value, this approach (10.15 and 10.16) was employed Gilg et al. (2003) in their study of lemming/predator interactions in Greenland, as well as by Turchin and Ellner (2000).

10.8
Parasitoid/host interactions and the Nicholson–Bailey models

A parasitoid is a special type of predator. Rather than killing and consuming the prey immediately, a female parasitoid lays its eggs on a particular stage of a host species (usually a larval stage of an insect) after stinging and paralyzing it. The egg hatches into a larva which consumes the host over a period of time, eventually killing it. The parasitoid emerges from the host ready to hunt for more prey. The parasitoids are usually very host-specific and must time their life cycles to those of their hosts. In addition, since there is a time delay between egg laying and emergence of the next generation, discrete time equations are used to model these interactions.

Nicholson and Bailey (1935) developed models for parasitoids and their hosts. These were the starting point for a series of more complex models developed by and expanded to book length by Hassell (1976, 1978).

The assumptions of the basic Nicholson–Bailey model include:
1. The number of encounters, N_e, between P_t parasitoids and a host or prey species is proportional to the host density N_t.
2. The encounters are randomly distributed among the hosts. This means that some host individuals will be encountered more than once, so that N_e can be larger than N_t.

Given the assumption of random distribution of encounters, the proportion of hosts *not* parasitized is given the zero term of the Poisson distribution:

$$p_0 = e^{\left(\frac{-N_e}{N_t}\right)} \tag{10.17}$$

The number of hosts actually parasitized is N_a:

$$N_a = N_t \left[1 - e^{\frac{-N_e}{N_t}} \right] \tag{10.18}$$

The number of encounters between P_t parasitoids and their hosts N_t, can be restated as:

$$N_e = aN_tP_t \tag{10.19}$$

Where a is constant called the parasitoid's rate of discovery. It is a measure of searching efficiency. It follows that $N_e/N_t = aP_t$, and the number of hosts that are parasitized can be rewritten as:

$$N_a = N_t[1 - e^{(-aP_t)}] \tag{10.20}$$

This expression implies that the rate of parasitism will reach a saturation level as parasitoids find fewer and fewer hosts not previously attacked. Nicholson–Bailey called this relationship a competition curve.

As mentioned above, because of the discrete seasonality of most arthropods and because there is no age structure in most populations, the equations defining host and parasitoid dynamics are written as difference equations. The number of progeny a female parasitoid leaves behind is based on the number of hosts attacked.

The prey equation is:

$$N_{t+1} = \lambda N_t e^{-aP_t} \tag{10.21}$$

Where λ = the finite rate of increase per generation and e^{-aP_t} represents the proportion of hosts escaping attack by the parasitoid.

If we assume that only one parasitoid emerges from each host, the parasitoid equation is:

$$P_{t+1} = N_t[1 - e^{-aP_t}] \tag{10.22a}$$

However, we can allow more than one parasitoid to emerge from each host by adding the parameter n to this equation. In equation 10.22b, the parameter n equals the number of parasitoids emerging from each host.

$$P_{t+1} = nN_t[1 - e^{-aP_t}] \tag{10.22b}$$

The number of parasitoids in the next generation, then, is the product of n times the number of host individuals times the proportion of host individuals successfully attacked.

Unfortunately, this model does not improve on the Lotka–Volterra model; the main difference is the built-in time lag associated with the use of difference equations. While the Lotka–Volterra model produces a neutrally stable cycle, the Nicholson–Bailey model predicts unstable, increasing oscillations (Figure 10.22)

However, the introduction of a density dependent host growth factor will stabilize the Nicholson-Bailey model (Hassell 1976) as in Figure 10.23 in which a carrying capacity term was added (equation 10.23). The result is a stable cycle, although the oscillations are very large.

$$N_{t+1} = \lambda N_t e^{\left[\left(1-\frac{N}{K_n}\right)-aP_t\right]} \tag{10.23}$$

If we look at these two interactions by graphing the host population on the x-axis and the parasitoid population on the y-axis, we get Figures 10.24 and 10.25.

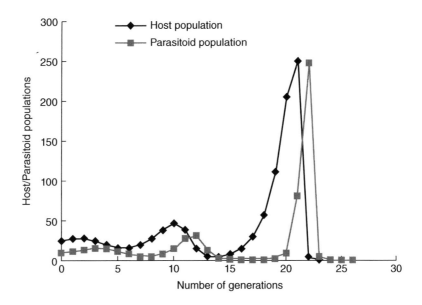

Fig. 10.22 Density independent host/parasitoid interaction. $\lambda = 2.0$, $a = 0.06$, and $n = 1.0$.

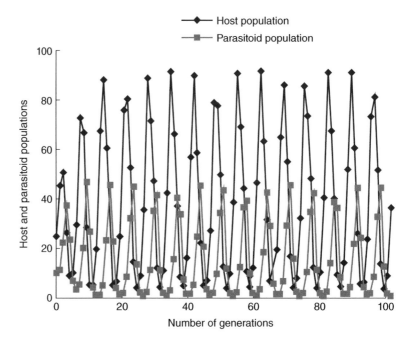

Fig. 10.23 Density dependent host/parasitoid interaction. Parameters as in Figure 10.22, but K = 50.

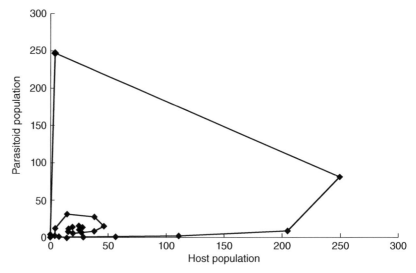

Fig. 10.24 Density independent host/parasitoid relationship showing mutual extinction. (Source: Based on data from Figure 10.22.)

Here we clearly see the mutual extinction of the two populations in Figure 10.24 and the ellipse in Figure 10.25, which illustrates the stable limit cycle more clearly than does Figure 10.23.

The introduction of the carrying capacity term for the prey produces a number of interesting results, depending on value of the other parameters. Under a large number of conditions the predator/prey interaction produces a stable cycle. For example, as we have shown in Figure 10.25, if $\lambda = 2.0$, $K_n = 50$, $n = 1.0$, $a = 0.060$, $N_0 = 25$, and $P_0 = 10$, the result is a stable cycle, whereas without the density dependent term, the result is mutual extinction (Figure 10.24). However, if we run simulations using a

Fig. 10.25 Host/
parasitoid relationship
showing stable limit
cycle. (Source: Based on
data from Figure 10.23.)

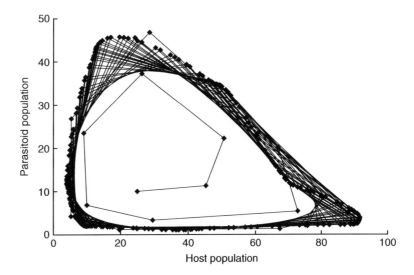

computer program like *Populus*, we find a wide range of results from mutual extinction
to a stable point. For example, if the initial prey population is too large or too small
relative to its carrying capacity, the result can be extinction of the parasitoid population.
If the host population has a finite rate of increase that is too small, it can also lead
to extinction of the parasitoid. But an extremely high host rate of increase does not
destabilize the interaction. A λ value of over 12 (= r of 2.5) simply leads to a stable
limit cycle. On the other hand, given the values shown above, lowering the carrying
capacity of the prey population to 25 leads to a stable point instead of a stable cycle.
But either lowering K_n to 10 or raising it to 75 leads to extinction of the predator.
The stability of this interaction is also affected by the searching efficiency parameter,
a. A highly efficient predator produces extinction of both the predator and the prey.
An *a*-value of about 0.09 results in extinction of only the predator, whereas values
between 0.03 and 0.06 result in a stable point rather than a limit cycle (Figures 10.26
and 10.27). If the parasitoid becomes even less efficient (values much less than 0.03),
it goes extinct.

Other factors that have been found to stabilize the parasitoid/host interaction include
Hassell (1978): (1) Interference competition among the parasitoids. If female parasitoids avoid each other or avoid laying eggs where others have laid theirs, the rate
of parasitism rises more slowly. (2) A refuge for the host. As in other predator–prey
models, if the prey has hiding spots or habitats that cannot be found by the parasitoid,
the interaction is more stable.

10.9 Section summary

As we indicated at the beginning of this chapter, simple predator/prey models
inevitably predict an oscillatory interaction between predators and their prey. However, by adding a density dependent factor for the prey population, and some version
of a functional response by the predator, depending upon the values of the parameters,
these models predict a range of responses, from extinction to cycles to stable points.
Adding a carrying capacity term for the predator, independent of the prey population,
may also help stabilize a predator–prey interaction. More realistic models would
include the ability of the predator to choose from among several prey species and/or

Fig. 10.26 Host/ parasitoid interaction with searching efficiency, $a = 0.003$. Result is a stable point.

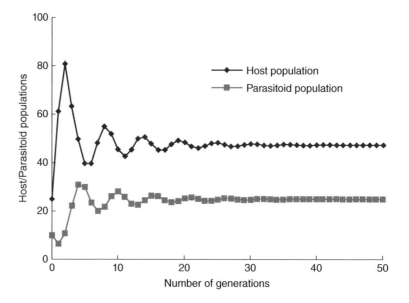

Fig. 10.27 Host/ parasitoid interaction with a searching efficiency, $a = 0.03$. Result is a stable point.

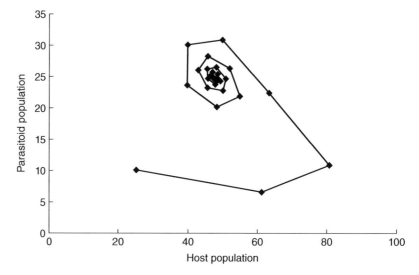

among age classes within a species. Readers interested in a more theoretical treatment should consult Case (2000) or Pastor (2008).

10.10 **Field studies**

10.10.1 *Fox/prey interactions in Sweden*

The role of predators in controlling bird and mammal populations has had a controversial history. As discussed earlier, based on his studies of rodent populations in England, and after reviewing data on snowshoe hare and lynx populations, Charles Elton (1924) believed that predator-prey interactions tend to be oscillatory (display a stable limit cycle) and that predators often control prey populations. At one point Peterson et al. (1984), based on their experiences with the wolf and moose populations on Isle Royal in Michigan seemed to have agreed. On the other hand, Errington (1946) and his followers thought that predators mostly took a "doomed surplus" of the prey population.

However, for 40 years following Errington's work, there had been few studies in which vertebrate predators were manipulated to discover their effect on vertebrate prey (but see NRC 1997).

Lindstrom et al. (1994) published data based on a "natural experiment" in which a Scandinavian red fox (*Vulpes vulpes*) population underwent a severe sarcoptic mange (*Sarcoptes scabiei*) epidemic. The mange mite appeared in Sweden in 1975 and had a dramatic effect on fox populations throughout the 1980s. The mite caused hair loss, skin deterioration and death, and the fox populations declined by 70%. This allowed Lindstrom et al. (1994) to survey the effects on the small game community of a dramatic drop in a top carnivore population.

The "alternate prey" hypothesis states that certain prey species are cyclic (3–4 years in voles and lemmings, 10 years in snowshoe hares, e.g.) and are mostly uncontrolled by predation. Such species are little affected by predation and the oscillations are likely due to interactions with the vegetation (Turchin and Batzli 2001), while other species are controlled by predation. In years when these cyclic prey numbers are in the low phase of their cycle, predators switch to "alternate prey." These alternate prey species *are usually limited* by predation especially during the years when the cyclic prey species are in the low phase. In Scandinavia, Lindstrom et al. (1994) proposed that while voles are largely uncontrolled, hare, grouse, and other small mammals and birds are alternate prey species subject to control by predation

Lindstrom et al. (1994), therefore, examined the effect of the reduced fox populations on voles, hares, grouse and red deer. Methods were as follows: (1) Rodent populations were surveyed by snap trapping; (2) the investigators checked red fox reproduction by entering dens and determining litter sizes; (3) populations of hare, grouse, and deer were checked by the pellet count method; (4) bird populations were estimated by using imitation birdcalls and recording responses; (5) hunters were given questionnaires concerning populations of all of the above and hunting records were consulted. Fawn/doe ratios were estimated from "incidental observations."

In the 1970s, previous to the mange introduction, red fox litter sizes followed the vole densities, increasing during the vole peaks. The fox population declined during the mange epidemic and then recovered after 1990 to the level before the mange appeared. Vole populations were not affected by fox numbers. They continued their normal cycles before, during, and after the fox mange epidemic. Lindstrom et al. hypothesized that these cycles were in response to variations in plant productivity and snow depth. Mountain hare and grouse populations went up 40–100% during the fox mange epidemic and their populations dropped as the fox populations recovered.

Previous to the mange epidemic, black grouse populations peaked one year after the peak of vole density. As foxes concentrated on consuming voles, predation on the hares dropped, and hare numbers increased accordingly. European hare and black grouse populations followed the same patterns, consistent with the alternate prey hypothesis.

For Roe deer the fox population affects the mortality of fawns. After the mange epidemic the number of fawns per doe increased 30% and the average deer density went up 64%. However, the production of fawns is also correlated with previous year's winter weather. With a delay of 1.5 years, snow depth explains 48% of the variation in fawn/doe ratios. Roe deer usually go into estrous in the second summer of life and, with delayed implantation, give birth at age two. A fawn undergoing a hard winter, however, would delay puberty by one year. Thus, there are fewer births 1.5 years after a hard winter, introducing a time lag into the deer population response. Still, fox density did have a significant effect on fawn production, even accounting for winter weather.

Summarizing: (1) All prey populations except voles increased in density as the mange struck the fox population. All but roe deer returned to previous population levels as the fox population returned to normal. (2) Consistent with the alternate prey hypothesis, whereas voles are not limited by fox predation, hare and grouse are. (3) Red fox is a "keystone" species in structuring the small game community in Scandinavia. It conveys the 3–4 year vole cycle to the hare and grouse populations. It shows both a functional response and a numerical response, since vole populations affect fox litter size. (4) The mange mite, since it affects red fox populations so drastically, could also be called a keystone species. (5) The role of humans cannot be ignored. First, by exterminating wolves from this area, the role of the fox has been enhanced. Second, by opening up the forest through clear-cutting practices, man has also created excellent fox habitat. In a closed forest the role of the fox is likely to be of much less significance.

10.10.2 Are population cycles caused by predation?

Other work, however, seems to indicate that vole cycles are caused by predation. In the paper by Lindstrom et al. (1994), although red fox predation was temporarily eliminated, weasel-type (Mustelidae) and avian predators were not controlled. Recent research suggests that population cycling by prey species is only eliminated when all potential predators or parasites are eliminated (May 1999).

For example, Hudson et al. (1998) studied population cycles in the red grouse (*Lagopus lagopus*). Red grouse have cycles of population abundance with an average period of 7 years. Moreover these cycles are synchronized over large portions of Northern England (Cattadori et al. 2005). These cycles are of special interest because grouse are a traditional English/Scottish shooting bird. About 75% of the British populations of red grouse undergo these cycles. Hypotheses for causation included vertebrate predators, food supplies, territory size, or parasite infections. According to Hudson et al. (1998), the answer is that the cycles are caused by parasites and the synchronization is set by climatic conditions affecting the transmission of the parasite (Cattadori et al. 2005).

The parasite is a nematode (*Trichostrongylus tenuis*). Hudson et al. (1998) began a study of six grouse populations, in which they predicted cyclic troughs in 1989 and 1993. In two of these populations birds were caught and treated with an antihelminthic medicine that eliminated the parasite. These treatments were in advance of the 1989 and 1993 predicted crashes. Two of the populations were not treated, and two of the populations were treated prior to 1993, but not prior to 1989.

The results were remarkable (May 1999). In the two untreated populations the predicted crashes occurred, on schedule, in 1989 and 1993. The magnitude of these crashes is a drop of more than 3 orders of magnitude. In the two populations treated twice, one population remained steady throughout the period 1988–1996, while the second showed a small drop (threefold) in 1993. The populations treated only in 1989 had a major crash in 1993, but remained relatively steady in 1989. For this vaccination program to work, however, at least 20% of the birds had to be injected. This is based on models of epidemiology, discussed in the previous chapter. These results have led to speculation as to whether removal of parasites could prevent population cycles in other species such as the mountain hare (*Lepus timidus*) in the United Kingdom (Townsend et al. 2009).

In a second example, Korpimaki and Norrdahl (1998) studied two species of field voles (*Microtus agrestis* and *M. rossiaemeridionalis*) and a species of bank vole (*Clethrionomys glareolus*) that show persistent three-year cycles in Finland. Korpimaki and Norrdahl set up six large study areas of 2–3 km². They reduced the numbers of weasels and stoats, the major Mustelid predators. They also removed kestrels and owls, the major avian predators. In 1992, prior to an anticipated population crash they were successful in removing most of the weasels but not the stoats, and they did not attempt to remove the birds of prey. In 1995, however, they were able to remove all of the predators. None of the predators were removed in the control areas. In 1992, with only the weasels removed, the vole populations crashed, as did the control populations in both 1992 and 1995. However, in 1995, when all the vertebrate predators were removed, the vole populations remained steady without a crash. In earlier work, Noordahl and Korpimaki (1995, 1998) had removed only avian predators. Here again, the voles crashed in both manipulated and control plots. We can conclude that vole cycles are indeed caused by predators; to eliminate cycles all of the predators must be removed. Removal of only a subset of the predator community leaves other predators to drive the cycle.

Turchin et al. (2000) drew a large distinction between vole and lemming populations. They agree that voles, such as (*Microtus agrestis*), are limited by their predators. They concluded, however, that lemming populations in Norway and other regions of Scandinavia are controlled by their food supply. They make the case that because voles feed on grasses and other herbs with the ability to quickly recover after defoliation by mobilizing underground food reserves, the vole-vegetation interaction is highly stable. According to Turchin et al. lemmings are primarily moss-eaters. Since mosses grow very slowly, the inherent time lag is highly destabilizing. Lemmings tend to deplete their forage in the arctic and alpine habitats where they live before predator populations begin to affect their dynamics. Lemmings are highly mobile during these population peaks, leading to the myths about lemming suicidal behavior (Chitty 1996). Thus, although both voles and lemmings undergo periodic oscillations, the cycles are driven by different ecological mechanisms, a predator-prey interaction in one case (voles) and a food-herbivore interaction in the other case (lemmings). This is a classic top-down (voles) versus bottom-up (lemmings) population control paradigm.

All of these works stress that cyclic prey species are actually preyed upon by a community of predators, again reminding us of the inadequacy of a single predator-single prey model. True cycles appear to be confined to northern temperate and sub-arctic ecosystems and are unknown in the tropical latitudes or southern hemisphere (Sinclair and Gosline 1997). Furthermore, it is widely believed that the amplitude of these cycles decreases at lower latitudes. However a review of 700 long-term (25+ years) animal population studies found that this expected latitudinal gradient in population cycles can only be confirmed for North American carnivores, British lagomorphs and Fennoscandian rodent populations (Kendall et al. 1998). The data base used by Kendall et al. (1998) did not include studies from the tropics or the southern hemisphere. Nevertheless they found periodicity in 33% of the mammal and 43% of the fish populations reviewed. Periodicity was less common among bird (13%) and insect (16%) populations, but was particularly common among grouse. Although we will discuss interactions between herbivores and plants in the next chapter, we note here that at least 18 species of forest Lepidoptera in North America and Europe have population cycles with average periodicities of 8–11, although outbreak periodicity is usually irregular (Myers 1988). Since 30% of all populations in the review by Kendall et al. (1998) showed regular cycles, this phenomenon cannot be dismissed.

One reason that cycles may be more common at high latitudes among mammals is the relative simplicity of the predator/prey community. For example, when Gilg et al. (2003) studied lemming cycles in the high Arctic tundra of Greenland, they found one major prey species, the collard lemming, and four predator species. Three species, as shown in Figure 10.11, were generalist predators (snowy owl, arctic fox, and long-tailed skua) and one was a specialist predator, the stoat (*Mustela erminea*), a member of the weasel (Mustelidae) family. Mustelids are prolific predators and prolific breeders, capable of a very strong numerical response. In their study, Gilg et al. (2003) have concluded that the lemming cycle in Greenland is driven by predation and there is no evidence of food or space (nest site) limitation. The specialist predator, the stoat, produces the cycle. There is a one-year delay in the numerical response of the stoat, resulting in a cyclical dynamic between lemmings and stoats. But the authors emphasize that generalist predators are necessary to stabilize the lemming-stoat interaction. The three generalist predators focus on lemmings only when the lemming population is very dense. As seen in Figure 10.11, all three have strong functional responses to lemming density. Snowy owls and arctic foxes, in addition to stoats, also display a strong numerical response. The generalist predators are thought to be necessary to slow down the growth rate of the lemming at its highest densities, thereby allowing the specialist predator to catch up and begin driving the lemming population downward.

Some Canadian lemming populations, however, do not cycle (Reid et al. 1997). The collard lemmings at Pearce Point in the Northwest Territories are preyed upon by a specialist predator, rough-legged hawks, a semi-specialist predator (red fox), and by several generalist predators (golden eagle, grizzly bear, arctic ground squirrel, peregrine falcon, and gyrfalcon). The interactions are governed by differences between summer and winter predation rates. The major stabilizing factor, however, is the presence of ground squirrels which become the primarily prey or principal alternative prey for the predators when lemming densities are low. Not only do ground squirrels provide a prey base to keep a diverse array of predators in the ecosystem, but they also prey on lemmings themselves. Without ground squirrels, Reid et al. (1997) hypothesize the loss of three of the generalist predator species. The complex guild of generalist predators plus the ground squirrels found in this ecosystem never allow the lemmings to go through an irruption.

10.10.3 *Snowshoe hare cycles*

As discussed earlier, snowshoe hare populations undergo 10 year cycles in North America, and based on pelts returned to the Hudson Bay Company, Charles Elton proposed that lynx controlled these hare populations. In spite of the popularity of this set of data, Gilpin (1973) pointed out long ago that the hare–lynx data might well by an artifact of the efforts of the trappers.

A more reliable set of data on this interaction has been produced by Krebs et al. (1995, 2001), who set up 1 km² plots in the boreal forests of the Yukon in Canada. Fertilizer was added to promote plant growth in some plots, and in some plots predators were excluded. To summarize this very complex study, they found that both food and predation help drive the snowshoe hare cycle, and that the two effects are likely linked. Although Krebs et al. (2001) emphasize that "the hare cycle is not driven primarily by plant–herbivore interactions," food limitation increased predation by forcing hares to search more extensively for food, by reducing their health, and by making them less likely to escape predation. This result reinforces the predator-sensitive food hypothesis described below for wildebeest (Sinclair and Arcese 1995).

When radio-collars were placed on hares, Krebs et al. (2001) found that 95% of the hares died as a result of predation. In the Yukon, predators included lynx, coyotes, goshawks and great horned owls. In this study, Krebs et al. (2001) could not identify a predominant role for any one predator species. As phrased by Krebs et al. (2001) "the hare cycle is not strictly a lynx-hare cycle, as many textbooks claim." The entire community of predators drives the hare cycles, not just lynx. In fact, in areas such as the Anticosti Island in Eastern Canada, where no lynx are found, the hare cycles continue.

Another interesting twist is that when snowshoe hares are in a down portion of their cycle, the predatory species turn on each other (O'Donoghue et al. 1995). These data also come from the Yukon, where hare populations cycle every 8–11 years. Prey species included red squirrels and voles. The primary predators included lynx, coyotes, great horned owls and goshawks, as well as red fox, wolverines and wolves. Since the primary predators appear to go through population cycles related to those of the hare, O'Donoughue et al. put radio-collars on lynxes and coyotes and radio-tagged owls, hawks and other birds of prey in order to determine their fate.

Snowshoe hare populations peaked at 450 per square mile in 1989–1990 and declined to 7 per square mile by the spring of 1992. The lynx population declined from 60 to 15 per square mile. Did the lynx leave the area? If they died what were the causes of death?

Of 15 radio-collared lynx in 1991–1992, by the spring of 1993 only seven remained. Six individuals emigrated and two deaths were recorded. Collars were found as far away as 500 miles in Alaska, British Columbia and the Northwest Territories.

In 1992–1993, of nine radio-collared lynx, all had died by April. One lynx was found healthy and well-fed, but had been killed by a wolf. A female lynx was killed by another lynx! One young male starved. A male and female pair was migrating into the mountains when a wolverine killed the female. The male was killed by either a wolf or a wolverine based on evidence from blood and tracks in the snow. The investigators actually witnessed a lynx being killed by a coyote. Other scientists have also reported witnessing lynx being killed by other lynx or by wolverines. All of this does not happen when hare are abundant. Furthermore, when hare populations are low, lynx themselves become very aggressive, routinely killing red fox.

Birds of prey followed the same pattern. In 1989 and 1990, when hare were abundant, 10 of 11 goshawk nests succeeded in fledging their young. In 1991, 50% nest failure was reported. All failed because of predation, evidently by great horned owls. The owls ate the adult female birds as well as their young.

Various authors have emphasized external factors such as predation, food shortages, fire and even sun spots as causative agents of the snowshoe hare cycle. Sinclair et al. (2003), however, raise the issue of intrinsic differences via different morphs, which could be different phenotypes or even genotypes, proliferating at different phases of the cycle. The authors define intrinsic as different sub-populations with differential sensitivity to crowding. These differences would be manifested through differences in birth and mortality rates at low versus high density. Sinclair et al. (2003) predicted that hares bred from the high phase of a cycle would have lower reproduction and/or survival compared to animals bred from the low phase. In their experiment they obtained hares from both high and low phases of the cycle and brought them to the

Animal Care Facility at the University of British Columbia. The results were consistent with the prediction that animals the high and low phases differ in reproduction and survival characteristics. The colonies derived from high-phase colonies declined to extinction while the low-phase colonies persisted until the experiment was terminated. For example, both high-phase colonies lost reproductive capacity (in terms of young born per year) while the low-phase animals retained theirs. There were also differences between the two populations in terms of longevity and senescence. Without going into further details here, the results are consistent with the hypothesis that population cycles, at least in snowshoe hares, are driven by intrinsic population mechanisms as well as by extrinsic factors.

In conclusion, snowshoe hare cycles are not simply driven by the lynx population as has been perpetuated in popular media as well as in text books. The vegetation, the entire community of predators and factors intrinsic to the population of hares are all involved in the cycles. Furthermore, these trophic interactions involve the community of predators itself. When prey populations are scarce, predatory animals turn on each other, and mortality of the predators due to starvation was, in fact, relatively rare.

10.10.4
Moose–wolf
interactions

Another interaction that receives regular treatment in the media and in textbooks is the moose–wolf relationship on Isle Royale. Isle Royale is a 574 km^2 island in Lake Superior that is a protected National Park. Moose (*Alces alces*) populations have been present on the island for about 100 years (Pastor et al. 1988). Shortly after 1900 moose arrived on the island from Minnesota or Canada. They found abundant food and no major predators. The Isle Royal ecosystem is, in fact, relatively simple in terms of large herbivores and carnivores. As described by Smith et al. (2003), when moose arrived, the only mammalian carnivores were coyote and red fox, and the large herbivores included red squirrel, snowshoe hare and beaver. Ravens were also present. The moose population rapidly increased to some 3000 individuals by the early1930s (Murie 1934). As they over-browsed their food supply the population declined. Several forest fires in the 1930s resulted in the regeneration of aspen and paper birch, which allowed the moose population to increase again since these are preferred foods. In the late 1940s, wolves (*Canis lupus*) arrived and evidently quickly extinguished the coyote population, maintaining a relatively simple food web. According to Mech (1966) predation limited the moose population below its food supply. More recent research (Peterson 1999), however, indicates that both vegetation and wolves play a role in moose population dynamics.

Between 1948 and 1950 four exclosures (each 101 m^2 in size) were set up to evaluate the effect of moose on the vegetation and other ecosystem properties. Pastor et al. (1988) concluded that intensive moose browsing lead to fewer deciduous trees and a forest dominated by spruce (*Picea glauca*) and balsam fir (*Abies balsamea*), which are less palatable to moose. In addition, lower levels of nitrogen remain in the soil. The net result was an ecosystem less able to support moose populations, since moose do less well on this diet.

Although there seems to be some agreement that moose populations are affected by wolves, the question remains, how much of an effect? Peterson et al. (1984), based on about twenty years of data on the wolf population at Isle Royal, concluded that moose and wolves would go through population cycles typical of predator–prey interactions involving rodent and hare populations. However, since the body mass (M) of moose is larger, they concluded that moose–wolf interactions would cycle with considerably longer period lengths. In fact, after noting that the intrinsic rate of increase of mammals scales as $M^{-0.26}$, they proposed that vertebrates cycle as the fourth root of the body mass, $M^{0.25}$. From this they predicted that moose populations should cycle with a period length of 38 ± 13 years. As we shall see, actual events were different.

Making predictions about prey and predator populations based on allometric relationships or on prey-induced vegetation changes also ignores the role of climate. Mech et al. (1987) showed that Isle Royale moose and North-eastern Minnesota deer populations were both significantly affected by the snow accumulation of the previous winter. In severe winters moose populations were not able to find browse and their physical condition deteriorated. Many of the moose and deer that were killed by wolves would have starved to death anyway. Therefore, although wolf predation was a direct mortality agent, Mech et al. (1987) believed it was secondary to winter weather in influencing deer and moose populations. In addition, snow accumulation during previous winters affects maternal nutrition to such a degree that fecundity and/or calf survivorship during the next growing seasons is seriously influenced.

The wolf population on Isle Royale decreased from 1981 through the mid-1990s, because of a canine parvovirus, probably introduced from the mainland on the hiking boots of visitors. During this period the moose population increased until it reached a record level in 1995 of 2,400. Before the wolf population could recover (there were approximately 50 in 1980 and 24 in 1995), the moose population crashed. Almost 80% died, mostly from starvation, but a severe winter tick infestation contributed to the crash (Peterson et al. 1998). Peterson (1999) concluded that although the moose population density is influenced by wolves, the population level is ultimately set by available food.

In any case, the end of this story may be nigh. By 2012–2013 the wolf population had declined to two packs rather than the historic three or four and only eight wolves were found during the annual count in January, 2013 (Mlot 2013). The cause seems to be the long term result of inbreeding among the wolves.

Smith et al. (2003) have pointed out how much more complex the food web is at Yellowstone National Park in Wyoming where wolves have recently been introduced, than is the case at Isle Royale. Whereas the only mammalian predators at Isle Royale are wolf and red fox, at Yellowstone coyote, mountain lion, grizzly bear, and black bear are found along with wolves. Human hunting on elk must also be factored in. Avian predators at Yellowstone include bald eagles, golden eagles, magpies and ravens, whereas only ravens are found on Isle Royale. Additional large prey species, beside moose, found at Yellowstone are: pronghorn antelopes, bighorn sheep, mule deer, elk and bison. As of 2003, only 26 instances of wolves killing moose have been recorded since the wolves were reintroduced at Yellowstone. Wolves mainly prey on elk, and are predicted to reduce elk populations. However, the elk herds had only declined by the amount expected (Smith et al. 2003; Ripple and Beschta 2003).

After reviewing 27 studies on moose–wolf interactions, Messier (1994) was able to generate both functional and numerical for wolves as a function of moose density. He concluded that when wolves are the single predator, a moose population will stabilize at 1.3 moose/km^2, whereas the equilibrium density with no predators is 2.0 moose/km^2. His analysis is consistent with a model proposing that under these circumstances, moose populations are regulated at low densities due to density-dependent wolf predation.

Sinclear and Arcese (1995) studied the interaction between food supply and predation in the regulation of large herbivore populations on the Serengeti Plain in East Africa. They distinguished among three hypotheses:

1. The *predation-sensitive food (PSF)* hypothesis states that both food and predation limit prey populations. As food becomes limiting, animals take greater risks to obtain food and become victims of predation.
2. The *predator regulation* hypothesis states that predators hold prey populations well below starvation levels.
3. The *surplus predation* hypothesis states that predators kill only those prey individuals that are excluded from optimal habitat and are already dying of starvation. This is similar to Errington's (1946) ideas about the "doomed surplus" of muskrat populations.

These three hypotheses were tested by examining the body condition of Serengeti wildebeest (*Connochaetes taurinus*) over a 24-year period. Two phases of population growth were examined: (1) 1968–1973 when food was abundant and prey populations were increasing; and (2) 1977–1991 when the wildebeest population was stationary and partially regulated by competition for food. Sinclear and Arcese examined live animals, predation kills, and non-predation deaths. Body condition was measured by an examination of bone marrow, the last reservoir of fat in these animals.

The predator regulation hypothesis predicts that bone marrow condition will be similar in predator killed and live samples, while the surplus predation hypothesis predicts that bone marrow will be similar in predator killed and non-predator deaths.

The PSF hypothesis predicts that bone marrow condition of animals killed by predators should be: (1) poorer than that of live animals; (2) better than that of the animals who die of causes, such as disease, unrelated to predation; (3) better when food is limiting than when it is abundant (because when food is abundant, predators only kill sick or young animals).

Analyses of the bone marrow from animals dying due to predation or from non-predation related causes showed that these animals were in poorer health than the live population. In both the increasing and stationary phases, the animals dying from predation were in better condition than the animals dying from non-predation causes. These results are consistent with the PSF hypothesis, and inconsistent with the other two hypotheses. Lions and hyenas killed animals in similar condition, but lions took younger animals.

The results suggest that: (1) body condition affects vulnerability of individual wildebeest to predation, and (2) predation jointly limits the population along with intraspecific competition for food resources.

Sinclair et al. (2003) have also analyzed the community wide patterns of predation in the Serengeti ecosystems of Tanzania and Kenya. Twenty eight species of ungulates and ten species of carnivores inhabit these areas, consisting mostly of open grassland (savannah). In any one habitat as many as seven species of canid and felid carnivores are present. The predators range from 8 kg (Golden jackal, *Canis aureus*) to 150 kg (lions, *Panthera leo*), while prey sizes range from small gazelles (*Gazella sp.*), which weigh 18–20 kg, to elephants (*Loxodonta africana*), rhinoceros (*Diceros bicornis*), and hippopotamus (*Hippopotamus amphibious*), which come in at over 3000 kg. Long term studies of the causes of ungulate mortality show that the proportion of adult mortality due to predation is above 80% in the smallest species such as orbi (*Ourebia ourebia*), impala (*Aepyceros melampus*), topi (*Damaliscus lunatus*), and wildebeest (*Connochaetes taurinus*). By contrast, in a heavier species such as zebra (*Equus quagga*), adult mortality due to predation is 70%. There is a "threshold" in body size of about 150 kg, after which deaths due to predation decline substantially. Only about 23% of adult buffalo and 5% of adult giraffe (*Giraffa camelopardalis*) mortality is caused by predation; rhinos, hippos and elephants (the "big three") suffer virtually no adult predation. While the smallest ungulates are preyed upon by as many as seven different predators, the number of potential predators falls off linearly as a function of the log of the herbivore weight to zero for the big three. Thus the smallest ungulates (less than 150 kg) are limited by the diverse array of predatory species, while the large ungulates are basically food limited. A "natural experiment" confirmed this generalization. In the northern Serengeti poaching and poisoning eliminated most of the carnivores, including lions, hyenas, and jackals, whereas in the nearby Mara Reserve the predator community remained intact. During the years when predator populations were eliminated five of the species below 150 kg in weight increased their populations conspicuously as compared to the Mara Reserve. Once the predators returned, their populations declined. By comparison, the 800 kg giraffes did not increase in the predator removal area. Sinclair et al. (2003) concluded that the mammalian herbivore populations in the Serengeti ecosystem are subject to top-down (predation) or bottom-up (food limitation) processes depending on their size.

10.11 The dangers of a predatory lifestyle

Ross (1994) has described the fate of solitary predators, such as cougars. Cougars, also known as pumas or mountain lions, are almost unique among solitary predators in that they consistently seek prey larger than themselves. African lions, hyenas, wild dogs, and wolves all practice cooperative hunting when attacking large prey. Most solitary hunters, such as weasels or foxes, generally prey on creatures smaller than themselves. Among the cats, leopards, cheetahs, jaguars, and lynxes also usually follow this pattern of attacking smaller animals.

Attacking large prey has the drawback that it can sometimes prove fatal for the hunter. Among Alberta cougars, there are some dramatic examples of what can go wrong in a violent struggle with a large prey animal. Aside from human hunters, these struggles are the main cause of death in this cougar population.

One young adult female suffered a broken back when the mule deer she was riding down a steep slope slammed into a pine tree. Another female was speared by a sharp branch when the elk that she eventually killed, tried to shake her loose from its throat. An adult male cougar that attacked a bighorn sheep lost his footing in the struggle, and both fell to their death over a 90-foot cliff.

10.12 Escape from predation

Let us now change our focus to an evaluation of predator–prey relationships from the perspective of the lower trophic level. What mechanisms have evolved allowing organisms to escape predation? These means of escape can be classified into the following four general categories: (1) Escape in time; (2) Escape in space; (3) Escape through behavior; (4) Escape through physical/chemical mechanisms.

10.12.1 *Escape in time or predator satiation*

Life cycles have evolved so as to minimize predation or herbivory. Usually one part of the life cycle is highly vulnerable to predation, such as newly produced offspring or adult stages involved in reproductive behavior. Examples include mast-fruiting in oaks, periodical cicadas (Williams et al. 1993; Karban 1997), bamboos that reproduce once a century (Janzen 1976), century plants, Mayflies, and the mating flights of ants and termites. In all of these cases reproduction is synchronized in a local population or in many populations over a large geographical area. So many potential prey organisms are produced in such a short time period, or in such an unpredictable manner (e.g., periodical cicadas reproduce once every 13 or 17 years), that potential predators can eat their fill (satiation) without seriously reducing the prey population. Since predator populations cannot build up due to the brevity of the vulnerable stage and/or the unreliability of the prey population, the vast majority of seeds, seedlings, larvae or other vulnerable stages "escape" predation.

10.12.2 *Escape in space*

In this case, the prey population is relatively rare and disperses readily, so that it is not abundant in any local habitat. Since the prey population is so rare, the predator may never develop a search image for it. Furthermore, predators that specialize on this species cannot become very common themselves, given the limited number of opportunities to feed. Such predators, once they find a prey population, however, may wipe it out completely. Although the prey population goes locally extinct, if it disperses rapidly enough it will not go extinct over a larger area. Such an interaction is consistent with the definition of a metapopulation that survives local extinctions as long as there are other local populations in the region. Examples of such interactions are the herbivorous and predatory mites studied in the laboratory by Huffaker (1958) and the *Opuntia stricta-Cactoblastis cactorum* relationship in Australia (Dodd 1940).

10.12.3 *Behavioral means*

Prey species have a variety of means by which they confuse predators. Some of these are combined with coloration schemes (see below). It is well known that among birds, adults will attempt to draw the attention of predators away from nests or vulnerable offspring by attacking and mobbing them, creating noisy disturbances, or pretending to be injured. Noteworthy also are herding (among mammals) and schooling (among fish) behaviors. In order to hunt effectively, a predator usually needs to focus on a given individual. Animals moving in herds or schools create confusion or disrupt the path of a predator that has focused his/her attention on a particular prey individual. In other cases, potential prey attempt to intimidate their predators through aggressive displays. The frilled lizard (*Chlamydosaurus kingii*) exposes a skin flap (a "frill") around its neck, opens its mouth widely, and makes aggressive lunges at a potential predator in an attempt to make it look large and fierce. Several species of fish and toads are known to fill themselves with air in order to look larger to potential predators.

10.12.4 *Physical/chemical mechanisms*

Physical toughness
In animals, spines, scales, quills and bony plates are used for protection. Among plants there exist spines (modified leaves), thorns (modified stems), epidermal hairs, thick cuticles, heavy cork layers, lignified or silicified cells in the epidermis or cortex, barbed and sticky trichomes (an outgrowth of the epidermis), and the general use of fibers, lignin, crystals, silicon, and cellulose itself to make plant tissues difficult to eat and to digest. When combined with noxious chemical defenses ("secondary compounds") produced by plants, the result is that most plant material including leaves, bark, stems, and roots constitute an unrewarding diet for most animals.

Resistance to chemical degradation
Cellulose is the most abundant compound in plants. Approximately 10 billion tons of carbon is transformed into cellulose per year (Albersheim 1965). Cellulose is a

linear polysaccharide, and the molecules are juxtaposed so as to form linear crystals or microfibrils. As such, cellulose is highly resistant to chemical and enzymatic degradation. Numerous bacteria, one termite species, and a few protistans have been shown to manufacture an enzyme, cellulase, capable of partially degrading cellulose. But cellulase has rarely been evolved in the animal kingdom. Actually two enzymes are necessary to degrade cellulose, cellulase, which yields the disaccharide cellobiose, and cellobiase that hydrolyzes the dimer to yield free hexose.

Wood contains large amounts of lignin. Conducting vessels and the fibers that strengthen plants are mostly lignified cell walls. Lignin is a close second to cellulose as the most abundant compound in nature (perhaps 60% of the amount of cellulose). Lignin is a polymer that contains a large number of aromatic side chains and is quite resistant to chemical degradation and almost impervious to enzymatic digestion. In fact, lignin is probably the most resistant compound to degradation by enzymes found in nature (Albersheim 1965).

Coloration

(a) *Cryptic or Camouflage Coloration*. Cryptic patterns, that is, blending in with the environment, are commonly found among arthropods, especially insects, and all groups of vertebrates. Even cephalopods such as octopuses have evolved cryptic color patterns. Among insects, larval and adult stages have evolved patterns that appear to be sticks, living leaves, dead leaves, flowers (Greene 1989), bark, lichens, and even bird droppings.

(b) *Confusing Patterns*. Several species of insects have evolved false "heads" or other strange appendages on their rear ends. A possible explanation for such traits is that they distract a predator, confuse it, and the result is an attack that is much less likely to be fatal than one directly on the head.

(c) *Startle Patterns*. Several tropical butterfly species have color patterns on the under-wings that are said to resemble owl's eyes, and many other Lepidopteran species have evolved concentric rings resembling vertebrate eyes (Blest 1957). When disturbed by a bird, the butterfly flashes these false eyes with the result that the bird may temporarily back away. Other insect larvae have evolved morphological patterns and/or colors that make them resemble snakes (Pough 1988).

(d) *Flash Patterns*. In this situation, a cryptic animal flashes a bright color pattern, which is normally hidden on the underside of the animal or by some other means. For example, white-tailed deer flash their white tails when alarmed. The white patch is highly visible as it runs away from its predator. Subsequently, the deer usually takes a sharp angle turn; at this point the white tail is no longer visible and the deer, which is highly cryptic otherwise, disappears from view. Having focused on the white tail, predators have a hard time visually locating where the deer turned and have trouble tracking it. Other examples include frogs, which are a cryptic green color except for a yellow underbelly. If discovered the frog jumps and the potential predator is "flashed" the bright yellow. This yellow is concealed as the frog settles again into the green vegetation.

(e) *Aposematic Patterns*. Here we are interested in the interaction of color patterns with chemical defenses (discussed below). When an animal is dangerous, toxic, poisonous, or odoriferous, it is adaptive if the potential predator has been warned of the consequences of dealing with such an animal. Thus, such chemically protected animals have usually evolved highly colorful patterns (red, orange or other bright colors) that warn predators with color vision (birds, primates etc.). Alternatively, they have evolved alternating stripes or spots of light and dark colors (e.g., striped and spotted skunks, and Hymenoptera such as bees and wasps) to warn animals that lack color vision. These are known as aposematic or warning color patterns.

Chemical defenses

Plants produce a multitude of compounds that are not part of the primary metabolism and are therefore called "secondary compounds." We will discuss these in the next chapter. At this point we will simply note that the great diversity of physical and chemical defenses among plants is one line of evidence suggesting that herbivores have greatly influenced the evolution of plants. The chemical diversity of plants has evidently resulted in a similar diversity in chemical defenses among animals. While some animals are able to synthesize their own chemical defenses, many acquire them from the plants they consume. Such animals have not only evolved mechanisms to detoxify or tolerate the chemicals found in their diet, but they are also able to sequester or modify these chemicals so as to provide them with protection from their own predators. Some animals have active chemical defenses, for example skunks, bombardier beetles (Coleoptera) (Eisner and Aneshansley 1982) and stinkbugs (Hemiptera). These animals are able to spray their potential predators with noxious compounds. Other animals have a more passive defense in that they taste bitter or are toxic if eaten.

Mimicry

The combination of poisonous host plants and chemically defended herbivores with aposematic coloration set the stage for the evolution of mimicry. In Batesian mimicry, the color pattern of an unpalatable model species is copied or mimicked by another species that is not itself chemically protected. The classic case of the monarch (*Danaus plexippus*) and the viceroy butterflies (*Limenitis archippus*) was described by Brower and his colleagues (Brower and Brower 1964, Brower, Brower and Corvino 1967). The larvae of monarch butterflies feed on milkweed (*Asclepius spp.*) plants that contain cardiac glycosides in a milky sap. The larvae are able to feed upon the glycosides without ill effect and store them in their tissues. The adult monarchs have an aposematic color pattern that is mimicked by the viceroy. In the classic interpretation, viceroys are palatable to birds (but see below). When a bird ingests a monarch, however, it often reacts violently to the cardiac glycosides, which cause vomiting and seizures. Once a bird has this experience it never eats either a monarch or a viceroy again.

In another type of mimicry (known as Muellerian mimicry) a group of species (again, often butterflies) all are chemically protected and converge on a common color pattern. The theory is that once a predator learns that a butterfly with this pattern is poisonous, it generalizes the lesson to all of the species that have this color pattern. Recent experiments have suggested that under some circumstances the viceroy is chemically protected, and that the viceroy-monarch story may actually be one of Muellerian rather than Batesian mimicry (Ritland and Brower 1991).

Although mimicry is most famous among members of the Lepidoptera, it should be noted that many other groups of insects can be mimics. Many flies (Diptera), for example, mimic bees and wasps (Hymenoptera) (Pough 1988).

Although discussions of mimicry usually focus on invertebrates, Pough (1988) described many cases of mimicry among vertebrates. Among amphibians, the unpalatable salamander, *Plethodon jordani*, is mimicked by palatable species of *Desmognathus*, and the noxious red salamanders *Pseudotriton ruber* and *P. montanus* are probably Mullerian mimics. Among frogs, the terrestrial, diurnal frogs of the genus *Phyllobates* (Dendrobatidae) are protected by curare secretions in the skin. There is evidence that several species of frogs in the family Leptodactylidae are Batesian mimics of *Phyllobates*.

Among reptiles, the most famous examples involve several species of snakes in the family Colubridae that are mimics of the coral snakes (genus *Micrurus*) (Greene and McDiarmid 1981). Pough (1988) suggests that snakes, especially venomous snakes, are such extremely noxious models that predators have generalized their characters very broadly. Additionally, if the model is sufficiently noxious, selection may act to produce innate avoidance of the model. Because of the extreme toxicity of poisonous snakebites, there is no opportunity for the trial-and-error learning by the predator that is the basis for mimicry in insects or other less toxic vertebrates. Selection must therefore favor predators that display an innate avoidance of the characteristics of model species, such as coral snakes.

Several studies have shown that, indeed, several vertebrate predators do have an innate fear of the coral snake pattern. Zoo-reared coatimundis (*Nasua narica*) and javelinas (*Tayassu pecari*) were presented with rubber snakes. The coral snake pattern startled both species and individuals fled. Rubber snakes painted green were attacked (Gehlbach 1972). Smith (1975, 1977, 1980) obtained similar results using hand-reared birds. When turquoise-browed motmots (*Eumomota superciliosa*) and great kiskadees (*Pitangus sulphuratus*) were presented with wooden dowels painted in various neutral colors, both species readily attacked. Both species reacted with alarms calls and fled from dowels painted yellow with red rings. They also fled from a coral snake pattern of yellow, black and red rings. On the other hand, birds from the temperate zone, such as sparrows, blackbirds, and jays, attacked all patterns presented, including the coral snake pattern. Thus, aversion to aposematic colors does not seem to be an innate behavior of all birds. Innate aversion to the coral snake pattern is, instead, a property of those species of birds potentially exposed to coral snakes.

10.13   Summary

What have we learned about predator–prey interactions? Combining what we know from theory, laboratory and field studies, we can conclude that predator–prey interactions are stabilized by: (1) Heterogeneity of the environment; (2) non-random hunting by the predator; (3) density dependence in the prey population unrelated to the predator population; (4) a prey refuge; (5) a less efficient predator; (6) providing the prey with less than optimal resources to slow its growth rate and thereby avoiding the "paradox of enrichment"; (7) a complex of predators including enough generalists to dampen the destabilizing effects of the specialist predators; and (8) the evolution of effective physical, chemical or behavioral means by the prey species to limit predation. On the other hand, simple predator/prey communities, time lags, and specialist predators that overwhelm prey defenses tend to promote cycles or extinctions. Furthermore, although generalist predators may stabilize some ecosystems, the introduction of either a specialist or a generalist predator into a "naïve" prey community may have devastating consequences. A functional response by the predator which includes a significant handling/digestion time (energy) component can stabilize a predator–prey interaction if the prey population is kept at low densities. Once the prey population "escapes" and becomes so large that the functional response of the predator has reached the "plateau," the effect is destabilizing. Unless there is also a numerical response by the predator population, the overall predation rate would become inversely density dependent; the more prey, the lower the proportion of prey being consumed. Finally, some prey species are controlled by predators, producing a top-down effect on the community. As we will explore in the next chapter, other herbivorous prey species are controlled by plant productivity resulting in a bottom-up control of the community.

References

Albersheim, P.. 1965. Biogenesis of the cell wall. Pages 298–321 in J. Bonner and J.E. Varner, Editors. *Plant Biochemistry*. Academic Press, New York, NY.

Baldwin, N.S. 1964. Sea lamprey in the Great Lakes. *Canadian Audubon Magazine*. November–December: 142–147.

Baumhover, A.H., A.J. Graham, B.A. Bitter, D.E. Hopkins, W.D. New, F.H. Dudley, R.C. Brushland. 1955. Screw worm control through the released of sterilized flies. *Journal of Economic Entomology* 48: 462–466.

Bergstedt, R.A. and M.B. Twohey. 2007. Research to support sterile-male-release and genetic alteration techniques for sea lamprey control. *Journal of Great Lakes Research* 33: 48–69.

Blest, A.D. 1957. The function of eyespot patterns in the Lepidoptera. *Behaviour* 11: 209–256.

Botkin, D.B. 1990. *Discordant Harmonies*. Oxford University Press, Oxford, UK.

Bozarth, C.A., F. Hailer, L. Rockwood, C. Edwards, and J. Maldonado. 2011. Coyote colonization of northern Virginia and admixture with Great Lakes wolves. *Journal of Mammalogy* 92: 1070–1080.

Brower, L.P. and J.V.Z. Brower. 1964. Birds, butterflies, and plant poisons: a study in ecological chemistry. *Zoologica* 49: 137–159.

Brower, L.P., J.V.Z. Brower and J.M. Corvino. 1967. Plant poisons in a terrestrial food chain. *Proceedings of the National Academy of Sciences* 57: 1059–1066.

Case, T.J. 2000. *An Illustrated Guide to Theoretical Ecology*. Oxford University Press, Oxford, UK.

Cattadori, I.M., D.T. Haydon, and P.J. Hudson. 2005. Parasites and climate synchronize red grouse populations. *Nature* 433: 737–741.

Caughey, G. 1970. Eruption of ungulate populations, with emphasis on Himalayan Thar in New Zealand. *Ecology* 51: 51–72.

Chitty, D. 1996. *Do Lemmings Commit Suicide? Beautiful Hypotheses and Ugly Facts*. Oxford University Press, Oxford, UK.

Crawley, M.J. 1992. Population dynamics of natural enemies and their prey. Pages 40–89 in M.J. Crawley, Editor. *Natural Enemies: The Population Biology Predators, Parasites and Diseases*. Blackwell Scientific, Oxford, UK.

Dodd, A.P. 1940. *The Biological Campaign Against Prickly Pear*. Commonwealth Prickly Pear Board, Government Printer. Brisbane, Australia.

Eisner, T. and D.J. Aneshansley. 1982. Spray aiming in bombardier beetles: jet deflection by the coanda effect. *Science* 215: 83–85.

Elton, C.S. 1924. Periodic fluctuations in the numbers of animals: their causes and effects. *Journal of Experimental Biology* 2: 119–163.

Elton, C.S. and M. Nicholson. 1942. The ten-year cycle in numbers of lynx in Canada. *Journal of Animal Ecology* 11: 215–244.

Errington, P.L. 1946. Predation and vertebrate populations. *The Quarterly Review of Biology* 21: 144–177, 221–245.

Fox, C.J. 2013. *The Status of Virginia's Cooperative Coyote Damage Control Program – Fiscal Year 2012*. USDA-APHIS-Wildlife Services, Christianburg, Virginia.

Frits, T.H. 1988. The brown tree snake, *Boiga irregularis*, a threat to Pacific Islands. U.S. Fish and Wildlife. Service, *Biology Reports* 88(31): 36 pp.

Gause, G.F. 1934. *The struggle for existence*. Williams and Wilkins, Baltimore. Reprinted in 1964 by Macmillan (Hafner Press), New York, NY.

Gehlbach, F.R. 1972. Coral snake mimicry reconsidered: the strategy of self-mimicry. *Forma et Functio* 5: 311–320

Glig, O., I. Hanski, and B. Sittler. 2003. Cyclic dynamics in a simple vertebrate predator-prey community. *Science* 302: 866–868.

Gilpin, M.E. 1973. Do hares eat lynx? *American Naturalist* 107: 727–730.

Greene, Erick. 1989. A diet-induced developmental polymorphism in a caterpillar. *Science* 243: 643–635.

Greene, H.W. and R.W. McDiarmid. 1981. Coral snake mimicry: does it occur? *Science* 213: 1207–1212.

Hassell, M.P. 1976. Arthropod predator-prey systems. Pages 71–93 *in* R.M. May, Editor. *Theoretical Ecology, Principles and Applications*. W.B. Saunders, Philadelphia, PA.

Hassell, M.P. 1978. *The Dynamics of Arthropod Predator-Prey Systems*. Princeton University Press, Princeton, NJ.

Holling, C.S. 1959. The components of predation as revealed by a study of small mammal predation of the European pine sawfly. *Canadian Entomologist* 91: 293–320.

Holling, C.S. 1961. Principles of insect predation. *Annual Review of Entomology* 6: 163–182.

Holling, C.S. 1966. The functional response of invertebrate predators to prey density. *Memoirs of the Entomological Society of Canada* 48: 1–86

Holling, C.S. 1973. Resilience and stability of ecological systems. *Annual Review of Ecology and Systematics* 4: 1–24.

Hudson, P.J., A.P. Dobson, and D. Newborn. 1998. Prevention of population cycles by parasite removal. *Science* 282: 2256–2258.

Hudson, P.J. and O.N. Bjørnstad. 2003. Vole stranglers and lemming cycles. *Science* 302: 797–798.

Huff, D.E. and J.D. Varley. 1999. Natural regulation in Yellowstone National Park's northern range. *Ecological Applications* 9: 17–29.

Huffaker, C.B. 1958. Experimental studies on predation: dispersion factors and predator-prey oscillations. *Hilgardia* 27: 343–383.

Janzen, D.H. 1976. Why bamboos wait so long to flower. *Annual Review of Ecology and Systematics* 7: 347–391.

Karban, R. 1997. Evolution of prolonged development: A life table analysis for periodical cicadas. *American Naturalist* 150: 446–461.

Kauffman, M.J., J.F. Brodie, and E.S. Jules. 2010. Are solves saving Yellowstone's aspen? A landscape-level test of a behaviorally mediated trophic cascade. *Ecology* 91: 2742–2755.

Kendall, B.E., J. Prendergast, and O.N. Bjørnstad. 1998. The macroecology of population dynamics: taxonomic and biogeographic patterns in population cycles. *Ecology letters* 1: 160–164.

Korpimaki, E. and K. Norrdahl. 1998. Experimental reduction of predators reverses the crash phase of small-rodent cycles. *Ecology* 79: 2448–2455.

Krebs, C.J., S. Boutin, R. Boonstra, A.R.E. Sinclair, J.N.M. Smith, M.R.T. Dale, K. Martin, and R. Turkington. 1995. Impact of food and predation on the snowshoe hare cycle. *Science* 269: 1112–1115.

Krebs, C.J., R. Boonstra, S. Boutin, and A.R.E. Sinclair. 2001. What drives the 10-year cycle of snowshoe hares? *Bioscience* 51: 25–35.

Leopold, A. 1943. Deer irruptions. *Wisconsin Conservation Department Publication* 321: 3–11.

Leopold, A. 1949. *A Sand County Almanac and Sketches Here and There*. Reprinted 1977. Oxford University Press, Oxford, UK.

Lindstrom, E.R., H. Andrén, P. Angelstam, G. Cederlund, B. Hörnfeld, L. Jäderberg, P.A. Lemnell, B. Martinsson, K. Sköld, and J.E. Swenson. 1994. Disease reveals the predator: Sarcoptic mange, red fox predation, and prey populations. *Ecology* 75: 1042–1049.

Lotka, A.J. 1925. *Elements of Physical Biology* Reprinted 1956. Dover, New York, NY.

Luckinbill, L.L. 1973a. Coexistence in laboratory populations of *Paramecium aurelia* and its predator, *Didinium nasutum*. *Ecology* 54: 1320–1327.

Luckinbill, L.L. 1973b. Coexistence in laboratory populations of *Paramecium aurelia* and its predator, *Didinium nasutum*. *Ecology* 54: 1320–1327.

May, R.M. 1976. Models for two interacting populations. Pages 49-70 *in* R.M. May, Editor. *Theoretical Ecology, Principles and Applications*. W.B. Saunders, Philadelphia, PA.

May, R.M. 1999. Crash tests for real. *Nature* 398: 371–372.

McCullough, D.R. 1997. Irruptive behavior in Ungulates. Pages 69–98 *in* W.J. McShea, H.B. Underwood, and J.H. Rappole, Editors. *The Science of Overabundance*. Smithsonian Institution Press, Washington, DC.

Meagher, M.M. and D.B. Houston. 1998. *Yellowstone and the Biology of Time: Photographs Across a Century*. University of Oklahoma Press, Norman, OK.

Mech, L.D. 1966. *The Wolves of Isle Royale*. Fauna of the National Parks of the United States, Fauna Series, Washington, DC.

Mech, L.D., R.E. McRoberts, R.O. Peterson, and R.E. Page. 1987. Relationship of deer and moose populations to previous winter's snow. *Journal of Animal Ecology* 56: 615–625.

Messier, F. 1994. Ungulate population models with predation: a case study with the North American moose. *Ecology* 75: 478–488.

Middleton, A. 2014. Is the wolf a real American hero? New York Times Monday March 10, 2014: A21.

Mlot, C.. 2013. Are Isle Royale's wolves chasing extinction? *Science* 340: 919–921.

Murie, A.1934. *The Moose of Isle Royale*. University of Michigan Museum of Zoology Miscellaneous Publication 25, Ann Arbor, MI.

Myers, J.H. 1988. Can a general hypothesis explain population cycles of forest Lepidoptera? *Advances in Ecological Research* 18: 179–242.

National Research Council (NRC), Board on Biology Commission on Life Sciences, Committee on Management of Wolf and Bear Populations in Alaska. 1997. *Wolves, Bears and Their Prey in Alaska: Biological and Social Challenges in Wildlife Management*. National Academy Press, Washington, D.C.

Nicholson, A.J. and V.A. Bailey. 1935. The balance of animal populations, part I. *Proceedings of the Zoological Society of London* 3: 551–598.

Norrdahl, K. and E. Korpimaki. 1995. Effects of predator removal on vertebrate prey populations: birds of prey and small mammals. *Oecologia* 103: 241–248.

O'Donoghue, M., E. Hofer, and F.E. Doyle. 1995. Predator versus predator. *Natural History* 104: 6–9.

Pastor, J.. 2008. *Mathematical Ecology of Populations and Ecosystems*. Wiley and Sons, Hoboken, New Jersey.

Pastor, J., R.J. Naiman, B. Dewey, and P. McInnes. 1988. Moose, microbes, and the boreal forest. *Bioscience* 38: 770–777.

Peterson, R.O. 1999. Wolf-moose interaction on Isle Royale: The end of natural regulation? *Ecological Applications* 9: 10–16.

Peterson, R.O., R.E. Page, and K.M. Dodge. 1984. Wolves, moose, and the allometry of population cycles. *Science* 224: 1350–1352.

Peterson, R.O., N.J. Thomas, J.M. Thurber, J.A. Vucetich, and T.A. Waits. 1998. Population limitation and wolves of Isle Royale. *Journal of Mammalogy* 79: 828–841.

Porter, K.G., J. Gerritsen, and J.D. Orcutt, Jr. 1982. The effect of food concentration on swimming patterns, feeding behavior, ingestion, assimilation, and respiration by *Daphnia*. *Limnology and Oceanography* 27: 935–949.

Porter, K.G., J.D. Orcutt, Jr. and J. Gerritsen. 1983. Functional response and fitness in a generalist filter feeder, *Daphnia magna* (Crustacea: Cladocera). *Ecology* 64: 735–742.

Pough, H. 1988. Mimicry of vertebrates: are the rules different? *American Naturalist* 131 (Supplement): S67–S102.

Pyne, S.J. 1997. *Fire in America: A Cultural History of Wildland and Rural Fire.* University of Washington Press, Seattle, WA.

Rasmussen, D.I. 1941. Biotic communities of Kiabab Plateau, Arizona. *Ecological Monographs* 3: 229–275.

Reid, D.G., C.J. Krebs, and A.J. Kenney. 1997. Patterns of predation on noncyclic lemmings. *Ecological Monographs* 67: 89–108.

Rhodda, G.H. and T.H. Fritts. 1992. The impact of the introduction of the Colubrid snake *Boiga irregularis* on Guam's lizards. *Journal of Herpetology* 26: 166–174.

Rhodda, G.H., T.H. Fritts, and P.J. Conry. 1992. Origin and population growth of the Brown Tree Snake, *Boiga irregularis*, on Guam. *Pacific Science* 46: 46–57.

Ricklefs, R.E. 1990. *Ecology*, 3rd Edition. W.H. Freeman, New York, NY.

Ripple, W.J. and R.L. Beschta. 2003. Wolf reintroduction, predation risk, and cottonwood recovery in Yellowstone National Park. *Forest Ecology and Management* 184: 299–313.

Ripple, W.J. and R.L. Beschta. 2008. Trophic cascades involving cougar, mule deer and black oaks in Yosemite National Park. *Biological Conservation* 141: 1249–1256.

Ripple, W.J. and R.L. Beschta. 2012. Large predators limit herbivore densities in northern forest ecosystems. *European Journal of Wildlife* 58: 733–742.

Ripple, W.J., J.A. Ester, R.L. Beschta, C.C. Wilmers, E.G. Ritchie, M. Hebblewhite, J. Berger, B. Elmhagen, M. Letnic, M.P. Nelson, O.J. Schmitz, D.W. Smith, A.D. Wallach, and A. J. Wirsing. 2014. Status and ecological effects of the world's largest carnivores. *Science* 343: 1241484-1–1241484-11.

Ritland, D.B. and L.P. Brower. 1991. The viceroy is not a Batesian mimic. *Nature* 350: 497–498.

Rosenzweig, M.L. 1969. Paradox of enrichment: destabilization of exploitation systems in ecological time. *Science* 171: 385–387.

Rosenzweig, M.L. and R.H. MacArthur. 1963. Graphical representation and stability conditions of predator-prey interactions. *American Naturalist* 97: 209–223.

Ross, I. 1994. Lions in winter. *Natural History* 103: 52–59.

Savidge, J.A. 1988. Food habits of *Boiga irregularis*, an introduced predator on Guam. *Journal of Herpetology* 22:275–282.

Sinclair, A.R.E. and P. Arcese. 1995. Population consequences of predation-sensitive foraging: the Serengeti Wildebeest. *Ecology* 76: 882–891.

Sinclair, A.R.E. and J.M. Gosline. 1997. Solar activity and mammal cycles in the Northern Hemisphere. *American Naturalist* 149: 776–784.

Sinclair, A.R.E., S. Mduma, and J.S. Brashares. 2003. Patterns of predation in a diverse predator-prey system. *Nature* 425: 288–290.

Smith, S.M. 1975. Innate recognition of coral snake pattern by a possible avian predator. *Science* 187: 759–760.

Smith, S.M. 1977. Coral-snake pattern recognition and stimulus generalization by native great kiskadees (Aves: Tyrannidae). *Nature* 265: 535–536.

Smith, S.M. 1980. Responses of naïve temperate birds to warning coloration. *American Midland Naturalist* 203: 346–352.

Smith, B.R. and J.J. Tibbles. 1980. Sea lamprey (*Petromyzon marinus*) in Lakes Huron, Michigan and Superior: history of an invasion and control, 1936-78. *Canadian Journal of Fisheries and Aquatic Science* 37: 1780–1801.

Smith, D.W., R.O. Peterson, and D.B. Houston. 2003. Yellowstone after wolves. *Bioscience* 53: 330–340.

Solomon, M.E. 1949. The natural control of animal populations. *Journal of Animal Ecology* 18: 1–35.

Southern, H.N. 1970. The natural control of a population of tawny owls (*Strix aluco*). *Journal of Zoology* 162: 197–285.

Townsend, S.E., S. Newey, S.J. Thirgood, L. Matthews, and D.T. Haydon. 2009. Co parasites drive population cycles in mountain hares? *Proceedings of the Royal Society B.* 276: 1611–1617.

Turchin, P. 2003. *Complex Population Dynamics: A Theoretical/Empirical Synthesis.* Princeton University Press, Princton, NJ.

Turchin, P., L. Oksanen, P. Ekerholm, T. Oksanen, and H. Henttonen. 2000. Are lemmings prey or predators? *Nature* 405: 562–565.

Turchin, P. and G. Batzli. 2001. Availability of food and the population dynamics of arvicoline rodents. *Ecology* 82: 1521–1534.

Turchin, P. and S.P. Ellner. 2000. Living on the edge of chaos: population dynamics of Fennoscandian voles. *Ecology* 81: 3099–3116.

Utida, S. 1957. Cyclic fluctuations of population density intrinsic to the host-parasite sysem. *Ecology* 38: 442–449.

Volterra, V. 1926. Fluctuations in the abundance of a species considered mathematically. *Nature* 118: 558–560.

Volterra, V. 1931. *Leçons Sur la Théorie Mathématique de la Lutte Pour la Vie.* Gauthiers-Vilars, Paris.

Williams, K.S., K.G. Smith, and F.M. Stephen. 1993. Emergence of 13-year periodical cicadas (Cicadidae: *Magicicada*): Phenology, mortality, and predator satiation. *Ecology* 74: 1143–1182.

11 Plant–herbivore interactions

11.1 Introduction

One of the most important fields of recent ecological inquiry has been that of plant–herbivore interactions. The reasons for the rapid developments in this field are as follows: (1) Herbivores are ecologically important. Globally, it is estimated that herbivores consume from 7% (Pimentel 1988) to 18% (Cyr and Pace 1993) of leaf area and 20% of annual net primary productivity in terrestrial ecosystems, (Agrawal 2011) and from 30% to 79% of net production in aquatic ecosystems (Cyr and Pace 1993). (2) Caught between "plants and predators" (Olff et al. 1999), herbivores must deal with the chemical and morphological defenses of plants while simultaneously defending themselves from their own predators. The study of plants and their herbivores leads to new understandings of interactions involving more than two trophic levels. (3) Herbivores have been, and still are, important to the evolution of plants and other animals. Through their activities, herbivores have evolutionarily shaped the plant community, partially determining the diversity, abundance, and life form of plants. This, in turn, affects what other types of animals are present and influences ecosystem processes such as energy flow and nutrient cycling. (4) Herbivores are economically important. In agricultural systems, herbivores may often consume 50% of net productivity and therefore depress yield. Under the worst of circumstances agricultural crops may be wiped out. Furthermore, the animals we raise for food and other products are almost all herbivores themselves. Thus, a better understanding of plant–herbivore relationships is crucial to the success of our agricultural endeavors. (5) Plant–herbivore interactions provide unique opportunities to investigate basic ecological and evolutionary processes such as co-evolution, food webs, plant chemistry, animal foraging, competition, and ecosystem processes such as nutrient cycling.

Introduction to Population Ecology, Second Edition. Larry L. Rockwood.
© 2015 John Wiley & Sons, Ltd. Published 2015 by John Wiley & Sons, Ltd.
Companion Website: www.wiley.com/go/rockwood/populationecology

The relationship between plant "secondary compounds" and herbivore–plant relationships was first outlined by Fraenkel (1959). Since the compounds under discussion are not part of the primary metabolism of plants, they were and still are known as "secondary compounds." Fraenkel (1959) suggested that plant secondary compounds have evolved as defenses against herbivores and herbivores have strongly affected plant evolution. Since the publication of his paper research has exploded in the fields of plant–herbivore relationships and chemical ecology. Ecologists such as Harborne (1993, 1997) and Hartley and Jones (1997) accept that the major functions of plant secondary compounds are as defenses against herbivores. Skeptics, such as Smith and Smith (2001), have asserted that there is little evidence that plants evolved secondary compounds for a defensive purpose. However, as Hartley and Jones (1997) put it, although "we are still not sure why plants have such a huge array of secondary compounds or how this came to be, we do know that these chemicals are important in keeping the world green." Furthermore, it is now known that some plant traits function exclusively as defense without any known function in primary metabolism (Agrawal 2011). For example, about 10% of plant species and plant families produce latex, which serves as physical barrier to feeding and contains large doses of defensive chemicals (Agrawal and Konno 2009).

The basic theory, which has guided ecological thinking concerning plant–herbivore interactions, was set forth by Ehrlich and Raven (1964). At that time, the great diversity of chemical compounds produced by plants was widely recognized. Many chemists and botanists referred to these compounds as waste products of plant metabolism, lacking any adaptive value. Ehrlich and Raven, in contrast, asserted that secondary compounds were the product of co-evolution with herbivores. Plants that produced secondary chemicals made their tissues unpalatable or toxic, lowered herbivore damage, and had a selective advantage since they could devote more energy to competition and reproduction. If less energy were lost to herbivores, then plants would grow more rapidly and enjoy competitive success. Any genes that allowed the plant to produce these chemicals would spread throughout the population. By the same argument, herbivores could be expected to adapt to plant defenses. Again, an herbivore that evolved a detoxification enzyme, or other adaptation to allow it to feed on protected plants, would enjoy competitive success compared to individuals unable to feed on these plants. As discussed earlier, some herbivores have turned defensive chemicals to their own advantage. That is, by modifying and storing plant toxins, the herbivore itself became unpalatable to predators. Thus, an herbivore, which evolved an adaptation for feeding on a chemically defended plant, could potentially enjoy both competitive success and protection from its own predators.

Faced with an increasing number of herbivores adapted to a particular chemical defense, plants have, according to Ehrlich and Raven's theory, counter-adapted by producing additional defensive chemicals. This process of evolution and counter evolution of chemical defenses is often referred to as the "evolutionary arms race" and is thought to have helped produce the great diversity of both angiosperms ("escape and radiate" hypothesis) and the insects that feed on them. Taken together, land plants (particularly Angiosperms) and insects make up more than half of all known terrestrial species on earth (excluding microorganisms).

The Ehrlich–Raven theory is based on the following assumptions:

1. Herbivore activity is harmful to plants. This would seem to be obvious, but this assumption cannot be accepted unequivocally.

2. Plants are able to evolve defenses that are effective in deterring feeding by herbivores. Note that, although the Ehrlich–Raven theory stresses herbivores, these plant chemical defenses could just as easily have been evolved to defend plants against attacks by fungi, bacteria, and other microorganisms. Those chemicals that evolved as antifungal defenses, for example, might also be effective against other microorganisms or herbivores.

3. Herbivore feeding activities, growth, reproduction, and evolution have been guided by the ability of plants to defend themselves, both physically and chemically.

4. Although there exist herbivores that feed on plants of many species, genera or families with seemingly little regard for the identity of the plants, these "generalists" are actually much more selective than they appear. Generalists engage in a broad "sampling program" in which they eat small amounts of material from many plant species. The majority of what they consume, however, comes from a much smaller species list (Rockwood 1976, 1977; Rockwood and Glander 1979). The careful manner in which generalists eat is thought to be consistent with the central importance of secondary compounds in plant–herbivore interactions.

5. The majority of herbivore species are not generalists but are specialists, feeding on just one plant species, one plant genus, or perhaps one plant family (e.g., the Cruciferae or mustard family). Such specialization is consistent with the idea that an herbivore that evolves a way of feeding on a particular plant type eventually loses the ability to feed on other plants.

For example, one simple hypothesis, based on these assumptions, is that plants with few specialized chemical defenses will be fed upon primarily by generalists herbivores, while plants with specialized, complex chemical defenses will by fed upon mostly by specialist herbivores. When Berenbaum (1981) examined the chemical defenses in the Umbelliferae (the carrot family), she found some species defended by a relatively simple phenolic (coumarin), some by linear furanocoumarins, and some by sophisticated angular furanocoumarins. As the defenses became more complex, the herbivores feeding on the plants changed from mostly generalists (defined as feeding on more than three plant families) to mostly specialists (those feeding on only one to three genera).

11.2 Classes of chemical defenses

Plants have evolved a wide variety of chemical defenses. Coley and Kursar (2014), for example, estimate that leaves of a single tropical tree contain hundreds of distinct chemical compounds that function as herbivore deterrents. Over 40 000 chemical compounds known as allelochemicals have been described from plants (Berenbaum 2002). Allelochemicals are characterized as having a negative impact on herbivores, disease organisms, or other plants and have been defined as non-nutritional metabolites that have stimulatory or inhibitory effects on other organisms (Haig 2008). Although plant biochemistry is extraordinarily diverse, Haig (2008) broke plant allelochemicals into six basic groups. For our purposes here, we further subdivide these six groups into compounds containing nitrogen and so-called carbon-based compounds that do not. Nitrogen-containing compounds include glucosinolates, alkaloids, and benzoxazinones. The other groups described by Haig (2008) are phenolics, terpenoids, and "other." We will treat glycosides separately.

The production of secondary compounds is metabolically expensive. To synthesize a gram of alkaloid requires 5 g of photosynthetic carbon dioxide, while a comparable figure would be 2.6 g for a phenolic (Gershenzon 1994). On the other hand, compounds such as alkaloids are effective at low concentrations such as 2%–4% dry weight, while others, such as tannins, require a higher concentration, such as 5% dry weight in leaves to be effective. Alkaloid concentration is sometimes only 10%–20% of the concentration of a comparable phenolic (Harborne 1997).

(a) Nitrogen-based defenses

Glucosinolates

Glucosinolates are sulfur- and nitrogen-containing compounds found mostly in the plant order Capparales. Every member of this group has a sulfate attached to an N=C. Within this order, glucosinolates are known from 16 plant families, mostly notably in the Brassicaceae, which is the home of the genus *Brassica* (the cabbage and mustard family). Over 130 different compounds of this type are known, and they help produce the bitter taste to plants in that genus.

Alkaloids

Alkaloids are a heterogeneous group of substances that have two common characteristics: they occur in plants, and they have an organic base containing a nitrogen atom. They were given their name by Meissner because of their alkali-like behavior when extracted with acid from plants (Haig 2008). They often have a carbon ring structure and are derived from biosynthetic pathways beginning with an amino acid. Therefore, they have at least one nitrogen atom. Alkaloids have evolved many times in a variety of plant taxonomic groups. Accordingly, alkaloid biosynthesis has no underlying unity, and there is no core fragment that can represent this entire group. Alkaloids impart a bitter taste to plants, some are very toxic, and some are hallucinogenic. When Levin (1976) reviewed the literature, he found that 16% of temperate zone plant species and over 35% of tropical plant species contain alkaloids. Approximately 12 000 types of alkaloids are known (Harborne 1988, 1997; Haig 2008).

Alkaloids are less important than the phenolics since they are found in only about 20% of angiosperm species worldwide, and they are generally absent from mosses, ferns, and gymnosperms (Harborne 1997). One reason why alkaloids and nitrogen-based defenses, in general, are less ubiquitous than carbon-based defenses, such as terpenoids and phenolics, is that these metabolites require protein amino acids. Amino acids are, in turn, dependent on a supply of nitrogen, either from the soil or fixed by bacterial mutualists, and nitrogen is usually limiting to plant growth. In legumes, although *Rhizobium*-type bacteria fix nitrogen from the air, the metabolic costs of fixation are high. The fact that nitrogen-based defenses potentially drain needed nitrogen from protein synthesis may have prevented such defenses from evolving more consistently. Nevertheless, there is a great diversity of alkaloids of great importance to humans, including nicotine, cocaine, morphine, codeine, quinine, strychnine, theobromine, and caffeine.

Benzoxazinones

Hydroxamic acids, part of a group called benzoxazinones, consist of a pair of benzene-like rings containing one N atom. These compounds are usually found in plant families such as the Acanthaceae, Poaceae, Ranunculaceae, and Scrophulariaceae. These compounds act against a wide range of insects and microbes (Haig 2008). Key enzymes of attacking organisms are inactivated.

(b) Carbon-based defenses

Phenolic compounds

Phenolic compounds are essentially ubiquitous in the plant kingdom (Swain 1965; Harborne 1988); they are found in ferns and gymnosperms, as well as in angiosperms. In fact, cinnamic acid is thought to be present in all higher plants, and benzoic acid has been found in all flowering plants (Haig 2008; Dalton 1999). Phenolics are defined as compounds with aromatic structures with one or more hydroxyl groups attached (Harbone 1997). They have a common biosynthetic origin from the amino acid phenylalanine. The total number of known phenolic compounds is estimated at 8000 (Harborne 1997). Phenolic compounds range in size from simple phenol, the essential oil in *Pinus sylvestris*, to condensed tannins with molecular weights of 20 000. Phenolics include flavanoids, derivatives of which give color to flowers and fruits, coumarins, and tannins. The tannins are substances present in vegetable extracts that are responsible for converting animal skins into leather by the tanning process. When leaves of many plants are boiled or decompose, tannins give the water its brown color. Tannins are defined as naturally occurring compounds of high enough molecular weight (500–3000 g/mol) and containing a sufficient number of phenolic hydroxyl groups to enable them to form crosslinks between proteins and other macromolecules. Phenolic compounds of lower molecular weight are apparently too small in size to form effective cross-links, while compounds of molecular weight beyond 3000 g/mol may be ineffective tanning agents because they are too large to penetrate between plant collagen fibers. Although tannins are among the most effective types of defensive compounds (Feeny 1976), simple phenols are also effective feeding deterrents. Phenolic compounds are broadly toxic to most forms of life. Due to their ability to form crosslinks with proteins and other polymers, phenols and tannins are capable of inhibiting enzymes and are also markedly astringent. That is, they cause a dry or puckery sensation in the mouth by reducing the lubricant action of the glycoproteins in the saliva.

There are two distinct types of vegetable tannins. The first class has a core poly-hydric alcohol such as glucose. Acids, bases, or enzymes can readily hydrolyze these tannins to yield a carbohydrate and a phenolic acid. Known as hydrolyz-able tannins, the main effect of these compounds is to inactivate the digestive enzymes of herbivores, especially insects.

Tannins of the second class contain only phenolic nuclei. These are called non-hydrolyzable or condensed tannins. Condensed tannins are bound to cellulose and proteins of the cell walls. They make it very difficult for herbivores to extract amino acids from plant material and also defend plants from microbial or fungal attack.

Furanocoumarins are another phenolic group (Harborne 1988) whose biosyn-thesis can be traced back to the amino acid phenylalanine (Berenbaum and Zangerl 1999). These allelochemicals are found primarily in two plant families, the Apiaceae (the carrot family) and the Rutaceae (the rue family), but thanks to the work of Berenbaum and her colleagues (Berenbaum 1991, 1995; Berenbaum and Zangerl 1996), we know a great deal about the biosynthesis and effects of the furanocoumarins in wild parsnip (*Pastinaca sativa*). Furanocoumarins have biological effects on a wide variety of organisms, from bacteria and fungi to insects and vertebrates. These chemicals, "act principally by binding covalently to pyrimidine bases in DNA and interfering with transcription, but these compounds are also capable of causing toxicity by binding to unsaturated lipids in membranes, inactivating enzymes and generating oxyradicals … " (Berenbaum and Zangerl 1999, p. 62). Furanocoumarin content of leaves, roots, and stems can be rapidly "induced" to increase by either biotic or abiotic damage. For example, furanocouramin content in leaves can increase over 200% due to mechanical damage. Furthermore, this induced response can be rapid. Mechanical damage can cause xanthotoxin concentration to reach a maximum level within 6 hours (Berenbaum and Zangerl 1999).

Terpenoids

Isoprenoids and their derivatives constitute a group of naturally occurring plant materials that have in common the fact that are derivatives of the 5-carbon compound isoprene, C_5H_8. This group includes essential oils, resins, steroids, carotenoids, and rubber. Terpenoids are classified according to the number of C_5 units they contain. For example, monoterpenoids are C_{10} compounds; sesquiterpenoids are C_{15} compounds and so on through tetraterpenoids, which are C_{40} compounds.

The 24 000 different compounds, produced by over 2000 species of plants in 60 plant families, monoterpenoids are usually pleasant smelling, as in the characteristic odor of the pine family (Pinaceae) (Haig 2008). Cells in the resin ducts of pine produce oleoresin, from which turpentine is made via steam distillation. Oils from citrus and *Eucalyptus* are also monoterpenoids. When resin ducts are disrupted by herbivores, the oleoresin, usually under pressure, spills out and is effective in "gumming up" the mouth parts of insects and even vertebrate herbivores.

The triterpenoids, C_{30} compounds, include the toxic and bitter cucurbitacins produced by the cucumber family (Curcurbitaceae), the cardiac glycosides discussed below and the saponins found in over 70 plant families (Harborne 1993).

(c) Glycosides

Glycosides are sugar derivatives that contain a nonsugar group bonded to an oxygen or nitrogen atom and that on hydrolysis yield a sugar (as glucose). They are biologically active forms of steroids, derived from oligosaccharides. Cardiac glycosides, known as cardenolides, are C_{23} steroids and belong to the terpenoid group. They are known from at least 11 plant families but are especially well known from the plant families Apocynanceae, Asclepiadaceae, and Scrophulariaceae. Often the youngest leaves have the greatest amount of these compounds. Cardiac glycosides are heart poisons, which usually provoke vomiting among animals that ingest them, and can cause death in livestock. The heart drug, digitalis, is derived from a cardiac glycoside. The cardiac glycosides in the milky sap of milkweeds (*Asclepias* spp.) are well documented.

Cyanogenic glycosides include a CN group such that hydrogen cyanide (HCN) is released on consumption of plant tissue. It is found in plant families such as the Fabaceae, Rosaceae, Linaceae, and Compositae (Vetter 2000).

A review of the role of cyanogenic glycosides in plant herbivore interactions is instructive (Gleadow and Woodrow 2002). Over 2500 plant species are able to release HCN on ingestion by an herbivore. When an herbivore chews a leaf, it disrupts tissues and cells that had previously separated the β-glucosidase enzymes from the cyanogenic glycoside compounds. The resultant degradation of the cyanogenic compound eventually releases HCN.

Examples of poisonings, some leading to death, of humans and animals abound. In addition, there are many cases in which animals have been shown to avoid foods containing cyanogenic glycosides. On the other hand, numerous studies have found that glycosides have little effect or even act as a feeding stimulant (Gleadow and Woodrow 2002).

As explained by Gleadow and Woodrow (2002), in attempting to understand a specific herbivore–plant interaction, four factors must be accounted for. These are: (1) concentration of the presumed toxin; (2) the status of the herbivore as a "specialist" versus a "generalist"; (3) the ability of an herbivore to monitor glycosides in its diet; and (4) the feeding mode of the animal.

In the first case, the cyanogenic compounds must be present at a threshold level in order to be effective. For example, Gleadow and Woodrow (2000) found a significant inverse relationship between concentrations of cyanogenic compounds in young leaves of *Eucalyptus cladocalyx* and herbivore damage. The age of the leaves and the health of the plant often determines the level of defensive compounds in plants and hence their level of toxicity.

There exist a number of studies that illustrate the principle that, while specialists have evolved the means to tolerate cyanogenic glycosides, plants containing these compounds are avoided by generalist herbivores. For example, Schappert and Shore (1999a, 1999b) found that the number of herbivore species feeding on the plant, *Turnera ulmifolia*, in Jamaica was inversely related to the cyanogenic glycoside content. Yet the total amount of herbivory was similar regardless of the glycosides. While the cyanogenic compounds deterred generalist herbivores, the specialists were not, and total leaf area lost was equivalent.

In many cases, an animal is able to sequester the cyanogens while preventing the release of HCN. These sequestered compounds are then used as a defense from predators. For example, larvae of *Eupitoiea hegesia* accumulate cyanogenic compounds from *Turnera ulmifolia* and become bad tasting to Anolis lizards (Schapper and Shore 1999b).

The third point is simply based on the fact that animals, including humans, are able to monitor the amount of glycosides in their diet due to their bitter taste. Thus, it is unlikely that animals will actually poison themselves while feeding on plant material containing cyanogenic glycosides. Hungry animals are more likely to overexpose themselves to plant material containing glycosides. On the other hand, exposure to cyanogenic glycosides leads to an increased ability to tolerate them through greater detoxification ability. For example, humans and other mammals can be induced to produce increased levels of the enzyme rhodanese, which detoxifies HCN as it is released (Gleadow and Woodrow 2002).

Finally, Gleadow and Woodrow (2002) point out that any mode of feeding that minimizes tissue disruption will allow insects and other animals to successfully feed on plants containing cyanogenic glycosides. The main successful tactic involves using sucking mouthparts (Hemiptera, Homoptera) as opposed to chewing mouthparts (Coleoptera).

Gleadow and Woodrow (2002) conclude that, although cyanogenesis is not a totally effective defense, it is toxic and distasteful enough to be effective against a wide variety of herbivores if available in sufficient concentration.

11.3 Constitutive versus Induced Defense

Plants usually produce a certain quantity of a chemical defense, a sort of background amount. This is called a constitutive defense. After a plant is attacked, however, the amount of these chemicals usually increases. In other cases, entirely new compounds are produced after an attack. Such a reaction, known as an induced defense, to herbivore attack can be thought of as a parallel to the immune system in animals.

According to Karban and Myers (1989), any change in a plant following herbivory can be thought of as an induced response. Such changes include not only allelochemical induction (Baldwin 1994) but also increases in physical defenses such as thorn length (Young 1987), spine density, production of trichomes, emission of volatiles that attract predators and parasites, reduction in plant nutritional quality for herbivores, and even increases in extrafloral nectar in plants protected by ants (Agrawal 2000a). Induced defenses are not limited to plants. Phytoplankton, rotifers, ciliated protozoans, cladocerans, and even carp are known to respond morphologically or behaviorally to the presence of herbivores or predators (Tollrian and Harvell 1999). Even marine bryozoans have been shown to have inducible physical defenses. Bryozoans produced spines, which were effective in reducing mortality, after being attacked by nudibranchs (Harvell 1984).

To qualify as an induced defense (or induced resistance), the response must result in a decrease in herbivore or predatory damage, and an increase in fitness must be demonstrated as compared to noninduced controls (Karban and Baldwin 1997). Harvell and Tollrian (1999) identified the following as conditions necessary for the evolution of an inducible defense: (1) The selective pressures should be variable and unpredictable but sometimes strong. If the inducing species is constant, then permanent, constitutive defenses should be present. (2) A reliable cue is necessary to activate the defense. (3) The defense must be effective. (4) The inducible defense must save energy as compared to a constitutive defense or no defense at all.

The basic hypothesis is that while defenses increase plant fitness when herbivores are present, the energy invested in such defenses results in lowered plant fitness when herbivores are absent (Agrawal et al. 1999). Inducible defenses allow an organism to invest in defense when necessary but avoid costly allocations to defense when the herbivore or predator is absent.

In the 1980s, Schultz and Baldwin (1982) and Rossiter et al. (1986) showed that defoliation by gypsy moth larvae stimulated oaks to increase the phenolic content of leaves. Rossiter et al. (1986) found a significant negative correlation between phenolic content of leaves and the size of female gypsy moth pupae. Assuming reproductive output of a female gypsy moth is positively correlated with size, the induction of a high phenolic content in oak leaves would be expected to eventually cause a decline in gypsy moth populations. However, a complication to this seemingly straightforward picture is the fact that another control agent, the gypsy moth nuclear polyhedrosis virus, is inhibited by high levels of phenolic compounds in oak leaves. Even though leaf phenolics depress gypsy moth reproduction, survivorship is enhanced due to suppression of the virus. The two potential control agents work in opposition to each other, making it increasingly difficult to predict the dynamics of gypsy moth populations (Foster et al. 1992). Furthermore, throughout the 1990s, gypsy moth populations declined to very low levels, apparently due to a fungus. The moth populations in the Eastern United States increased during the period 2000–2002 but did not reached the population levels common in the early 1990s.

Agrawal (2000a) exposed the leaves of the herbaceous plant *Lepidium virginicum* (Brassicaceae or mustard family) to herbivory by the larvae of *Pieris rapae* (Pieridae). Induced plants responded by producing more trichomes per leaf and increasing the glucosinolates (mustard oil) content of the leaves. While induction did not affect the feeding behavior of the specialist *P. rapae,* feeding by generalist caterpillars in the Noctuid family was reduced.

In a related field study, Agrawal (2000a) again induced a response by allowing leaves to be consumed by larvae of *P. rapae.* However, in addition to undamaged controls, he damaged leaves by clipping with scissors. In the field, aphids were important herbivores. Controlled herbivory by *P. rapae* induced resistance to attack by the aphids, but clipping with scissors did not. The number of aphids feeding on control and clipped plants averaged five per plant, while the average number on induced plants was just three per plant. Furthermore, plant survivorship was lowest in the clipped plants. Previous studies have suggested that the saliva of herbivores is necessary for successful defensive induction (Bodnaryk 1992, Mattiacci et al. 1995). Plants therefore may respond differently to leaf losses from herbivores as opposed to leaf damage from storms or other physical causes.

Other research has focused on the ability of plants to produce proteinase inhibitors that inhibit the major digestive enzymes of insects. For example, when attacked by herbivores, sagebrush (*Artemisia tridentate*) produces a compound known as methyl jasmonate or jasmonic acid. Under the influence of methyl jasmonate, tobacco, tomato (Farmer and Ryan 1990, 1992), and alfalfa plants all were induced to produce proteinase inhibitors. More recently, it was found that injury to a plant tissue causes the production of a peptide hormone. The hormone stimulates the release of linolenic acid, a fatty acid common to plant cell membranes. Linolenic acid is then converted to jasmonic acid that, in turn, stimulates proteinase inhibitors (Chen 1990). In another study, jasmonic acid stimulated the production of nicotine in tobacco plants (Olhnmeiss and Baldwin 2000).

Some induced responses to wounds are considered "systemic." This means that damaged plant tissue may transmit a signal to other areas of the plant, resulting in the induced reaction. Karban and Baldwin (1997) have outlined the requirements for a hypothetical signal to be taken seriously.

1. The signal must be rapidly generated at the wound site.
2. The inducer must be known.
3. The signal must travel through the plant in a time course consistent with the induced response.
4. The signal must stimulate the induced response at concentrations consistent with those known from damaged plants.

Although many signal compounds have been proposed, what Agrawal (2011, p. 427) has termed the "master regulator of … defensive mechanisms," is jasmonic acid and its volatile cousin, methyl jasmonate. Jasmonic acid meets the requirements listed above

(Karban and Baldwin 1997), has been shown to stimulate everything from alkaloid to phenolic production, but also may have a role in stimulating the production of extrafloral nectar (Agrawal 2011; Heil et al. 2004).

Jasmonic acid is derived from common fatty acids, and it, along with its methyl ester relative (methyl jasmonate), is commonly found in plants. Both compounds elicit a multitude of responses in plants. Mechanical wounding increases the levels of jasmonic acid, which then moves rapidly through the phloem. Moreover, methyl jasmonate is very volatile. Minute concentrations of gaseous methyl jasmonate can induce the synthesis of proteinase inhibitors, as described earlier. Other chemicals involved in plant defense such as ethylene, systemin, and several alkaloids increase after plants are exposed to methyl jasmonate. The possibility that the gas, methyl jasmonate, is a signaling compound provides a potential mechanism for communication among plants (Karban and Baldwin 1997).

11.4 Plant communication

Damage to a plant may result in an induction of chemical defenses in neighboring plants. Several experiments have shown that volatile chemicals are released when plants are damaged. Experiments by Rhoades (1983) and Baldwin and Schultz (1983) suggested that plants, which share the same air space with damaged plants, increase the production of defensive chemicals even though they are not damaged themselves. In these early experiments, however, the effect was temporary, and alternative explanations have been offered. Many scientists have questioned the results, and, in other experiments, the evidence for communication among plants was negative (Karban and Baldwin 1997). As outlined earlier, methyl jasmonate was proposed to be the gaseous carrier, and a number of plant species have responded to this signal by producing proteinase inhibitors. Evidence is increasing, however, that volatile cues from damaged plant tissues may be used as cues for both herbivores and the predators and parasites of these herbivores, to locate these plants. However, it remains to be seen whether signals released by plants that are damaged are used by unattacked, neighboring plants, to induce defensive responses (Karban and Baldwin 1997).

11.5 Novel defenses/herbivore responses

Many plants store resins, latexes, gums, and mucilages under pressure in networks of canals. When these plants are damaged, the fluids are released from the injured tissues. In many cases the canals contain defensive compounds. The secretions also mechanically deter herbivores by interfering with their feeding since the secretions solidify when exposed to air. Sometimes insects are trapped in the secretions. Several species of Lepidoptera, Coleoptera, and Orthoptera circumvent the secretions by cutting trenches in the leaves, biting the veins, or even cutting the petioles. These species often have morphological adaptations for trenching or vein-cutting. For example, Dussourd and Denno (1994) showed that the cabbage looper (*Trichoplusia ni*) is able to cut a trench in leaves of *Lactuca serriola*, thereby disrupting the flow of latex to areas of the leaf distal to the trench. Not only is the cabbage looper able to feed on leaves distal to the trench, but so also is the yellow-striped armyworm (*Spodoptera ornithogalli*), a generalist herbivore. Lacking the ability to trench on its own, however, the armyworm is unable to successfully feed on plants with intact canals.

Becerra (1994) has reported on a squirt-gun defense of the succulent, deciduous shrub, *Bursera schlechtendalii*, found in Mexican deserts. This plant has simple leaves that produce terpenes distributed in a network of resin canals located in the stem cortex and throughout the leaves. When a leaf is cut, a syringe-like squirt of terpenes is released. The squirt travels 5–150 cm and persists for several seconds. Some leaves release large amounts of terpenes without the squirt response. This "rapid bath response" covers the entire surface of the leaf. Other leaves do not release resins when damaged.

A specialized beetle (genus *Blepharida*, Chrysomelidae) feeds only on the leaves of this species. The larvae are able to sever the resin canals. They bite the mid-vein to eliminate the squirt or bath response. If larvae are bathed or squirted, they withdraw from the leaf and try to clean themselves, remaining inactive for many hours.

Becerra (1994) showed that there is a great deal of variation in terpene response within the *Bursera* population. In most plants, 80% of the leaves released resins when damaged. Some had extremely high responses, with resin squirting from mid-veins and from lateral veins. The number of plants with high response increased throughout the season. Larvae growing on plants with high rates of response had higher mortality rates. Early instar larvae are unable to sever veins and feed by leaf mining. Sometimes, they inadvertently rupture canals. When this happens they die, covered with resins. Larvae also grow at a slower rate on plants with a higher frequency of leaf response. High frequency of leaf response is correlated with leaf scars, probably indicating an induced response from a prior attack. The frequency of vein cutting behavior is also associated with frequency of leaf response.

In summary, canals can be effective barriers against nonspecialized insects. Even specialized trenching insects have higher mortality on highly responsive plants. It can take 1.5 hours for a larva to deactivate a resin canal of a leaf. The consumption of a leaf after disarming takes only 10–20 minutes. Thus, there is high handling cost. Meanwhile, the plant's response is not fixed but depends on water potential and plant health.

11.6 Detoxification of plant compounds by herbivores

Animals are by no means helpless in the face of plant secondary compounds. Both invertebrate and vertebrate herbivores have evolved detoxification mechanisms. There are basically three parts to detoxification. Since these toxins are usually attracted to lipids (lipophilic), the first step is usually for the herbivore to convert the compound to a water soluble form, usually an alcohol, so that it can be excreted. So-called phase I enzymes are found in the livers of vertebrates and involve cytochrome P-450. These enzymes are capable of acting on a wide variety of chemicals and are known as polysubstrate mixed-function oxidases (PSMOs) or simply mixed-functional oxidases. PSMOs are known from many groups of animals. In insects, they are found in the gut, fat bodies, and Malpighian tubules. In addition to oxidation, other reactions such as reduction and/or hydrolysis may occur (Harborne 1993).

In the second stage of detoxification, the metabolite from stage I is conjugated with a sugar or sulfate anion. This means that the compound from stage I is united with another molecule to form an inactive, nontoxic product. These reactions require energy. Therefore, although an herbivore may not suffer damage from ingesting plant material containing plant toxins, the detoxification process is an energy drain. In the last stage, the conjugated chemical is excreted through bile or urine. There is a great deal of variation in the detoxification process, both between and within species. For example, Millburn (1978) found that benzoic acid is excreted as a conjugate of glycine in mammals, amphibians, fish, and insects, as an ornithine conjugate in birds and reptiles and as an arginine conjugate by arachnids. Furthermore, as pointed out by Harborne (1993), 14% of the British people cannot metabolize the alkaloid betanin, the red pigment in beet roots. The result is red-colored urine after ingesting beets (beeturia). The major point of this section is that herbivores are capable of detoxifying defensive chemicals, but there are high metabolic costs in doing so.

11.7 Plant apparency and chemical defense

A general theory developed by Feeny (1976) stated that the amount and the type of defense a plant has evolved is dependent on the likelihood it would be found and consumed by herbivores. His hypotheses are as follows:

1. Plants that are large, woody, persistent, have clumped distributions, and that are abundant are "bound to be found" by herbivore populations. Feeny called such species "apparent," and asserted that they would have a relatively large investment in chemical defenses. Such plants produce chemicals such as phenolics that reduce the ability of all herbivores to digest the plant parts that contain them. Such "digestibility reducing compounds" often bind to proteins in such a way that herbivores have great difficulty eating, swallowing, and digesting this vegetation. These compounds are more effective in larger doses and take more time and energy to produce than some other defensive compounds. Thus, Feeny (1976) termed them "quantitative defenses."

2. "Unapparent" plants are small, herbaceous, and ephemeral, have highly dispersed distributions, and/or are rare. If toxins are present, they are effective at low concentrations, are rapidly produced, and require less total energy in their formation. Alkaloids and glycosides are examples of such compounds, termed "qualitative defenses" by Feeny (1976). These compounds are effective in deterring generalist or nonadapted herbivores. They are subject to counter-adaptation and often are consumed by host-specific herbivorous insects. These r-selected plants, which exist in temporary environments, where growth rates can be very high due to lack of competition, primarily use the qualitative defenses, while K-selected plants should favor the quantitative defenses. K-selected plants normally use carbon-based defenses (especially phenolics). This may be because "apparent" plants are often members of a community in which nutrients eventually become limiting to growth. On the other hand, defenses found among r-selected species include nitrogen-based alkaloids, glycosides, and carbon-based coumarins.

3. Defenses vary with time in plant parts. New leaves are ephemeral and somewhat unapparent, in that their characteristics change. Young leaves are less tough, higher in water and protein content, and lower in defenses of the quantitative type (Feeny 1970; Rockwood 1976). According to Coley and Kursar (2014), 75% of lifetime leaf damage by herbivores in the tropics occurs in the first few weeks after a leaf expands. As a leaf ages, it increases its quantitative defenses, becomes tougher through changes in fiber or lignin content, and its protein and moisture contents drop. At the other extreme, as leaves age, they often become senescent and more susceptible to herbivore or fungal attack (e.g., lilac leaves). As explained in the section on optimal defense theory, plants invest chemical defenses in the most valuable tissues. Nitrogen-based defenses are withdrawn from less valuable tissues. Thus, new leaves have higher amounts of qualitative defenses than older leaves. For example, in *Coffea arabica* leaves, purine alkaloids are present at those periods when the plant is most susceptible to herbivory. During leaf development, the concentration of the alkaloid increases to 4% of dry weight in new leaves. The rate of synthesis decreases as the leaf matures. At senescence, the leaf contains only traces of the alkaloids (Harborne 1993).

11.8 Soil fertility and chemical defense

Janzen (1974) observed that tropical plant communities growing on nutrient poor, white-sand soils always had fewer herbivores than nearby plant communities based on fertile soils. He hypothesized that plants growing on infertile soils would have a difficult time replacing tissues lost to herbivores due to their slow growth rates. Therefore, the permanent or "apparent" plants found on such soils should be heavily protected by quantitative defenses. Indirect evidence in favor of this hypothesis is that rivers draining such white-sand environments contain tea-colored waters. These brown or black water rivers contain large amounts of phenolic compounds that have leached from the surrounding plant communities. The famous Rio Negro of the Amazon basin drains such a white-sand area.

McKey et al. (1978) tested this idea in Africa. Plants from a white-sand region in Uganda had twice the phenolic compounds compared with plants from a nutrient-rich soil in Cameroon.

Finally, Coley and others (Coley 1980; Coley et al. 1985) have shown that plants not limited by a physiological stress of some kind are fast growing and have higher leaf protein contents, shorter leaf lifetimes, higher herbivory rates, lower amounts of defensive metabolites, and higher turnover rates of defenses compared with plants that grow slowly. Furthermore, herbivores, especially generalist herbivores, have a definite preference for the fast-growing plant species.

11.9 Modeling plant–herbivore population dynamics

Most plant–herbivore models are for grazers. As pointed out by Turchin (2003, p. 112), a grazer is a "consumer that scores low on both intimacy and lethality." That is, models are usually based on generalist herbivores, which do not usually kill their resources, and which consume small amounts from many different individuals and species of plants. Specialist herbivores, especially sucking insects, are functionally parasites and score high on the "intimacy" scale. As we will see, grazer-vegetation models tend to be similar to predator–prey models when they focus on the quantity of vegetation harvested by grazers. In these models, we assume that plant quality does not change in spite of attack by herbivores. In another approach, the assumption is made that plant quality is the variable, while quantity is held constant.

Crawley (1997) summarized several possible models for vegetation–herbivore population interactions of the first type (the quantity of the vegetation is the variable). To begin, we can use the basic ideas gleaned from predator–prey models described in the previous chapter. Density-independent plant growth can be described as $r_v V$, where r is maximum growth (equivalent to the intrinsic rate of increase), and V represents plant abundance or biomass. For density-dependent plant growth, we can simply use the logistic equation $dV/dt = r_v V \frac{K_v - V}{K_v}$, where K_v is a carrying capacity term for the plant population. We then subtract plant losses due to herbivore feeding. This is equivalent to the herbivore functional response and can be of several forms. Since herbivores are not usually killing the vegetation, we measure the grazer's functional response in terms of amounts of biomass removed. We also have to be careful how we define V for the herbivore since most herbivores cannot eat entire plants but specialize on leaves, stems, buds, and so on. Recall from the last chapter that we used E as a coefficient measuring the efficiency of predation. Here F will be used to measure the efficiency of herbivore removal of vegetation biomass. If the herbivore functional response is simply modeled as FNV, where N is the number of herbivores, we have the equivalent of the Lotka–Volterra equation for prey. This is essentially a type I functional response, linear and unlimited, meaning that there is no handling component, and that each herbivore has an unlimited appetite. However, we can add either a type II or type III functional response as we did in Chapter 10. The type II functional response would be $\frac{FNV}{1+Fh_2V}$, where h_2 is the handling time component. The type III functional response is $\frac{FNV^2}{1+Fh_3V^2}$, and finally, the functional response with a half-saturation constant for the herbivore–plant interaction is $\frac{fNV}{b+V}$. The term "f" equals the maximum consumption or grazing rate when the search and capture components have been minimized, and b equals half of the maximum consumption rate.

Using the functional response with the half-saturation constant, the plant growth equation becomes

$$\frac{dV}{dt} = r_vV\left(\frac{K_v - V}{K_v}\right) - \frac{fNV}{b + V} \tag{11.1}$$

Before we continue by modeling the herbivore or grazer population, we should stop and consider a point raised by Turchin (2003) and others. Using the logistic for growth of the vegetation assumes that when V is close to zero, growth rate is rather slow at first, then accelerates until $V = K_v/2$, before slowing down to zero at K_v. The problem with this is that many plants, especially grasses, store 80%–90% of their biomass under ground (Wielgolaski 1975). When we are modeling V, we are usually only measuring above ground vegetation (e.g., leaves and stems). As Turchin (2003) points out, when re-growth occurs after an herbivore attack, the initial re-growth pattern is more likely to follow a linear path to K, as opposed to a logistic pattern (Figure 11.1).

Using this logic, the equation describing re-growth in the absence of herbivore losses simplifies to equation 11.2:

$$\frac{dV}{dt} = u_0\left(1 - \frac{V}{K_v}\right) \tag{11.2}$$

where u_0 represents plant growth rate when V is close to zero, and V specifically represents above ground biomass. Originally proposed by Turchin and Batzli (2001), this

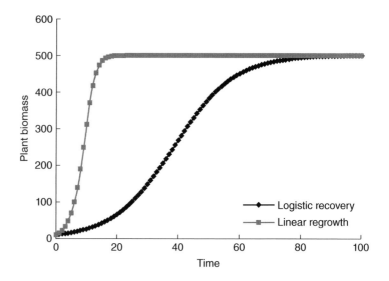

Fig. 11.1 Vegetation recovery from grazing; logistic recovery versus linear re-growth.

is known as the linear initial re-growth or re-growth model. In contrast to the logistic model, V increases linearly to K with no initial slow growth phase. Turchin (2003) points out that both models are simplistic, but different vegetation systems are reasonably characterized by one of these two models. For example, when lemmings feed on mosses, the logistic best describes the interaction since most of the moss biomass is vulnerable to direct consumption. On the other hand, grasses and sedges are best modeled with the linear re-growth equation, in that most of the biomass is underground and escapes herbivory. The exception would be if herbivores such as brown lemmings consume rhizomes and other underground parts (Pitelka 1957; Tast 1974).

The herbivore population is modeled with growth based on a positive numerical response to increases in plant net production, coupled with a density-independent or density-dependent mortality rate. The numerical response can be of several forms. Following the Lotka–Volterra format for the predator population, with no limits based on a functional response, the numerical response is simply, $\chi_h f N V$, where f is maximum grazing rate and χ_h is the herbivore's assimilation rate. The product $\chi_h f$ describes the maximum rate at which plant material is turned into herbivores and is the equivalent of $\chi_p c$ in equation 10.16. We can again add either a type II or type III functional response to the numerical response. Alternatively, we add the numerical response based on the half-saturation constant (equation 10.16): $\frac{\chi_h f N V}{b+V}$

The herbivore death rate can simply be a density-independent constant, m_h. Or, we can add a coefficient, θ, which is density independent when equal to one but which increases the herbivore death rate at high densities if $\theta > 1$. The resultant equation (11.3) describes the dynamics of the herbivore population with a functional response and a density-dependent death rate:

$$\frac{dN}{dt} = \frac{\chi_h f N V}{b+V} - m_h N^\theta \tag{11.3}$$

Crawley (1997) makes the interesting point that functional responses often work more in terms of the amount of food per herbivore as opposed to the amount of food per area. This point has also been made for predator–prey relationships. Therefore, the functional response might be more appropriately modeled as ratio dependent. Using equation 11.3, if we assume $\theta = 1$, and the functional response is dependent on the ratio of V/N, we have equation 11.4:

$$\frac{dN}{dt} = \chi_h f N \left(\frac{V}{N}\right) - m_h N \tag{11.4}$$

If we assume both populations of herbivores and plants have stopped growing, they reach an equilibrium when $dN/dt = 0$ in equation 11.4. By canceling the Ns, the result $\chi_h f V = m_h N$ can be solved for V^* (equation 11.5), which is the equilibrium plant abundance. V^* turns out to be directly proportional to the herbivore equilibrium population size, N^*, and the herbivore death rate, but inversely proportional to herbivore foraging and assimilation efficiency. Obviously, high plant abundance is favored when herbivore foraging efficiency is low and/or the herbivore population has a high death rate. Less intuitively obvious is the positive relationship between herbivore equilibrium population size and plant abundance. In fact, if we solve equation 11.5 for N^*, we have equation 11.6, which implies that an increase in vegetation results in an increase in herbivores. But Equation 11.5 seems to say that an increase in herbivores leads to an increase in vegetation biomass. As pointed out by Turchin (2003), these equations, which are the equivalent of the Rosenzweig and MacArthur (1963) model in chapter 10, lead to the so-called paradox of enrichment in which the system is highly unstable when there is an abundance of vegetation:

$$V^* = \frac{m_h N^*}{\chi_h f} \tag{11.5}$$

$$N^* = \frac{\chi_h f V^*}{m_h} \tag{11.6}$$

If we replace the logistic equation for vegetation growth with the linear re-growth equation, we have the herbivore/re-growth model of Turchin and Batzli (2001) (11.7 and 11.8). The herbivore equation is based on equation (10.16b) in which χ_h is again the assimilation efficiency of the herbivore, f represents the maximum grazing rate, and d is the half maximal grazing rate by the herbivores. The parameter μ_h is the minimum food intake necessary for the herbivore to survive, and u_0 represents plant growth rate when V is close to zero:

$$\frac{dV}{dt} = u_0 \left(1 - \frac{V}{K_v} \right) - \frac{f N V}{b + V} \tag{11.7}$$

$$\frac{dN}{dt} = \chi_h N \left(\frac{f V}{b + V} - \mu_h \right) \tag{11.8}$$

According to Turchin (2003), this model produces stability, mainly because the logistic model has an inherent time lag built into it, while the linear model does not. Also, the fact that most of the biomass of the plants is protected below ground means that the vegetation has a refuge, invulnerable (normally) to grazing. As we saw in chapter 10, the presence of a refuge has a stabilizing effect on a consumer-resource system.

To this point, we have exclusively focused on the quantity of the vegetation available to grazers, but based on the theory of inducible defenses discussed earlier, the dynamics of herbivore–plant interactions include changes in the quality of the resource for the grazer. By preferentially grazing on better-quality plants and plant parts, and by inducing defenses, heavy grazing by an herbivore population results in lower quality resources for itself.

Edelstein-Keshet (Edelstein-Keshet 1984; Edelstein-Keshet and Rausher 1989) developed a model that includes plant quality as a variable. She assumed that, without herbivory, plants devote less energy to defenses and overall plant quality increases. By contrast, grazing tends to decrease plant quality for herbivores. A simplified form of her models is presented in equations 11.9 and 11.10 (from Tuchin 2003):

$$\frac{dQ}{dt} = q - f Q N (N - n) \tag{11.9}$$

$$\frac{dN}{dt} = r_h N \left(\frac{\frac{Q}{q} - N}{\frac{Q}{q}} \right) = r_h N \left(1 - \frac{qN}{Q} \right) \tag{11.10}$$

In this model, Q is average plant quality. N is herbivore density, n is a specific density of herbivores, and r_h is the maximum growth rate of the herbivore population.

Equation 11.9 assumes that in the absence of grazers, Q increases at the constant rate, q. The second term represents the loss of quality due to the presence of herbivores. The phrase $(N - n)$ describes a tipping point. When $N < n$, the plant quality is increased faster than in the absence of herbivores (see later). When $N > n$, plant quality is decreased by grazing. Equation 11.10 is simply the logistic equation for the herbivore, but the carrying capacity of the herbivore depends on average plant quality (Q/q).

This model predicts that an increase in grazing leads to dampened oscillations leading to a stable point. Furthermore, the interactions produce lower average plant quality when grazing intensity is high.

Turchin (2003) developed a discrete time model for herbivore–plant interactions that encompasses plant quality. He then applied this model to the interaction between larch trees (*Larix sp.*) and the larch budworm (LBM) (*Zeiraphera diniana*) in the Swiss Alps. The moth population displays regular oscillations, which cover five orders of magnitude. Time series data are available from 1959 through 1977 for this moth population. As Turchin (2003) noted, plant quality is currently the "reigning" explanation for the LBM oscillations. When he analyzed the time-series data, however, he found that plant quality explained only 31% of the variance in the LBM populations. Second, he found that parasitoids (Ichneumonid wasps) likely played a role in determining LBM rates of change. Finally, Turchin (2003) has found that a tritrophic model, combining both plant quality and parasitism, is superior to any single explanation. His conclusion mirrors what we have found previously: a model involving three trophic levels, combining plant quality and parasitism, did the best job in matching the observed dynamics of the LBM populations.

In summary, (1) the addition of self-limitation terms adds stability to both plant–herbivore and predator–prey systems and (2) modeling producers with the re-growth, rather than the logistic model, also produces a more stable outcome (Turchin 2003).

11.10 Summary: the complexities of herbivore–plant interactions

Herbivores can affect individual plants, plant populations, and plant communities in subtle, complex, and unexpected ways.

11.10.1 *Distribution*

Herbivores have the ability to limit the distribution of plants geographically or by habitat. Reef fishes limit the distribution of turtle grass to areas away from the protection provided to the fish by the reefs. Randall (1965) observed a conspicuous band of bare sand averaging 10 m in width that separated coral reefs from beds of sea grasses (*Thalassia testudinum* and *Cymodocea manatorum*). The Klamath weed, *Hypericum perforatum,* introduced to California from Europe, became so wide spread and common that it was considered a pest. The beetle, *Chrysolina quadrigemina,* which feeds upon it in Europe, was introduced. This introduction, a successful case of biological control, limited the Klamath weed so well that it is now found only in shady moist areas where the beetles do not reproduce (Holloway 1964).

11.10.2 *Attack by other organisms*

Although attack by herbivores may result in an increase in induced defenses, heavy attacks can leave the plant in such a weakened condition that it becomes subject to further attack. Rockwood (1974) described one variation on this theme. The calabash tree, *Crescentia alata*, was defoliated by hand in order to examine the effects of defoliation on reproduction (see later). At the time of defoliation, it was noted that mature leaves had very little herbivore damage, but a beetle was consuming newly produced leaves at the top of the tree. After the hand defoliation, the entire tree produced a crop of new leaves. Beetles flew in from a wide geographic area and began feasting on the new leaves. The tree was completely defoliated again, while neighboring control trees, full of mature leaves, were untouched by the beetles. The trees actually flushed a crop of leaves again and were once again defoliated by the beetles. This cycle of destruction ended when the beetles went dormant in the dry season.

11.10.3 *Effects of herbivores on productivity and reproduction*

One of the tenets of plant–herbivore theory is that herbivore activity has a negative effect on plant growth and reproduction. Such an obvious relationship is, however, more complex than it appears. Several authors have proposed that herbivory, up to a certain level, may be beneficial to plant growth and reproduction. Let us begin with the common recommendation that roses and fruit trees are kept productive by a constant pruning. Next, we should note that grasslands are kept at high productivity by either allowing grazing by herbivores or a consistent mowing. Finally, Paige and Whitham (1987) and Whitham et al. (1991) found that when the scarlet gilia (*Ipomopsis aggregata*), an herbaceous plant from the southwestern United States, was browsed such that the apical meristem was destroyed, it responded by producing multiple inflorescences and up to three times as many flowers, fruits, and seeds. In all of these studies, the plants appear to compensate for herbivore damage through re-growth. In some cases, the re-growth appears to be great enough to surpass the growth in the un-browsed plants. This is known as overcompensation.

Is overcompensation possible and is it real? In theory, low levels of herbivory can be beneficial to plants such as monocots, in which the meristems are near ground level, for the following reasons: (1) Herbivores remove old, nonproductive tissues while allowing light to penetrate to the ground where new leaves are found; (2) herbivores speed up the cycling of mineral nutrients (old leaves are consumed and deposited on the ground as manure, which is quickly broken down by fungi and bacteria, releasing mineral nutrients); and (3) the saliva of herbivores contains growth-promoting substances or encourages an induced defense that protects the plant from further herbivore attacks. McNaughton (1986) endorsed overcompensation, asserting that plants have the capacity to compensate for herbivory and overcompensate for damage so as to increase fitness. Crawley (1997), however, stated that there was actually no evidence that herbivory can increase Darwinian fitness, and Belsky et al. (1993) stated that neither life history theory nor any plausible models provide support for overcompensation.

However, Agrawal (2000b) has reviewed the literature and found a number of both field and greenhouse studies documenting overcompensation by plants in response to herbivory. He noted that overcompensation may be environmentally based, and the following factors must be explored: (1) Timing of herbivory. If a plant is browsed early in the growing season, there is ample evidence that many plants are able to recover. (2) Nutrient, water, and light availability. Overcompensation might only occur if the plants are not stressed by a lack of resources needed for photosynthesis. (3) Competition. If there are few competitors in the area, then a plant can recover, again assuming adequate resources are available. Under heavy competitive stress, however, a plant that is damaged by an herbivore will lose its position in the community to other plants that are not damaged. (4) Type of tissue lost. As seen in the paper by Ohnmeiss and Baldwin

(2000), damage to certain tissues, such as new leaves, is of much greater importance to plant productivity than is damage to other tissues. When overcompensation is environmentally based, as in the above examples, Bronstein (1994) and Agrawal (2000b) have termed it conditional mutualism.

Another potential benefit of herbivory is release from apical dominance, or "phenological escape." Release from apical dominance can result in more stems, more flowers and therefore more seeds (Agrawal 2000b).

There are at least three different models that try to explain the response of plants to defoliation (Bagchi and Ritchie (2011). The *compensatory continuum* hypothesis proposes that plants can more easily show compensation to herbivore damage in resource-rich environments. However, there have been studies seeming to document the opposite; that is, an inverse relationship between resource availability and herbivore tolerance (Hawkes and Sullivan 2001). The *growth rate hypothesis*, by contrast, proposes a greater tolerance under resource-limiting conditions (Hilbert et al. 1981). The argument here is that under stressful conditions plants maintain a slow growth rate relative to their maximum potential growth rate. After defoliation, they accelerate their growth rate as compensation for tissue lost to herbivores. Finally, the third theory, the resource *limiting resource model* (LRM), considers whether herbivory specifically affects the availability of a key limiting resource versus that of another resource that does not limit plant production (Wise and Abrahamson 2008). Differentiating among these three models is obviously challenging, and they may not actually be mutually exclusive. Without going into great detail here, Bagchi and Ritchie (2011) found in their experiment that after simulating herbivore damage via clipping, plants were nutrient limited whereas without clipping they were water limited. They concluded that the results were partially explained by the last model (LRM) but not consistent with either of the other two models.

In spite of the above, high levels of herbivore damage severely limit reproduction in many species. Rockwood (1973) showed that heavy defoliation virtually eliminated reproduction in six species of tropical shrubs and small trees. Subsequently, Marquis (1992) and Rockwood and Lobstein (1994) have demonstrated a graduated response to differing levels of herbivory. At low levels of herbivory, reproduction is little reduced. However, the timing of the defoliation, its local intensity, and the amount of competition from other plants all modify the reproduction responses to defoliation. Less than 50% defoliation had little effect on herbaceous plants in Northern Virginia (Rockwood and Lobstein 1994), and the effects of defoliation were often expressed a year later. Marquis (1992) discovered that a given branch of the tropical shrub, *Piper arieianum*, suffered an 80% reduction in seed production from a mere 10% leaf removal when it was concentrated on that one branch. Finally, when the plant, *Abutilon theoophrasti*, was grown at low densities, up to 75% defoliation had no effect on reproductive fitness. At high densities, however, the same amount of defoliation reduced reproduction by 50% (Lee and Bazzaz 1980).

In conclusion, herbivores have had significant effects on the evolution of plants and continue to exert considerable selective pressure on plants today. Yet herbivores themselves have been "under the gun" from their own predators, so to speak, in both contemporary and evolutionary time. Several lines of evidence tell us that the most effective way to investigate herbivores is to evaluate both their food resources and their predators. The roles of herbivores in an ecosystem are also undoubtedly affected by the local climate and soil regime. Opportunities for research on plant–herbivore relationships are almost unlimited. Only well designed and replicated long-term field experiments will begin to address the multitude of questions remaining (Crawley 1997).

References

Agrawal, A A. 2000a. Benefits and costs of induced plant defenses for Lepidium virginicum (Brassicaceae). *Ecology* 87: 1804-1813.

Agrawal, A.A. 2000b. Overcompensation of plants in response to herbivory and the by-product benefits of mutualism. *Trends in plant science* 5: 309-313.

Agrawal, A.A. 2011. Current trends in the evolutionary ecology of plant defence. *Functional Ecology* 25: 420-432.

Agrawal, A.A., S.Y. Strauss, and M.J. Stout. 1999. Costs of induced responses and tolerance to herbivory in male and female fitness components of wild radish. *Evolution* 55: 1093-1104.

Agrawal, A.A. and K. Konno. 2009. Latex: a model for understanding mechanisms, ecology, and evolution of plant defense against herbivory. *Annual Review of Ecology Evolution and Systematics* 40: 311-331.

Bagchi, S. and M.E. Ritchie. 2011. Herbivory and plant tolerance: experimental tests of alternative hypotheses involving non-substitutable resources. *Oikos* 120: 119-127.

Baldwin, I.T. 1994. Chemical changes rapidly induced by folivory. Pages 1-23 in E. Bernays, editor. *Insect Plant Interactions*. V. Chemical Rubber Company Press, New York, NY.

Baldwin, I.T. and J.C. Schultz. 1983. Rapid changes in tree leaf chemistry induced by damage: evidence for communication among plants. *Science* 221: 277-279.

Becerra, J.X. 1994. Squirt-gun defense in Bursera and the Chrysomelid counter ploy. *Ecology* 75: 1991-1996.

Belsky, A.J., W.P. Carson, C.L. Jensen, and G.A. Fox. 1993. Overcompensation by plants: herbivore optimization or red herring? *Evolutionary Ecology* 7: 109-121.

Berenbaum, M.R. 1981. Patterns of furanocoumarins distribution and insect herbivory in the Umbelliferae: Plant chemistry and community structure. *Ecology* 62:1254-1266.

Berenbaum, M.R., 2002. Post-genomic chemical ecology: from genetic code to ecological interactions. *Journal of Chemistry Ecology* 28: 873-896.

Berenbaum, M.R. and A.R. Zangerl. 1996. Phytochemical diversity: adaptive or random variation? *Recent Advances in Pytochemistry* 30: 1-24.

Berenbaum, M.R. and A.R. Zangerl. 1999. Coping with life as a menu option: inducible defenses of the wild parsnip. Pages 10-32 in R. Tollrian and C.D. Harvell, Editors. *The Ecology and Evolution of Inducible Defenses*. Princeton University Press, Princeton, NJ.

Bodnaryk, R.P. 1992. Effects of wounding on glucosinolates in the cotyledons of oilseed rape and mustard. *Phytochemistry (Oxford)* 31: 2671-2677.

Bronstein, J.L. 1994. Conditional outcomes in mutualistic interactions. *Trends in Evolutionary Ecology* 9: 214-217.

Chen, I. 1990. Plants bite back. *Science News* 138: 408-410.

Coley, P.D. 1980. Effects of leaf age and plant life history patterns on herbivory. *Nature* 284: 545-546.

Coley, P.D., J.P. Bryant, and F. S. Chapin, III. 1985. Resource availability and plant antiherbivore defense. *Science* 230: 895-899.

Coley, P.D. and T.A. Kursar. 2014. On tropical forests and their pests. *Science* 343: 35-36.

Crawley, M.J. 1997. Plant herbivore dynamics. Pages 401-474 in M.J. Crawley, Editor. *Plant Ecology*. Blackwell Scientific, Oxford, UK.

Cyr, H. and M.L. Pace. 1993. Magnitude and patterns of herbivory in aquatic and terrestrial ecosystems. *Nature* 361: 148-150.

Dalton, B.R. 1999. The occurrence and behavior of plant phenolic acids in soil environments and their potential involvement in allelochemicals interference interactions: methodological limitations in establishing conclusive proof of allelopathy. Pages 57-74 in, Inderjit, K.M.M.

Dakshini and C.L. Foy, Editors, *Principles and practices in plant ecology: allelochemicals interactions.* CRC Press, Boca Raton, LA.

Dussourd, D.E. and R.F. Denno. 1994. Host range of generalist caterpillars: trenching permits feeding on plants with secretory canals. *Ecology* 75: 69-78.

Ehrlich, P.R. and P.H. Raven. 1964. Butterflies and plants: a study in co-evolution. *Evolution* 18: 586-608.

Farmer, E.E. and C.A. Ryan. 1990. Interplant communication: airborne methyl jasmonate induces synthesis of proteinase inhibitors in plant leaves. *Proceedings of the National Academy Science of the United States of America* 87: 7713-7716.

Farmer, E.E. and C.A. Ryan. 1992. Octadecanoid precursors of jasmonic acid activate the synthesis of wound-inducible proteinase inhibitors. *Plant Cell* 4: 129-134.

Feeny, P.P. 1970. Seasonal changes in oak leaf tannins and nutrients as a cause of spring feeding by winter moth caterpillars. *Ecology* 51: 565-581.

Feeny, P.P. 1976. Plant apparency and chemical defense. Pages 1-40 in J. Wallace and R.L. Mansell, Editors. *Biochemical Interactions Between Plants and Insects.* Volume 10 of Recent Advances in Phytochemistry. Plenum Press, New York, NY.

Foster, M.A., J.C. Schultz, and M.D. Hunter 1992. Modelling gypsy moth-virus-leaf chemistry interactions: implications of plant quality for pest and pathogen dynamics. *Journal of Animal Ecology* 61: 509-520.

Edelstein-Keshet, L. 1984. *Mathematical theory for plant-herbivore systems.* Lefschetz Center for Dynamical Systems, Providence, RI.

Edelstein-Keshet, L. and M.D. Rausher. 1989. The effects of inducible plant defenses on herbivore populations. I. Mobile herbivores in continuous time. *American Naturalist* 133: 787-810.

Fraenkel, G.S. 1959. The raison d'être of secondary plant substances. *Science* 129: 1466-1470.

Gershenzon, J. 1994. The cost of plant chemical defense against herbivory: a biochemical perspective. Pages 105-173 in E.A. Bernays, Editor. *Insect-Plant Interactions*, Vol. 5. CRC Press, Boca Raton, LA.

Gleadow, R.M. and I.E. Woodrow. 2000. Temporal and spatial variation in cyanogenic glycosides in Eucalyptus cladocalyx. *Tree Physiology* 20: 591-598.

Gleadow, R.M. and I.E. Woodrow. 2002. Constraints on effectiveness of cyanogenic glycosides in herbivore defense. *Journal of Chemical Ecology* 28: 1301-1313.

Haig, T. 2008. Allelochemicals in plants. Pages 63-104 in R.S. Zeng, A.U. Mallik, and S. Luo, Editors. *Allelopathy in Sustainable Agriculture and Forestry. Springer*, New York, NY.

Harborne, J.B. 1988. *Introduction to Ecological Biochemistry* 3rd Edition. Academic Press, London, UK.

Harborne, J.B. 1993. *Introduction to Ecological Biochemistry* 4th Edition. Academic Press, London, UK.

Harborne, J.B. 1997. Plant Secondary Metabolism. Pages 132-155 in M.J. Crawley, Editor. *Plant Ecology.* Blackwell Scientific, Oxford, UK.

Hartley, S.E. and C.G. Jones. 1997. Plant chemistry and herbivory, or why the world is green. Pages 284–324 in M.J. Crawley, Editor. *Plant Ecology.* Blackwell Scientific, Oxford, UK.

Harvell, C.D. 1984. Predator-induced defense in a marine bryozoan. *Science* 224: 53–54.

Harvell, C.D. and R. Tollrian. 1999. Why inducible defenses? Pages 3–9 in R. Tollrian and C.D. Harvell, Editors. *The Ecology and Evolution of Inducible Defenses.* Princeton University Press, Princeton, NJ.

Hawkes, C.V. and J.J. Sullivan. 2001. The impact of herbivory on plants in different resource conditions: a meta-analysis. *Ecology* 82: 2045–2058.

Heil, M., S. Greiner, H. Meimberg, R. Kruger, J.L. Noyer, G. Heubl, K.E. Linsenmair, and W. Boland. 2004. Evolutionary change from induced to constitutive expression of an indirect plant resistance. *Nature* 430: 205–208.

Hilbert, D.W., D.M. Swift, J.K. Detling, and M.I. Dyer. 1981. Relative growth rates and the grazing optimization hypothesis. *Oecologia* 51: 14–18.

Holloway, J. K. 1964. Projects in biological control of weeds. Pages 650–670 in P. DeBach, Editor, *Biological Control of Insect Pests and Weeds*. Reinhold Publication Corporation, New York.

Janzen, D.H. 1974. Blackwater rivers, animals, and mast fruiting by the Dipterocarpaceae. *Biotropica* 6: 69–103.

Karban, R and J.H. Myers. 1989. Induced plant responses to herbivory. *Annual Review on Ecology and Systematics* 20: 331–348.

Karban, R. and I.T. Baldwin. 1997. *Induced Responses to Herbivory*. University of Chicago Press, Chicago, IL.

Lee, T.D. and F.A. Bazzaz. 1980. Effects of defoliation and competition on growth and reproduction in the annual plant Abutilon theophrasti. *Journal of Ecology* 68: 813–821.

Levin, D.A. 1976. The chemical defenses of plants to pathogens and herbivores. *Annual Review on Ecology and Systematics* 7: 121–159.

Marquis, R.J. 1992. A bite is a bite? Constraints on response to folivory in *Piper arieianum* (Piperaceae). *Ecology* 73: 143–152.

Mattiacci, L., M. Dicke, and M.A. Posthumus. 1995. Betaglucosidase: an elicitor of herbivore-induced plant odor that attracts host-searching parasitic wasps. *Proceedings of the National Academy of Science of the United States of America* 92: 2036–2040.

McKey, D.B., P.G. Waterman, C.N. Mbi, J.S. Gartlin, and T.T. Strusaker. 1978. Phenolic content of vegetation in two African rainforests: ecological implications. *Science* 202: 61–64.

McNaughton, S.J. 1986. On plants and herbivores. *American Naturalist* 128: 765–770.

Millburn, P. 1978. Biotransformation of xenobiotics by animals. Pages 35-76 in J.B. Harborne, editor, *Biochemical Aspects of Plant and Animal Coevolution*. Academic Press, London

Olff, H., V.K. Brown, and R.H. Drent. 1999. *Herbivore: Between Plants and Predators*. Blackwell Science Ltd., Oxford, UK.

Ohnmeiss, T.E. and I.T. Baldwin. 2000. Optimal defense theory predicts the ontogeny of an induced nicotine defense. *Ecology* 81:1765–1783.

Paige, K.N. and T.G. Whitham. 1987. Overcompensation in response to mammalian herbivory. *American Naturalist* 129: 407–416.

Pimentel, D. 1988. Herbivore population feeding pressure on plant hosts: feedback evolution and host conservation. *Oikos* 53: 289–302.

Pitelka, F.A. 1957. Some characteristics of micotine cycles in the Arctic. Pages 73–88 in H.P. Hansen, Editor. *Arctic Biology*. Oregon StateUniversity Press, Corvalis, OR.

Randall, J.E. 1965. Grazing effect of sea grasses by herbivorous reef fishes in the West Indies. *Ecology* 46: 255–260.

Rhoades, D.F. 1983. Responses of alder and willow to attack by tent caterpillars and weborms: Evidence for pheromonal sensitivity of willows. Pages 55–58 in P.A. Hedin, Editor. *Plant Resistance to Insects*. Symposium series 208. American Chemical Society, Washington, D.C.

Rockwood, L.L. 1973. The effect of defoliation on seed production of six Costa Rican tree species. *Ecology* 54: 1363–1369.

Rockwood, L.L. 1974. Seasonal changes in the susceptibility of Crescentia alata leaves to the flea beetle, Oedionychus, sp. *Ecology* 55: 142–148.

Rockwood, L.L. 1976. Plant selection and foraging in two species of leaf-cutting ants, *Atta*. *Ecology* 57: 48–61.

Rockwood, L.L. 1977. Foraging patterns and plant selection in Costa Rican leaf-cutting ants. *Journal of New York Entomological Society* 85: 222–233.

Rockwood, L.L. and K.E. Glander. 1979. Howling monkeys and leaf-cutting ants: comparative foraging in a tropical deciduous forest. *Biotropica* 11: 1–10.

Rockwood, L.L. and M.B. Lobstein. 1994. The effects of experimental defoliation on reproduction in four species of herbaceous perennials from Northern Virginia. *Castanea* 59: 41–50.

Rosenzweig, M.L. and R.H. MacArthur. 1963. Graphical representation and stability conditions of predator-prey interactions. *American Naturalist* 97: 209–223.

Rossiter, M., J.C. Schultz, and I.T. Baldwin. 1986. Relationships among defoliation, red oak phenolics, and gypsy moth growth and reproduction. *Ecology* 69: 267–277.

Schappert, P.J. and J.S. Shore. 1999a. Cyanogenesis, herbivory, and plant defense in *Turnera ulmifolia* on Jamaica. *Ecoscience* 6: 511–520.

Schappert, P.J. and J.S. Shore. 1999b. Effects of cyanogensis in polymorphism in Turnera ulmifolia on Euptoieta hegesia and potential Anolis predators. *Journal of Chemical Ecology* 25: 1455–1479.

Schultz, J.C. and I.T. Baldwin. 1982. Oak leaf quality declines in response to defoliation by gypsy moth larvae. *Science* 217: 149–150.

Smith, R.L. and T.M. Smith. 2001. *Ecology and Field Biology*. 6th Edition. Benjamin Cummings, San Francisco, CA.

Swain, T. 1965. The tannins. Pages 552–580 in J. Bonner and J.E. Varner, Editors. *Plant Biochemistry*. Academic Press, New York, NY.

Tast, J. 1974. The food and feeding habits of the root vole, Microtus oecnomus, in Finnish Lapland. *Aquilo Seriae Zooligica* 15: 25–32.

Tollrian, R. and C.D. Harvell. 1999. *The Ecology and Evolution of Inducible Defenses*. Princeton University Press, Princeton, NJ.

Turchin, P. 2003. *Complex Population Dynamics: A Theoretical/Empirical Synthesis*. Princeton University Press, Princton, NJ.

Turchin, P. and G. Batzli. 2001. Availability of food and the population dynamics of arvicoline rodents. *Ecology* 82: 1521–1534.

Vetter, J. 2000. Plant cyanogenic glycosides. *Toxicon* 38: 11–36.

Whitham, T.G., J. Maschinski, K.C. Larson, and K.N. Paige. 1991. Plant responses to herbivory: the continuum from negative to positive and underlying physiological mechanisms. Pages 227–256 in, P.W. Price, T.M. Lewinsohn, G.W. Fernandes, and W.W. Benson, Editors. *Plant-Animal Interactions: Evolutionary Ecology in Tropical and Temperate Regions*. Wiley-Interscience, New York, NY.

Wielgolaski, F.E. 1975. Productivity of tundra ecosystems. Pages 1–12 in D.E. Reichle, J.F. Franklin, and D.W. Goodall, Editors. *Productivity of World Ecosystems*. National Academy of Sciences, Washington, DC.

Wise, M.J. and W.G. Abrahamson. 2008. Applying the limiting resource model to plant tolerance of apical meristem damage. *American Naturalist* 172: 635–647.

12 Multi-trophic interactions

Jonathan W. Witt

12.1	Introduction

Except for ecosystems that are very resource limited, most natural systems contain at least three trophic levels (plants, herbivores and predators). In previous chapters we only considered interactions between adjacent trophic levels – between plants and herbivores or between predators and prey. In this chapter, we will examine relationships among three trophic levels, species or groups: trophic cascades, *intraguild predation* (*IGP*), and cannibalism. Each can be identified by the presence of an intermediate consumer situated between resources and a top-tier predator (Figure 12.1). Trophic cascades connect three agents linearly, where consumers deplete resources and predators eat consumers (Figure 12.1a). IGP is similar, except that the predator can also deplete the resource (Figure 12.1b). With cannibalism, the consumer is replaced by young or small conspecific predators (Figure 12.1c).

These multi-trophic interactions are important to study for a number of reasons. Natural food webs can be incredibly complex and full of interactions that span the food web. Studying simplified multi-trophic interactions provides a glimpse of the connectedness that exists in natural communities. In these models, the three trophic levels or agents are connected such that changes in the population abundance of one species, especially at the highest level, can dynamically affect the behaviors of the other populations. Extensive study of multi-trophic interactions has informed our understanding of many community processes. For example, they have provided insights into factors that facilitate predator coexistence, and help explain ecosystem changes after apex predator removal or mesopredator release. In particular, these interactions demonstrate how predators can influence lower trophic levels, and provide compelling arguments for proper predator management and conservation (Sergio et al. 2008; Prugh et al. 2009; Soulé 2010; Estes et al. 2011).

Introduction to Population Ecology, Second Edition. Larry L. Rockwood.
© 2015 John Wiley & Sons, Ltd. Published 2015 by John Wiley & Sons, Ltd.
Companion Website: www.wiley.com/go/rockwood/populationecology

Fig. 12.1 A trophic
cascade (a), where
predators (*P*) indirectly
benefit resources (*N*) by
consuming consumers
(*C*). Intraguild predation
(b), where consumers
and predators also
compete for a common
resource. Cannibalism
(c), where larger or older
predators (*P*₂) consume
smaller or younger
conspecifics (*P*₁).

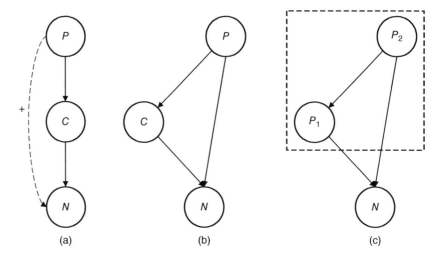

(a)   (b)   (c)

As with two species models, those presented in this chapter provide a simplified
mathematical framework for understanding population growth. They do not, however, always capture the complexity of natural systems, and their outcomes depend on
the ecological context. Because most ecologists need to integrate conceptual theory
and real-world complexities in understanding population dynamics, this chapter provides case studies that either have applications in conservation biology and ecosystem
management or demonstrate the importance of context in determining the outcome of
multi-trophic interactions.

12.2 Trophic cascades

The simplest multi-trophic interaction is a trophic cascade. The development of trophic
cascade theory can be traced to Hairston et al. (1960), who argued that because the
world is not devoid of vegetation, herbivores are not resource-limited and must be controlled in part by predation. Their idea became known as the Green World Hypothesis.
Although simplistic, this hypothesis was significant because it asserted that plants are
indirectly benefited by predators, and a full understanding of plant communities must
involve both herbivores and predators.

The idea that predators indirectly benefit plants by suppressing herbivores became
formally known as a trophic cascade (Paine 1980), and the definition now generally
encompasses all indirect effects of predators on non-adjacent trophic levels (Estes
et al. 2001). The classic example by Paine (1966) demonstrated that predation by the
starfish, *Piaster*, in the rocky intertidal zones of the western North America, indirectly
benefits invertebrate communities and kelp (reviewed in Menge 1995, Robles and
Desharnais 2002). When *Piaster* was removed, dominant competitors in the mussel
genus *Mytilus*, once kept in check by *Piaster*, depleted kelp populations. Through a
combination of kelp loss and competition with *Mytilus*, more than half of invertebrate
species in communities where *Piaster* was removed disappeared. Conversely, at sites
where *Piaster* was allowed to return, mussel beds receded and the community slowly
recovered (Paine 2010).

A trophic cascade is conceptualized as a linear connection of three species, where
predators eat consumers and consumers deplete resources (Figure 12.1a). For simplicity, in most of the examples presented, consumers are herbivores and resources
are plants. However, any group of linearly connected species that span three trophic

levels (e.g., an apex predator, mesopredator and herbivore) can create a trophic cascade. The cascade occurs if predators are efficient or numerous enough to keep consumer densities low, which results in higher amounts of resources than if predators were absent. Consumers in this tri-trophic situation are wedged between two opposing forces, resource availability and mortality from predation, and their abundance depends on the balance between these forces.

As a way to introduce a basic trophic cascade model and the feedbacks that occur between resource availability and predation, consider the consumer-resource (predator–prey) models discussed in Chapter 10. First, consider the impact of predation on resources using equations 10.4b and 10.5. If resource growth between two time periods depends on consumer population size, any predation event that lowers consumer abundance will result in relatively more resources in the next time step. Likewise, any increase in predator abundance will lead to fewer consumers in the next time step. If we combine these two models, the implication is that higher predator abundance will lead to higher plant abundance. Second, consider the impact of resources on predators using the same two sets of equations. Higher resource abundance supports more consumers, and more consumers support more predators. Combining these two examples, it suggests that high resource abundance leads to both more predators and more intense trophic cascades (Oksanen et al. 1981, Forkner and Hunter 2000, Carpenter et al. 2001).

We can formally introduce the trophic cascade model by adding a third trophic level to equations 10.4b and 10.5. The resulting sets of equations are very similar, except the consumer level has an added predator mortality term ($E_{pc}CP$) and unique conversion efficiencies and encounter rates for predators and consumers ($\chi_{pc}E_{pc}$ and $\chi_{cn}E_{cn}$, respectively). E_{cn} represents the searching efficiency of the consumer for the resource, and E_{pc} represents the searching efficiency of the predator for the consumer. χ_{pc} represents the efficiency by which consumers are turned into new predators, and χ_{cn} represents the efficiency by which resources are turned into new consumers. Depending on how one wishes to model species interactions, more complex functional responses from chapter 10 or saturation constants can be added. The models presented here use simple Type I functional responses (Equations 12.1-12.3):

$$\frac{dN}{dt} = r_n N \left(\frac{K_n - N}{K_n} \right) - E_{cn}CN \tag{12.1}$$

$$\frac{dC}{dt} = \chi_{cn}E_{cn}CN - E_{pc}CP - m_c C \tag{12.2}$$

$$\frac{dP}{dt} = \chi_{pc}E_{pc}PC - m_p P \tag{12.3}$$

This model can lead to four outcomes: (1) all populations eventually decline to zero, (2) resources persist but consumers and predators eventually decline to zero, (3) resources and consumers persist but predators eventually decline to zero, or (4) coexistence of all three populations. Under conditions where resources, consumers, and predators coexist, predators in this model will always create a trophic cascade. For example, Figure 12.2 shows consumer and resource populations oscillating towards equilibrium with (solid lines) and without (dashed lines) the predator present. Here, even a small consistent number of predators results in more resources and fewer consumers.

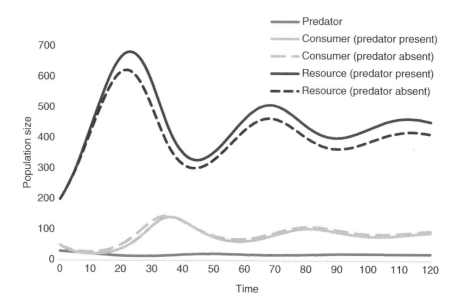

Fig. 12.2 Predator, consumer and resource abundance with damping oscillations using Equations 12.1–12.3. Here $N_0 = 200$, $r_n = 0.15$, $K_n = 1000$, $C_0 = 50$, $E_{cn} = 0.001$, $\chi_{cn} = 0.5$, $m_c = 0.2$, $P_0 = 30$, $E_{pc} = 0.001$, $\chi_{pc} = 0.6$, $m_p = 0.05$. Dashed lines show consumer and resource abundance when the predator is absent. The presence of a predator in the model creates a trophic cascade.

The behavior of this model can be explored by allowing carrying capacity and predator encounter rates to vary while fixing all other parameters. Figure 12.3 shows (a) which trophic levels are present when the model is close to equilibrium (here at $t = 5000$), (b) the final predator abundance, and (c,d) the difference in consumer and resource abundance compared to a simple consumer and resource model. Here, as the carrying capacity and encounter rate increase, the predator is more likely to be present (Figure 12.3a). Likewise, more predators are supported at higher resource carrying capacities (Figure 12.3b), which results in more intense trophic cascades (Figure 12.3c,d). At low resource abundances, not enough energy passes through the consumer level to support predators (Figure 12.3a,b).

This model, however, is sensitive to perturbation, and predators are predicted to persist over a smaller range of parameter values than consumers (reviewed in Terborgh et al. 2010). This sensitivity occurs in part because both consumer and predator populations are not self-limited, which can lead to large magnitude oscillations. In other words, self-limitation restricts how many new consumers and predators can be added in one time step, which adds stability to the model. To account for this, Turchin (2003) proposed adding predator self-limiting terms to a tri-trophic model based on his herbivore-plant regrowth equations (Eqns. 11.7 and 11.8), which are presented below. Here, the consumer growth rate is diminished by the term $(cCP/d + C)$, as described in Chapter 10 (Eqns. 10.8b and 10.11), and the predator growth rate is reduced by (sP^2/K_p), where s is an intraspecific competition coefficient and K_p is a predator carrying capacity related to territoriality (Turchin 2003). Here r_n and r_c are the maximum growth rates of the resource and consumer:

$$\frac{dN}{dt} = r_n \left(1 - \frac{N}{K_n}\right) - \frac{fCN}{b + N} \tag{12.4}$$

Fig. 12.3 Tri-trophic model behavior (equations 12.1–12.3) at $t = 5000$, where K_n varies from 300 to 1000 and E_{pc} from 0.0005 to 0.002. All other parameters are held constant: $N_0 = 200$, $r_n = 0.15$, $C_0 = 50$, $E_{cn} = 0.001$, $\chi_{cn} = 0.5$, $m_c = 0.2$, $P_0 = 30$, $\chi_{pc} = 0.6$, $m_p = 0.05$. Final outcome (a): (1) resources persist but predators and consumers decline to zero, (2) consumers and resources persist but predators decline to zero, and (3) all three persist. Under these parameter values, the outcome where all populations decline to zero does not occur. To illustrate the trophic cascade, (b) the final predator population size is shown (P_{5000}, and (c,d) the differences in consumer population size ($C_{5000,\ predator\ present} - C_{5000,\ predator\ absent}$) and resource population size ($N_{5000,\ predator\ present} - N_{5000,\ predator\ absent}$) with and without the predator present.

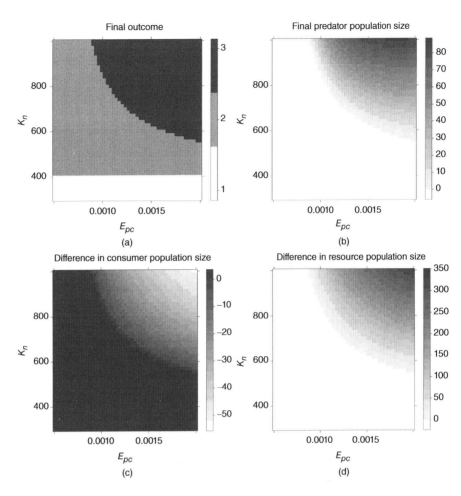

$$\frac{dC}{dt} = r_c C \left(\frac{\chi_{cn} f N}{b + N} - \mu_c \right) - \frac{cCP}{d + C} \tag{12.5}$$

$$\frac{dP}{dt} = \chi_{pc} P \left(\frac{cC}{d + C} - \mu_p \right) - \frac{sP^2}{K_p} \tag{12.6}$$

Trophic cascades have been found in all environments, but are thought to be more common in aquatic systems (Shurin et al. 2002). Within specific ecosystems, their strength and prevalence varies, depending on the environmental and community context (Shurin et al. 2002, Borer et al. 2005). Generally strong links between resources and consumers, and consumers and predators must exist to create strong cascading effects. In particular, highly efficient, but predation-vulnerable, herbivores create situations where predators can have strong indirect effects on plants (Polis and Strong 1996). Analogously, we can expect strong cascades in cases where highly competitive herbivores are subsequently preyed upon by a dominant or keystone predator (Estes and Duggins 1995, Estes et al. 2010). Keystone species are those that have an exceptionally high and wide-spread impact on their communities, such that changes in keystone species abundance typically lead to strong cascading effects. The *Piaster*-induced trophic cascade is a prime example of interactions between a competitively dominant herbivore and a keystone predator (Paine 1966, 2010).

Factors that alter the intensity of a trophic cascade often do so by modifying the interaction strength between predators and consumers or consumers and resources. Predator and prey behavior, nonlinear trophic interactions, community diversity, interactions with the physical environment, and system productivity can all alter how species interact, and thus, the outcome of a trophic cascade. Changes in prey behavior, such as reduced foraging or resource switching in the presence of a predator, may reduce the impact of predation on consumers (Schmitz et al. 2004, Janssen et al. 2007). Omnivores and mutualists, which form non-linear food chains, can confuse the roles of predators and consumers (Thompson et al. 2007, Hunter 2009). For example, if birds consume both invertebrate herbivores and pollinators (mutualists), any indirect links between birds and plants could be obscured because these consumers have contrasting effects on plant populations (Knight et al. 2006). Increases in herbivore diversity are thought to lower the intensity of trophic cascades by diluting the impact of predators on herbivores and herbivores on plants (Polis and Strong 1996, Schmitz et al. 2000, Stachowicz et al. 2007), whereas changes in predator diversity can raise or lower trophic cascade intensity, depending on the degree of antagonism between predators (Finke and Denno 2004, 2006, Stachowicz et al. 2007).

Physical and biological habitat characteristics, which vary across prey and predator ranges, can alter how species in a tri-trophic structure interact. For example, Frank et al. (2006, 2007) found that the impact of predators decreased with increasing water temperature and fish diversity. In this case, increased diversity diluted predator impacts across species, while higher water temperature increased prey growth rates, which further blunted the impact of predators. Habitat heterogeneity and changes in habitat complexity, which often occur at local or small scales, can provide refuges to consumers and lower predator efficiency (Finke and Denno 2002, 2006, Witt et al. 2013). In cases where herbivores benefit from complexity, weaker predator-herbivore relationships may occur, while in cases where small predator species benefit, stronger predator–herbivore relationships may occur (Janssen et al. 2007). For example, Power (1992) found that roach (*Hesperoleucas symmetricus*) and steelhead trout (*Oncorhynchus mykiss*are) were effective at capturing invertebrate predators on bedrock substrates, which allowed chironomids to suppress algae. However, these fish were relatively ineffective in gravel substrates, which allowed invertebrate predators to suppress chironomid densities, leading to higher algal stocks.

So far, we have discussed trophic cascades in terms of density-mediated effects, which is implicit in our understanding of populations and the models in this book. Nevertheless, changes in population abundance or density, although conceptually simple, are infrequently the only driver of trophic cascades. Rather, many of the indirect effects seen in cascades are facilitated by anti-predator behavior, or are a result of both behavioral and density-mediated changes (Box 12.1, Schmitz et al. 2004, Steffan and Synder 2010). Under predation risk, consumers are expected to adopt behaviors that balance their foraging needs against the additional risk of mortality. Although predator avoidance strategies vary, they often include reduced movement rates or foraging duration, increased refuge use, resource switching, and changes in foraging locations and times to minimize encounter rates (Lima and Dill 1990, Lima 1998).

Conversely, semi-social animals may herd or school together, which may increase detection rates of the group, but ultimately lowers individual mortality risk (Lima and Dill 1990, Hebblewhite and Pletscher 2002, Magurran 2005). All of these predator-induced non-lethal effects can alter the amount and type of resources consumed by prey without affecting mortality directly. This has commonly been referred to as the "ecology of fear" (Brown et al. 1999, Ripple and Beschta 2003, 2004, Berger 2010).

Box 12.1 Behavior mediated trophic cascades in old field invertebrates

Both density and behaviorally-mediated trophic cascades can occur within the same system and create unique cascading effects, depending on how prey respond to different predators. In an experimental old field system, Schmitz and Suttle (2001) and Schmitz (2008) examined how differences in hunting behavior between two spiders influence the density and behavior of the common grasshopper, *Melanoplus femurrubrum*. Here, the nursery web spider, *Pisaurina mira*, is a sit-and-wait predator, while the jumping spider, *Phidippus rimator*, is a roaming active hunter.

In their grassland system, plants can be broken into three dominant groups: a nutritious and common grass, *Poa pratensis*, a competitively dominant herb, *Solidago rugosa*, and a variety of competitively weaker herbs. Jumping spiders roam the plant-scape in search of unsuspecting prey; but, because the spiders have unpredictable hunting patterns, grasshoppers do not show a strong anti-predator behavioral response. As a result, these spiders cause a density-mediated trophic cascade and release the grass and dominant herb from herbivory, resulting in an 18% increase in above net primary plant production. However, this allows *S. rugosa* to suppress competitively weaker herbs, lowering plant diversity by 7%.

Conversely, nursery web spiders leave stronger cues due to their predictable (sit and wait) hunting patterns, and elicit strong anti-predator behavioral responses in grasshoppers. In their presence, grasshoppers change their foraging strategy to avoid the spider, and relegate themselves to refuges within *S. rugosa*. In this case, nursery web spiders cause a behaviorally-mediated trophic cascade by facilitating herbivory on *S. rugosa*, decreasing herbivory on *P. pratensis* and releasing other herbs from competition. Lower *S. rugosa* biomass results in a 28% decrease in above ground net primary plant production and a marginal increase in plant diversity.

Despite causing cascades through different pathways, both predators have an indirect positive effect on the grass, *P. pratensis*, which is the most nutritious plant available to grasshoppers. Without either predator present, declines in *P. pratensis* biomass occurr.

12.3 Trophic cascades and antropogenic change

Trophic cascade theory has been particularly useful in providing plausible, mechanistic explanations for the often drastic community and ecosystem changes observed after removal or introduction of large-bodied consumers or predators (Estes et al. 2011). Over the last 200 years, hunting and persecution of large predators, such as wolves, bears, sharks, and predatory cats have resulted in either drastic range reduction or the extinction of many species (Prugh et al. 2009, Estes et al. 2011). Due to their positions at the top of food chains, these animals are often keystone predators, and they exert a number of direct and indirect effects on lower trophic levels. Accordingly, removal of large predators can fundamentally alter the ecosystems they govern; some striking examples include the extirpation and reintroduction of wolves to Yellowstone National Park (but see Middleton et al. 2014), maintenance of kelp forests by sea otters, and ecosystem collapse following overfishing and the extirpation of sharks (Box 12.2a, Estes and Duggins 1995, Jackson et al. 2001, Ripple and Beschta 2003, 2004, Myers et al. 2007). Alternately, in freshwater systems, water managers have exploited trophic cascades between fish, zooplankton, and phytoplankton to improve water quality (Box 12.2b, Hansson et al. 1998, Carpenter et al. 2001, Lathrop et al. 2002).

Box 12.2a Sea otter maintenance of kelp forests in the northern Pacific

Sea otters (*Enhydra lutris*) once had a range that covered the northern Pacific Rim from Japan to Baja California (Reinman and Estes 1990). Exploitation of sea otters for their fur began over 250 years ago, and by the time the species gained protection under the Fur Seal Treaty in 1911, only 13 small remnant colonies remained. The patchy reintroduction and recovery of the species (as well as documented declines over the last 15 years) in the United States and British Columbia have provided one of the more compelling natural experiments in trophic cascades. By examining sites with and without sea otters, Estes and colleagues (Estes and Palmisano 1974, Estes et al. 1978) found that otter indirectly benefit kelp through a trophic cascade by reducing urchin populations.

Otters are a keystone predator in these coastal reef systems, and reefs with otters typically contain 90% fewer urchins than those without (Estes and Duggins 1995). In the presence of otter, urchin herbivory is minimized and kelp forests form. Conversely, when otters are removed, urchins are released and create habits denuded of kelp (urchin barrens).

Many extensive and indirect effects of the otter-urchin-kelp cascade have been documented (reviewed in Estes et al. 2004, Estes et al. 2010). The creation of complex kelp habitat fosters communities of barnacles and fish (Duggins et al. 1989, Reisewitz et al. 2005). Increased prey diversity shifts Glaucous-winged gull (*Larus glaucescens*) diets from predominantly macroinvertebrates in urchin barrens to fish in kelp forests, and bald eagle (*Haliaeetus leucocephalus*) diets from predominantly sea birds in urchin barrens to a mix of birds, marine mammals and fish in kelp forests (Irons et al. 1986, Anthony et al. 2008).

In the 1990s, however, otter populations at many Alaskan sites declined sharply and collapsed, reverting the affected locations to urchin barrens (Estes et al. 1998, Doroff et al. 2003). It has been hypothesized that killer whale predation caused these recent declines (Estes et al. 1998, Williams et al. 2004, Estes et al. 2009), but what portion of the decline and the underlying reason for increased killer whale exploitation remain in question (Wade et al. 2007, Estes et al. 2009).

Box 12.2b Piscivore control of primary production

Eutrophication of freshwater systems, caused primarily by increased anthropogenic nitrogen and phosphorus inputs, facilitates algal and bacterial production. This increased production can lead to the collapse or degradation of aquatic food webs, light limitation, toxic cyanobacteria blooms, and periods of summer hypoxia. Bio-manipulation of freshwater lakes, by stocking large piscivores, has provided one mechanism for improving the quality of degraded waters through a complex trophic cascade (Hansson et al. 1998, Carpenter et al. 2001, Lathrop et al. 2002).

In this trophic cascade, large piscivores release zooplankton from predation by consuming planktivorous fish; the zooplankton in turn graze on algae, which suppresses algal biomass and improves water clarity and quality (Carpenter et al. 2001). When piscivores are absent, planktivores consume zooplankton and algal growth accelerates. Extensive study of this system has revealed numerous complications to this relatively simple dynamic. The trophic cascade is enhanced and maintained by several factors, including the interaction of piscivores and lake hydrology, piscivore-induced changes in light and nutrient availability, changes in plankton community composition, behaviorally mediated effects, and the availability of allochthonous inputs.

First, shallow lakes and nutrient enrichment can amplify the effect of the cascade (Carpenter et al. 2001, Benndorf et al. 2002, Mehner et al. 2002). In relatively shallow lakes, cascade-induced improvements in water clarity allow macrophytes to recolonize littoral zones and compete for nutrients, which makes them unavailable to algae and which amplifies the strength of the cascade (Benndorf et al. 2002, Mehner et al. 2002). Although nutrient enrichment leads to higher algal abundance, it also increases zooplankton grazing, reproductive, and nutrient sequestration rates (Sarnelle 1992, Carpenter et al. 2001, Mehner et al. 2002). These stored nutrients are similarly unavailable to algae. In situations when phosphorous cannot be limited by zooplankton or macrophytes, such as in very large eutrophic lakes, these relationships break down and weaken the cascade (Benndorf et al. 202, Mehner et al 2002).

Second, zooplankton and algae composition change in this cascade, which further alters water quality and clarity. Planktivores selectively choose large bodied zooplankton, such as *Daphnia*, which have higher per capita grazing efficiencies and are active for a longer portion of the year (Shapiro 1990, Carpenter et al. 2001, Lathrop et al. 2002). In the presence of just planktivores, a larger portion of the zooplankton community becomes comprised of smaller, ephemeral species, and zooplankton impacts on algae decrease even when zooplankton biomass stays relatively constant (Carpenter et al. 2001).

Third, in the presence of piscivores, planktivores will shift from the pelagic zone to more complex littoral habitats to avoid predation (Schlosser 1987, Eklöv and VanKooten 2001). These shifts reduce planktivore foraging rates, cause shifts in planktivore diets, and compel zooplankton to move offshore to avoid planktivores, which exposes them to more algae (Hölker et al. 2002, Romare and Hansson 2003).

Fourth, allochthonous inputs into lake systems can subsidize and stabilize fish populations involved in the trophic cascade. Detritus from terrestrial systems supports many invertebrates consumed by fish, and terrestrial prey can be eaten by fish directly (Polis and Strong 1996, Hodgeson and Hansen 2005, Shurin et al. 2006). In support of this, Cole et al. (2006) found that in some lakes up to 40% of piscivorous largemouth bass (*Micropterus salmoides*) diets were made up of terrestrial prey.

12.4 Intraguild predation

Analyses of multi-trophic interactions are further complicated by the fact that many animals do not belong to a single trophic level (Polis and Strong 1996, Thompson et al. 2007, Hunter 2009). Many predators do not just eat herbivores, but will also consume other predators, members of their own species (cannibalism), and plants. Acquiring resources from more than one trophic level is called omnivory. The lack of clarity in trophic distinction often results in situations where predators and prey compete for resources, which creates complex population dynamics and alters the expected results of simple trophic cascades.

Intraguild predation (IGP), which is defined as predation that occurs between competitors for a common resource, is thought to be ubiquitous in nature and, like trophic cascades, more common in aquatic systems (Polis and Holt 1992). The interaction is so common that in an examination of over 100 food webs, Arim and Marquet (2004) found that 50-80% of all species participated in some form of IGP. These relationships are usually conceptualized as asymmetric, where a predator (or intraguild predator, or apex predator) consumes both a consumer (or intraguild prey, or mesopredator) and a shared resource (or common prey), while the consumer only utilizes the shared resource (Figure 12.1b). The most basic effect of IGP is to spread out the impact of predators between both consumers and resources (Polis and Strong 1996). IGP is thought to add stability to predator populations, because not only do predators have an alternate resource, but eating consumers also eliminates competition for resources (Polis and Strong 1996). Although simplistic, IGP is predicted to weaken the strength of trophic cascades, because it results in more consumers and fewer resources than would be expected by a simple linear relationship. Furthermore, predator removal can have multiplicative effects on consumers because resource levels increase at the same time consumer mortality decreases (Holt and Polis 1997, Brashares et al. 2010).

We can modify Equations 12.1–12.3 to incorporate IGP by extending predator influence down to the resource level. In this model, however, the predator-consumer assimilation coefficient (χ_{pc}) and encounter rate (E_{pc}) are confounded with competition, so they are replaced with β and α, respectively. Beta should be thought of as scaling the benefit a predator receives by causing predation or competitive mortality in the consumer, while alpha is a competition-predation functional response (Holt and Polis 1997). Again, these models can be modified to include more complex functional responses or saturation constants. The equations below again show Type I functional responses (Equations 12.7–12.9):

$$\frac{dN}{dt} = r_n N \left(\frac{K_n - N}{K_n} \right) - E_{cn} CN - E_{pn} PN \tag{12.7}$$

$$\frac{dC}{dt} = \chi_{cn}E_{cn}NC - \alpha PC - m_cC \qquad (12.8)$$

$$\frac{dP}{dt} = \beta\alpha PC + \chi_{pn}E_{pn}PN - m_pP \qquad (12.9)$$

This model can lead to five outcomes: (1) all populations eventually decline to zero, (2) resources persist but consumers and predators eventually decline to zero, (3) resources and consumers persist but predators eventually decline to zero, (4) resources and predators persist but consumers eventually decline to zero, or (5) coexistence of all three populations. Since IGP involves competition and predation, both of which can lead to the exclusion of one species or another, a major focus of research in this area has been to understand under what circumstances consumers and predators (or multiple predators) coexist.

In this and other more advanced IGP models, coexistence only occurs at intermediate resource levels and when the consumer is the superior competitor (Holt and Polis 1997, Mylius et al. 2001, Borer et al. 2003, van de Wolfshaar et al. 2006). When the consumer is the superior competitor, resource availability determines the relative importance of competition to coexistence. At low resource levels, consumers can persist because they are superior competitors, while at high resource levels, competition is relaxed and predator abundance becomes sufficiently high to drive down consumers. Intermediate levels of resources balance these two factors. In situations where the predator is the superior competitor, the consumer is always excluded by a combination of predation and competitive losses.

Again, these concepts can be illustrated by exploring the behavior of Equations 12.7–12.9, by allowing resource carrying capacity and the encounter rate between predators and resources to vary while fixing all other parameters (Figure 12.4). Figure 12.4a shows which trophic levels present when the model is close to equilibrium (here at $t = 5000$). Notice that if resource availability is too high or if the predator is too competitive on the common resource (Figure 12.4a,d), the consumer is excluded. Similarly, if the predator cannot sufficiently acquire resources, either because of resource limitation or poor competitive ability, the predator is excluded (Figure 12.4a,b). Only at intermediate levels does the model predict species coexistence (Figure 12.4a,c). In this particular case, the parameter values in Figure 12.4 are the same as those presented in Figure 12.3, except $\alpha = E_{pc} = 0.0005$ and $\beta = \chi_{pc} = 0.6$. Note that in the trophic cascade model, when $E_{pc} = 0.0005$, the predator is excluded at all resource capacities (Figure 12.3a). Here, the additional dimension introduced by omnivory (E_{pn}) leads to a variety of different outcomes and subsequent ways for the predator to persist (Figure 12.4a).

Although some systems adhere to these model predictions (e.g. Borer et al. 2003), many do not. For example, the diverse food webs examined by Arim and Marquet (2004) all contain intraguild predators, and Amarasekare (2007, 2008) found diverse parisitoids can coexist in highly productive systems. Similar to trophic cascades, factors that influence the outcome and intensity of IGP often modify the interaction strength between trophic levels. Predator and consumer behavior, alternate prey, interactions with the physical environment, and predator diversity can again play a role in predator-consumer (or multi-predator) outcomes. Under certain circumstances, the presence of these complex interactions can considerably expand the coexistence space of the species considered (Mougi and Kondoh 2012).

Fig. 12.4 Intraguild predation model behavior (equations 12.4–12.6) at $t = 5000$, where K_n varies from 300 to 1000 and E_{pn} from 0.0001 to 0.0003. All other parameters are held constant: $N_0 = 200$, $r_n = 0.15$, $C_0 = 50$, $E_{cn} = 0.001$, $\chi_{cn} = 0.5$, $m_c = 0.2$, $P_0 = 30$, $\chi_{pn} = 0.4$, $\alpha = 0.0005$, $\beta = 0.6$, $m_p = 0.05$. Final outcome (a): (1) resources persist but consumers and predators decline to zero, (2) consumers and resources persist but predators decline to zero, (3) predators and resources persist but consumers decline to zero, and (4) all three persist. For example, (b) at $K = 500$ and $E_{pn} = 0.00017$, predators decline to zero due to resource limitation, (c) at $K = 1000$ and $E_{pn} = 0.00017$ both consumers and predators coexist, and (d) at $K = 1000$ and $E_{pn} = 0.00025$ consumers decline to zero because of a combination of competition and predation. Under these parameter values, the outcome where all populations decline to zero does not occur.

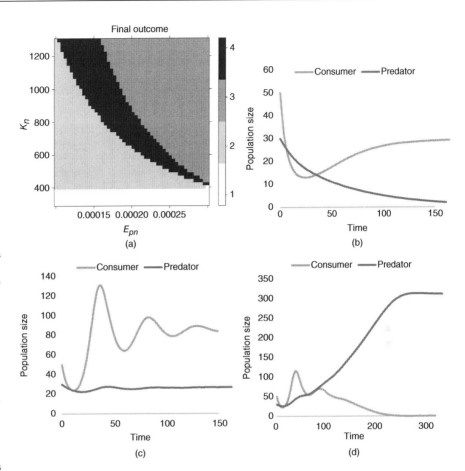

As with trophic cascades, consumers of all types frequently shift habitat use and foraging behaviors to reduce IGP risk (Crumrine and Crowley 2003, Rudolf and Armstrong 2008, Janssen et al. 2007). Lower predator efficiency in complex habitats leads to predators shifting their focus to shared resources (Finke and Denno 2006, Janssen et al. 2007). Furthermore, consumers and predators may not consistently appear together in all habits or times of the day or year (see Section 10.12, Witt et al. 2013). If consumers appear earlier in the year, for example, they may build up sufficient numbers or have completed important aspects of their life cycle and will not be excluded by predators (Morin 1984, Amarasekare 2007).

12.5 Intraguild predation and prey suppression

IGP has important ramifications for the management of pest populations in agricultural ecosystems. Biological control programs often seek to induce trophic cascades through the application of natural enemies (Box 12.3, Denoth et al. 2002, Finke and Denno 2004, Letourneau et al. 2009). In this setting, the goal is to predict under what conditions changing the diversity of natural enemies will increase the ability of the predator guild to suppress pests, and thus increase plant biomass through a trophic cascade. If the different predator species complement or facilitate each other, increasing predator diversity can increase prey suppression (Cardinale et al. 2003, Wilby et al.

2005, Casula et al. 2006, Letourneau et al. 2009). Predators may complement each other by consuming different prey or utilizing the same prey uniquely for example, by seeking prey in different habitats, on different parts of plants, or by attacking prey in different life stages. Predators can facilitate each other when the presence of one predator species increases the effectiveness of another, for example, by compelling prey to change habitats or foraging routines that expose them to new predators (Eklöv and VanKooten 1995, Losey and Denno 1998, Sih et al. 1998).

As the number of predator species increases, however, the opportunity for antagonism between predators also increases, which can lead to weaker effects on prey species. In communities where the number of generalist predators and IGP risk is high, increased predator diversity can release prey from predator pressure (Rosenheim 1998, Sih et al. 1998, Finke and Denno 2004, Vance-Charlcraft et al. 2007). In these circumstances, external factors, such as the availability of mesopredator refuges, habitat complexity, and the degree of spatial and temporal overlap between predators, become increasingly important for reducing IGP (Janssen et al. 2007, Amarasekare 2007).

Box 12.3 Habitat complexity and predator diversity determine the strength of trophic cascades in salt marshes

Habitat complexity may reduce the impacts of IGP and facilitate a trophic cascade (Finke and Denno 2006, Janssen et al. 2007). One such example is in North American Atlantic salt marshes. In these marshes, the amount of leaf litter (thatch) associated with *Spartina* grasses varies along an elevation gradient, with higher elevation habitats containing more thatch and higher complexity (Finke and Denno 2002). This system contains a diverse predator guild, which commonly includes the mirid bug, *Tytthus vagus*, the web building spider, *Grammonota trivittata*, and the wolf spider, *Pardosa littoralis*, all of which regularly consume common plant hoppers (*Prokelisia spp.*). Of these, *Pardosa* is a strong intraguild predator, while *Grammonota* and *Tytthus* primarily consume plant hoppers.

Finke and Denno (2002, 2006) experimentally manipulated the presence of these predators and thatch to examine their individual and combined influences on both plant hopper and Spartina density and biomass. The results hinged strongly on the interaction between predator diversity and habitat complexity. The introduction of a predator guild to mesocosms with *Spartina* and plant hoppers created a trophic cascade, which was amplified by the presence of complex thatch. Compared to predator-free treatments, the presence all three predators decreased plant hopper abundance by 60% in thatch-free treatments and 90% in thatch treatments, and increased *Spartina* biomass by 40% in thatch-free treatments and 300% in thatch treatments (Finke and Denno 2006).

To demonstrate why the predator guild was less effective in simple (thatch-free) treatments, they experimentally manipulated the presence of *Pardosa* with *Grammonota* and *Tytthus*. They found that plant hopper abundance between the thatch and thatch-free treatments depended largely on the intensity of IGP by *Pardosa*. When *Pardosa* was present, *Grammonota* and *Tytthus* densities decreased by approximately 60% and 90% in thatch-free treatments and 35% and 25% in thatch treatments, respectively (Finke and Denno 2002, 2006). This shift in consumption patterns occurs because capture and encounter rates between predators decrease in these complex habitats (Finke and Denno 2006, Janssen et al. 2007), which leads to relatively higher predator abundance and a shift to shared prey.

12.6 Intraguild predation and mesopredator release

Another lens through which to examine IGP and trophic cascades is mesopredator release. Here the goal is to understand how mesopredators affect lower trophic levels, following changes in top-tier predator abundance or their introduction into predator-free environments. Mesopredators are small or mid-sized predators that occupy mid-level trophic positions (consumers in our models). If predators and mesopredators share common prey, a wide range of potential outcomes may occur, depending on the interaction strength between between species. For example, Berger et al. (2008) found that pronghorn (*Antilocapra americana*) fawn survival was four times higher in areas where wolves had returned.. This counter-intuitive result occurred because wolves limited a mesopredator, solitary coyotes, that had a relatively larger impact on pronghorn populations.

Mesopredator release has been documented on all continents, except Antarctica, and negative effects on prey have been documented for a wide range of taxa, from insects, rodents, and lizards to birds, turtles, and deer (Brashares et al. 2010). In North America, universal declines in apex predators have coincided with approximately a 60% increase in mesopredator ranges and up to a four-fold increase in mesopredator abundance (Prugh et al. 2009, Richie and Johnson 2009). In 90% of studied cases, biodiversity losses accompanied mesopredator release (Richie and Johnson 2009). Mesopredator release is of particular concern to conservationists because mesopredators are thought to have broader diets, reproduce more quickly, and forage more efficiently than their predators, which allows them to more completely exploit habitats (Box 12.4, Brashares *et al.* 2010). Like other interactions we have explored, environmental context plays an important role in mediating the effects of mesopredator release (Elmhagen and Rushton 2007, Rayner et al. 2007, Elmhagen et al. 2010).

Box 12.4 Introduction of two predators, feral cats and rats, and subsequent mesopredator release effects on Cook's Petrel

Introduction of feral cats (*Felis catus*) to islands, where the cats often assume the role of a released mesopredator, have contributed to at least 33 of 238 known global bird, small mammal, and reptile extinctions on the IUCN Red List; of those 33 species, 22 were birds (Medina et al. 2011, IUCN 2013). Furthermore, more birds and mammals are lost to domestic and feral cats than are hit by cars, killed by pesticides, or any other examined anthropogenic effect (Loss et al. 2013). However, cats are not the only released mesopredator tied to bird decline. Rodents, red foxes, badgers, mongoose, raccoons, and opossum have all contributed to bird declines in different circumstances (Brashares et al. 2010).

Cook's petrel (*Pterodroma cookii*), a small nest-burrowing bird, which was historically found across New Zealand, is now only found on three islands on the edges of its former range (Imber et al. 2003). The introduction of feral cats is considered a major factor in the decline of Cook's petrel, as well as the decline or extinction of 17 other endemic New Zealand species (Veitch 2001). On Little Barrier Island, feral cats coexisted with another mesopredator, the introduced Pacific rat (*Rattus exultans*), where the cat assumed the role of top-tier predator. In an attempt to restore petrels on the island, feral cats were eradicated in 1980 (Veitch 2001). Despite high rates of feral cat predation on other islands and documented cat predation on Little Barrier Island, the bird populations did not recover. Instead the removal of cats released rats from predation, which resulted in further petrel declines (Imber et al. 2003, Rayner et al. 2007). Breeding success dropped from 32% with both mesopredators present to 9% with just rats present (Rayner et al. 2007). Following eradication of rats in 2004, breeding success on the island increased to 59%.

The effect was also habitat-specific. After cat eradication, petrel breeding success at upland sites was only 5%, but surpassed 65% at lowland sites. Following rat removal, breeding success in upland sites jumped to roughly 60%. Rayner et al. (2007) hypothesize that habitat specific rat diets created the observed patterns. After rats were released from predation, resource limitation in upland sites may have drove them to consume more petrel eggs and chicks. This suggests that IGP between cats and rats benefited petrels in upland areas, but that cat predation was likely the dominant source of petrel mortality in lowland areas before 1980.

12.7 Cannibalism

Another common form of omnivory that occurs within predator populations is cannibalism, which is defined as both the killing and eating of conspecifics (Fox 1975, Woodward and Hildrew 2002, Claessen et al. 2004). Like IGP, cannibalism involves both aspects of competition and predation (Figure 12.1c, Persson et al. 2000, 2003, Finke and Denno 2006). In many cases different age or size classes of cannibals share resources, but even if they do not, all potential cannibal victims eventually be come competitors (Claessen et al. 2000, 2004). Cannibalism is largely a density-dependent interaction, so it can also be thought of as an extreme form of density regulation. Despite this, its effects often differ from simple density-dependence, because it benefits certain age classes of individuals, harms others, lowers overall predator abundance, and can create complex feedback loops within predator populations (Box 12.5a, Persson et al. 2003, Claessen et al. 2004).

The types of feedback loops cannibalism creates depend on resource availability and competitive interactions. Similar to IGP, cannibalism can stabilize predator populations by removing present or future competition for shared resources and by providing an alternative resource (Claessen et al. 2004, Rudolf 2007). For example, cannibalism can create a "lifeboat" effect for a species in low resource situations, if long term reproductive benefits outweigh short term demographic losses. Similarly, if cannibals are also intraguild predators, shifts to conspecifics at low consumer densities can increase the likelihood of consumer and predator coexistence (Rudolf 2007). Cannibalism is not always a stabilizing mechanism, however, and can potentially hinder the recovery of exploited or conserved species, create chaotic population dynamics, or cause populations to go through large population cycles or demographic phase shifts (Box 12.5a,b, Persson et al. 2003, Claessen et al. 2004, Zohar et al. 2008).

Like all the trophic interactions we've considered before, cannibal behavior and environmental context can alter the intensity of cannibalism. Cannibal victims often behave differently in the presence of large conspecifics by reducing their movement rates or shifting habitats to reduce risk (Van Buskirk 1992). For example, Rudolf (2006) found that increased cannibalism risk to juvenile spring salamanders (*Gyrinophilus porphyriticus*) caused them to alter their microhabitat use and reduce their foraging rates, which incidentally reduced IGP rates on the smaller two-lined salamander, *Eurycea cirrigera*. Finke and Denno (2006) found that habitat complexity not only lowered IGP rates between predators, but also lowered cannibalism rates among *Pardosa*, which contributed to a larger trophic cascade (Box 12.3).

Box 12.5a Demographic phase shifts and trophic cascades in cannibalistic perch

Complex population dynamics induced by cannibalism can have community-wide effects. Persson et al. (2003) considered interactions among three age classes of perch (*Perca fluviatilis*) – adults at least two years old, one year old juveniles, and *young of the year* (*YOY*) – as well as interactions between perch, macroinvertebrates, zooplankton, and phytoplankton. They found that perch experienced population phase shifts between a "dwarf" stage that consisted of numerous small adults and a "giant" phase that consisting of few large adults. These demographic phases are associated with distinct perch consumption patterns, which cascade down to lower trophic levels.

During the dwarf phase, adults compete and suppress the YOY through cannibalism and juvenile prevent recruitment. Low YOY and juvenile abundance release zooplankton from predation, which in turn suppress phytoplankton. Because there are many adults and few energy poor YOY, adults also consume large quantities of macroinvertebrates.

The transition to the giants phase and back to the dwarf phase happens incrementally. As adult abundance declines over time, more YOY initially escape cannibalism and grow into juveniles. However, these juveniles are eventually cannibalized by adults, which begin to grow rapidly from consuming larger cannibal victims. During this transition, birth rates from these increasingly large adults eventually outpace losses from cannibalism. High fecundity marks the beginning of the giant phase, during which so many YOY are created by a dwindling number of gigantic adults that they outcompete juveniles and cause them to starve. Since there are so many YOY during this phase, they suppress zooplankton populations and release phytoplankton from predation.

As the remaining adults continue to decline from background mortality, fecundity eventually decreases, producing fewer YOY. At another transition point, the reduced YOY and adult abundances allow a cohort of juveniles to escape competition and cannibalism. This new cohort of adults suppresses the YOY and prevents juvenile recruitment, marking a return to the dwarfs phase.

Box 12.5b Cannibalism's role in blue crab declines and recovery

Since cannibalism is a density-dependent interaction, it is usually not a factor in species conservation. In captive breeding or stocking programs, however, where juveniles are raised at high densities, cannibalism has the potential to reduce efficiency and hinder recovery (Zohar et al. 2008)

The blue crab, *Callinectes sapidus*, is an economically and culturally important fishery in the Chesapeake Bay on the east coast of North America. Although this stock naturally fluctuates and has declined and recovered three times in the 20th century, it has experienced unusually severe declines in adult abundance and occasional recruitment failure since the mid-1990s (Lipcius and Stockhausen 2002). Overfishing is an major component of population declines, but concurrent loss of sea grass and bivalve rich habitats, eutrophication, and subsequent summer hypoxia have furthe hindered stock recovery. Blue crabs are also highly cannibalistic, and cannibalism can account for between 75% and 97% of juvenile mortality (Hines and Ruiz 1995, Lipcius et al. 2005).

These high cannibalism rates are exacerbated by habitat loss. Sea grass and bivalve habitats, which are associated with high blue crab settlement rates (Engel and Thayer 1998), increase juvenile survival largely by limiting cannibalism (Hovel and Lipcius 2001; 2002).

In addition to fishing limitations, augmenting wild populations with hatchery raised blue crabs (stock enhancement) has been a particularly popular idea. As such, a substantial amount of research has been conducted to determine the viability of hatchery raised juveniles (Zohar et al. 2008). Between 2002 and 2008, researchers have released a total of 290 000 juveniles into the Chesapeake Bay to test the survival of hatchery raised individuals, and raised a total of 570 000 juvenile blue crabs on a year round cycle in captivity. Controlling cannibalism has been one obstacle in making the hatcheries efficient and cost effective. Crabs are raised in mass production tanks, which allow crabs to interact with each other at high densities (Zmora et al. 2005). Since cannibalism is a density and size-dependent interaction, synchronizing early molting and finding ideal densities have been goals of the stocking program (Zohar et al. 2008). As such, simple changes in rearing tanks, such as the type of substrate and availability of shelters in the tanks, can lower interaction rates between crabs and reduce cannibalism (Zmora et al. 2005). In addition, choosing the right release size, season, and location are important for reducing cannibalism rates between hatchery and wild crabs (Hines et al. 2008, Johnson et al. 2008). In particular, experiments on release survival have shown spring and fall releases result in lower cannibalism rates for all sized hatchery crabs (Johnson et al. 2008).

Chapter and book summary

In this chapter we were introduced to three basic and common multi-trophic interactions: trophic cascades, IGP, and cannibalism. Exploring these interactions highlights the indirect and complex interactions that characterize natural systems. Each interaction demonstrated how changes in abundances at each trophic level influence dynamics between species, often in unintuitive ways. Predators, in particular, have wide reaching and diverse influences on their ecosystems, which alter the balance between consumers and resources.

Modeling three trophic levels brought to our attention unexpected issues that did not arise when comparing just two trophic levels. Species frequently feed at more than one trophic level, and models which place these species into linear relationships run the risk of making incorrect inferences (Polis and Strong 1996, Hunter 2009). Since omnivory is common in natural systems, accounting for it becomes crucial to predicting species population dynamics. Even models that incorporate ominvory simplify and can misrepresent natural processes. For IGP in particular, Holt and Polis (1997) drew attention to the apparent discrepancy between the models and empirical data; that strong IGP "raises a puzzle of species coexistence." Due to this disconnect between model expectations and empirical data, it has been argued that simple multi-trophic models do not adequately capture the range of situations under which coexistence can occur (Janssen et al. 2007). More recent modeling efforts that attempt to incorporate additional interactions have improved these outcomes (Holt and Huxel 2007, Rudolf 2007, Mougi and Kondoh 2012, Ohlberger et al. 2012).

Understanding of the context surrounding predation as well as the assumptions implicit in population models often provides insights into why natural systems deviate from model expectations. In particular, it is important to consider how interaction strength between species varies along a gradient of resource availability or across habitats, and subsequently, how the relative importance of top-down and bottom-up controls on consumers change along those gradients. For instance, predators may not be able to initiate a trophic cascade when they have weak control over consumers that refuge in complex habitats. The models that we have seen so far only include a few species, and assume that only the species in the models interact. They also assume individuals are uniform in age and size, have fixed behaviors, and are well mixed (but see Mylius et al. 2001, van de Wolfshaar et al. 2006). These assumptions are rarely true. Habitat preferences and temporal disjunction violate the assumption of well mixed prey, while the availability of alternate prey means population dynamics depend on more than just the species modeled. Different age and sized individuals frequently have different behaviors and may consume different resources (Sinclair et al. 2003, Chapter 4).

Before reading this chapter, you may not have considered how multi-trophic interactions apply to conservation or ecosystem management. Given the pervasiveness of human influence on the world's ecosystems, understanding how apex predator removal or mesopredator release cascades through these systems is essential for restoring or preserving the species and services offered by ecosystems. The loss of apex predators has inadvertently created "trophically downgraded" landscapes where the loss of predator control has altered everything from food supplies, species diversity and abundance, and nutrient dynamics (Estes et al. 2011).

In this book we began with simple growth models for single populations. These models became increasingly complex as we added time lags, stochastic effects and age structures. In the second half of the book we began with relatively simple competitive interactions involving only two species. We moved on to consider interactions between different trophic levels, again starting with simple two-species models (one predator, one prey. But you should now realize that all interspecific interactions must be analyzed in the context of the entire community of organisms, not simply in terms of one competing or one predatory species. In this final chapter we have looked at a number of variations on the theme of multiple trophic level interactions. Thus plants are attacked by a multitude of herbivores, while herbivores exist in a community of both competitors and predators, and both herbivores and predators must contend with parasites, and competition with their own species (or predation by members of their own species in the case of cannibalism). All of this takes place on a complex physical landscape and an ever changing climate. Attempting to understand, much less model, these complexities should keep population, community and landscape ecologists challenged for a long time to come.

References

Amarasekare, P. 2007. Trade-offs, temporal variation, and species coexistence in communities with intraguild predation. *Ecology* 88: 2720–2728.
Amarasekare, P. 2008. Coexistence of intraguild predators and prey in resource-rich environments. *Ecology* 89: 2786–2797.
Anthony, R.G., J.A. Estes, M.A. Ricca, A.K. Miles, and E.D. Forsman. 2008. Bald eagles and sea otters in the Aleutian archipelago: indirect effects of trophic cascades. *Ecology* 89: 2725–2735.

Arim, M. and P.A. Marquet. 2004. Intraguild predation: a widespread interaction related to species biology. *Ecological Letters* 7: 557–564.

Benndorf, J., W. Böing, J. Koop, and I. Neubauer. 2002. Top-down control of phytoplankton: the role of time, scale, lake depth and trophic state. *Freshwater Biology* 47: 2282–2295.

Berger, J. 2010. Fear mediated food webs. Pages 241–253 *in* J. Terborgh and J.A. Estes, editors, *Trophic Cascades: Predators, Prey, and the Changing Dynamics of Nature*. Island Press, Washington, DC.

Berger, K.M., E.M. Gese, and J. Berger. 2008. Indirect effects of traditional tropic cascades: a test involving wolves, coyotes, and pronghorn. *Ecology* 89: 818–828.

Borer, E.T., C.J. Briggs, W.W. Murdoch, and S.L. Swarbrick. 2003. Testing intraguild predation theory in a field system: does numerical dominance shift along a gradient of productivity. *Ecology Letters* 6: 929–935.

Borer, E.T., E.W. Seabloom, J.B. Shurin, K.E. Anderson, C.A. Blanchette, B. Brottman, S.D. Cooper, and B.S. Halpern. 2005. What determines the strength of a trophic cascade? *Ecology* 86: 528–537.

Brashares, J.S., L.R. Prugh, C.J. Stoner, and C.W. Epps 2010. Ecological and conservation implications of mesopredator release. Pages 221–240 *in* J. Terborgh and J.A. Estes, editors, *Trophic Cascades: Predators, Prey, and the Changing Dynamics of Nature*. Island Press, Washington, DC.

Brown, J.S., J.W. Laundre, and M. Gurung. 1999. The ecology of fear: optimal foraging, game theory and trophic interactions. *Journal of Mammalogy* 80: 385–399.

Cardinale, B.J., C.T. Harvey, K. Gross, and A.R. Ives. 2003. Biodiversity and biocontrol: emergent impacts of a multi-enemy assemblage on pest suppression and crop yield in an agroecosystem. *Ecology Letters* 6: 857–865.

Carpenter, S.R., J.J. Cole, J.R. Hodgson, J.F. Kitchell, M.L. Pace, D. Bade, K.L. Cottingham, T.E. Essington, J.N. Houser, and D.E. Schindler. 2001. Trophic cascades, nutrients, and lake productivity: whole-lake experiments. *Ecological Monographs* 71: 163–186.

Casula, P., A. Wilby, and M.B. Thomas. 2006. Understanding biodiversity effects on prey in multi-enemy systems. *Ecology Letters* 9: 995–1004.

Claessen, D., A.M. de Roos, and L. Persson. 2000. Dwarfs and giants: cannibalism and competition in size-structured populations. *American Naturalist* 155: 219–237.

Claessen, D., A.M. de Roose, and L. Persson. 2004. Population dynamic theory of size-dependent cannibalism. *Proceedings of the Royal Society B* 271: 333–340.

Cole, J.J., S.R. Carpenter, M.L. Pace, M.C. Van de Bogert, J.F. Kitchell, and J.R. Hodgson. 2006. Differential support of lake food webs by three types of terrestrial organic carbon. *Ecology Letters* 9: 558–568.

Crumrine, P.W., and P.H. Crowley. 2003. Partitioning components of risk reduction in a dragonfly-fish intraguild predation system. *Ecology* 84: 1588–1597.

Denoth, M., L. Frid, and J.H. Myers. 2002. Multiple agents in biological control: improving the odds? *Biological Control* 24: 20–30.

Doroff, A.M., J.A. Estes, M.T. Tinker, D.M. Burn, and T.J. Evans. 2003. Sea otter population declines in the Aleutian archipelago. *Journal of Mammalogy* 84: 55–64.

Duggins, D.O., C.A. Simenstad, and J.A. Estes. 1989. Magnification of secondary production by kelp detritus in coastal marine ecosystems. *Science* 245: 170–173.

Eklöv, P., and T. VanKooten. 2001. Facilitation among piscivorous predators: effects of prey habitat use. *Ecology* 82: 2486–2496.

Elmhagen, B., G. Ludwig, S.P. Rushton, P. Helle, and H. Linden. 2010. Top predators, mesopredators and their prey: inferences about ecosystems along bioclimatic productivity gradients. *Journal of Animal Ecology* 79: 785–794.

Elmhagen, B. and S.P. Rushton. 2007. Trophic control of mesopredators in terrestrial ecosystems: top-down or bottom-up? *Ecology Letters* 10: 197–206.

Engel, D.W., and G.W. Thayer. 1998. Effects of habitat alteration on blue crabs. *Journal of Shellfish Research* 17: 579–585.

Estes, J.A., K. Crooks, and R.D. Holt. 2001. Predators, ecological role. Pages 857–878 *in* S. Levin, editor, *Encyclopedia of Biodiversity*, Vol. 4. Academic Press, San Diego, CA.

Estes, J.A., E.M. Danner, D.F. Doak, B. Konar, A.M. Springer, P.D. Steinberg, M.T. Tinker, and T.M. Williams. 2004. Complex trophic interactions in kelp forest ecosystems. *Bulletin of Marine Science* 74: 621–638.

Estes, J.A., D.F. Doak, A.M. Springer, and T.M. Williams. 2009. Causes and consequences of marine mammal declines in southwest Alaska: a food-web perspective. *Philosophical transactions of the Royal Society B*: 364: 1647–1658.

Estes, J.A., and D O. Duggins. 1995. Sea otters and kelp forests in Alaska: generality and variation in a community ecological paradigm. *Ecological Monographs* 65: 75–100.

Estes, J.A., and J.F. Palmisano. 1974. Sea otters: their role in structuring nearshore communities. *Science* 185: 1058–1060.

Estes, J.A., C.H. Peterson, R.S. Steneck. 2010. Some effects of apex predators in higher-latitude coastal oceans. Pages 37–53 *in* J. Terborgh and J.A. Estes, editors, *Trophic Cascades: Predators, Prey, and the Changing Dynamics of Nature*. Island Press, Washington, DC.

Estes, J.A., N.S. Smith, and J.F. Palmisano. 1978. Sea otter predation and community organization in the western Aleutian Island, Alaska. *Ecology* 59: 822–833.

Estes, J.A., J. Terborgh, J.S. Brashares, M.E. Power, J. Berger, W.J. Bond, S. R. Carpenter, T.E. Essington, R. D. Holt, J.B.C. Jackson, R.J. Marquis, L. Oksanen, T. Oksanen, R.T. Paine, E.K. Pikitch, W.J. Ripple, S.A. Sandin, M. Scheffer, T.W. Schoener, J.B. Shurin, A.R.E. Sinclair, M.E. Soulé, R. Virtanen, and D.A. Wardle. 2011. Trophic downgrading of planet earth. *Science* 333: 301–306.

Estes, J.A., M.T. Tinker, T.M. Williams, and D.F. Doak. 1998. Killer whale predation on sea otters linking oceanic and nearshore ecosystems. *Science* 282: 473–476.

Finke, D.L., and R.F. Denno. 2002. Intraguild predation diminished in complex-structured vegetation: implications for prey suppression. *Ecology* 83: 643–652.

Finke, D.L., and R.F. Denno. 2004. Predator diversity dampens trophic cascades. *Nature* 429: 407–410.

Finke, D.L., and R.F. Denno. 2006. Spatial refuge from intraguild predation: implications for prey suppression and trophic cascades. *Oecologia* 149: 265–275.

Fox, L.R. 1975. Cannibalism in natural populations. *Annual Review of Ecology, Evolution, and Systematics* 6: 87–106.

Frank, K.T., B. Petrie, N.L. Shackell. 2007. The ups and downs of trophic control in continental shelf ecosystems. *Trends in Ecology and Evolution* 22: 236–242.

Frank, K.T., B. Petrie, N.L. Shackell, and J.S. Choi. 2006. Reconciling differences in trophic control in mid-latitude marine ecosystems. *Ecology Letters* 9: 1096–1105.

Forkner, R.E., and M.D. Hunter 2000 What goes up must come down? Nutrient addition and predation pressure on oak herbivores. *Ecology* 81: 1588–1600.

Hairston, N.G., F.E. Smith, and L.B. Slobodkin. 1960. Community structure, population control, and competition. *American Naturalist* 94: 421–425.

Hansson, L.A., H. Annadotter, E. Bergman, S.F. Hamrin, E. Jeppesen, T. Kairesalo, E. Luokkanen, P.A. Nilsson, M. Søndergaard, and J. Strand. 1998. Biomanipulation as an application of food-chain theory: constraints, synthesis, and recommendations for temperate lakes. *Ecosystems* 1 558–574.

Hebblewhite, M. and D.H. Pletscher. 2002. Effects of elk group size on predation by wolves. *Canadian Journal of Zoology* 80: 800–809.

Hines, A.H., E.G. Johnson, A.C. Young, R. Aguilar, M.A. Kramer, M. Goodison, O. Zmora, and Y. Zohar. 2008. Release strategies for estuarine species with complex migratory life cycles: stock enhancement of Chesapeake blue crabs (*Callinectes sapidus*). *Reviews in Fisheries Science* 16: 175–185.

Hines, A.H., and G.M Ruiz. 1995. Temporal variation in juvenile blue crab mortality: nearshore shallows and cannibalism in Chesapeake Bay. *Bulletin of Marine Science* 57: 884–901.

Hodgeson, J.R. and E.M. Hansen. 2005. Terrestrial prey items in the diet of largemouth bass, *Micropterus salmoides*, in a small, north temperate lake. *Journal of Freshwater Ecology* 20: 793–794.

Hölker, F., S.S. Haertel, S. Steiner, T. Mehner. 2002. Effects of piscivore-mediated habitat use on growth, diet and zooplankton consumption of roach: an individual based modelling approach. *Freshwater Biology* 47: 2345–2358.

Holt, R.D., and G.R. Huxel. 2007. Alternate prey and the dynamics of intraguild predation: theoretical perspectives. *Ecology* 88: 2706–2712.

Holt, R.D., and G.A. Polis. 1997. A theoretical framework for intraguild predation. *American Naturalist* 149: 745–764.

Hovel, K.A., and R.N. Lipcius. 2001. Habitat fragmentation in a seagrass landscape: patch size and complexity control blue crab survival. *Ecology* 82: 1814–1929.

Hovel, K.A., and R.N. Lipcius. 2002. Effects of seagrass habitat fragmentation on juvenile crab survival and abundance. *Journal of Experimental Marine Biology and Ecology* 271: 75–98.

Hunter, M.D. 2009. Trophic promiscuity, intraguild predation and the problem of omnivores. *Agricultural and Forest Entomology* 11: 125–131.

Imber, M.J., J.A. West, and W.J. Cooper. 2003. Cook's petrel (*Pterodroma cookie*): historic distribution, breeding biology and effects of predators. *Notornis* 50: 221–230.

Irons, D.B, R. G. Anthony, and J.A Estes. 1986. Foraging strategies of glaucous-winged gulls in rocky intertidal communities. *Ecology* 67: 1460–1474.

IUCN. 2013. The IUCN Red List of Threatened Species. Version 2013.2. http://www.iucnredlist.org. Downloaded on 3 March 2013.

Jackson, J.B.C., M.X. Kirby, W.H. Berger, K.A. Bjorndal, L.W. Botsford, B.J. Borque, R.H. Bradbury, R. Cooke, J. Erlandson, J.A. Estes, T.P. Hughes, S. Kidwell, C.B. Lange, H.S. Lenihan, J.M. Pandolfi, C.H. Peterson, R.S. Steneck, M.J. Tegner, and R.R. Warner. 2001. Historical overfishing and the recent collapse of coastal ecosystems. *Science* 293: 629–638.

Janssen A., M. Sabelis, S. Magalhães, M. Montserrat, and T. van der Hammen. 2007. Habitat structure affects intraguild predation. *Ecology*, 88, 2713–2719.

Johnson, E.G., A. H. Hines, M.A. Kramer, and A.C. Young. 2008. Importance of season and size of release to stocking success for the blue crab in Chesapeake Bay. *Reviews in Fisheries Science* 16: 243–253.

Knight, T.M., J.M. Chase, H. Hillebrand, and R.D. Holt. 2006. Predation on mutualists can reduce the strength of trophic cascades. *Ecology Letters* 9: 1173–1178.

Lathrop, R.C., B.M. Johnson, T.B. Johnson, M.T. Vogelsang, S.R. Carpenter, T.R Hrabik, J.F. Kitchell, J.J. Magnuson, L.G. Rudstam, and R.S. Stewart. 2002 Stocking piscivorus to improve fishing and water clarity: a synthesis of the Lake Mendota biomanipulation project. *Freshwater Biology* 47: 2410–2424.

Letourneau, D.K, J.A. Jedlicka, S.G. Bothwell, and C.R. Moreno. 2009. Effects of natural enemy biodiversity on the suppression of arthropod herbivores in terrestrial ecosystems. *Annual Review of Ecology, Evolution, and Systematics* 40: 573–592.

Lima, S.L., 1998. Nonlethal effects in the ecology of predator–prey interactions. *Bioscience* 48: 25–34

Lima, S.L. and L.M. Dill. 1990. Behavioral decisions made under the risk of predation: a review and prospectus. *Canadian Journal of Zoology* 68: 619–640.

Lipcius, R.N., and W.T. Stockhausen. 2002. Concurrent decline of the spawning stock, recruitment, larval abundance, and size of the blue crab, *Callinectes sapidus*, in Chesapeake Bay. *Marine Ecology Progress Series* 226: 45–61.

Lipcius, R.N., R.D. Seitz, M.S. Seebo, and D. Colon-Carrion. 2005. Density, abundance and survival of the blue crab in seagrass and unstructured salt marsh nurseries of Chesapeake Bay. *Journal of Experimental Marine Biology and Ecology* 319: 69–80.

Losey, J.E., and R.F. Denno. 1998. Positive predator-predator interactions: enhanced predation rates and synergistic suppression of aphid populations. *Ecology* 79: 2143–2152.

Loss, S.R., T. Will, and P.P. Marra. 2013. The impact of free-ranging domestic cats on wildlife in the United States. *Nature Communications* 4: 1–7.

Magurran, M.E. 2005. *Evolutionary Ecology: The Trinidadian Guppy*. Oxford University Press, New York, NY.

Medina, F.M., E. Bonnaud, E. Vidal, B.R. Tershy, E.S. Zavaleta, C.J. Donlan, B.S. Keitt, M. Le Corre, S.V. Horwath, and M. Nogales. 2011. A global review of the impacts of invasive cats on island endangered vertebrates. *Global Change Biology* 17: 3503–3510.

Mehner, T., J. Benndorf, P. Kasprzak, and R. Koschel. 2002. Biomanipulation of lake ecosystems: successful applications and expanding complexity in the underlying science. *Freshwater Biology* 47: 2453–2465.

Menge, B.A. 1995. Indirect effects in marine rocky intertidal interaction webs: patterns and importance. *Ecological Monographs* 65: 21–74.

Middleton, A.D., M.J. Kauffman, D.E. McWhirter, M.D. Jimenez, R.C. Cook, J.G. Cook, S.E. Albeke, H. Sawyer, and P. J. White. 2013. Linking antipredator behavior to prey demography reveals the limited risk effects of an actively hunting large carnivore. *Ecology Letters* 16: 1023–1030.

Morin, P.J. 1984. Odonate guild composition: experiments with colonization history and fish predation. *Ecology* 65: 1866–1873.

Mougi, A., and M. Kondoh. 2012. Diversity of interaction types and ecological community stability. *Science* 337: 349–351.

Myers. R.A., J.K. Baum, T.D. Shepherd, S.P. Powers, and C.H. Peterson. 2007. Cascading effects of the loss of apex predatory sharks from a coastal ocean. *Science* 315: 1846–1850.

Mylius, S.D., K. Klumpers, A.M. de Roos, L. Persson. 2001. Impact of intraguild predation and stage structure on simple communities along a productivity gradient. *American Naturalist* 158: 259–276.

Oksanen, L., S. D. Fretwell, J. Arruda, and P. Niemelä. 1981. Exploitation ecosystems in gradients of primary productivity. *American Naturalist* 118: 240–261.

Ohlberger, J., Ø. Langangen, N.C. Stenseth, and L.A. Vøllestad. 2012. Community level consequences of cannibalism. *American Naturalist* 180: 791–801.

Paine, R.T. 1966. Food web complexity and species diversity. *American Naturalist* 100: 65–75.

Paine, R.T. 1980. Food webs: linkage, interaction strength, and community infrastructure. *Journal of Animal Ecology* 49: 667–685.

Paine, R.T. 2010. Food chain dynamics and trophic cascades in intertidal habitats. Pages 21–35 *in* J. Terborgh and J.A. Estes, editors, *Trophic Cascades: Predators, Prey, and the Changing Dynamics of Nature*. Island Press, Washington, DC.

Persson, L., P. Byström, and E. Wahström. 2000. Cannibalism and competition in Eurasian Perch: population dynamics of an ontogenic omnivore. *Ecology* 81: 1058–1071.

Persson, L., A.M. de Roos, D. Claessen, P. Byström, J. Lövgren, S. Sjögren, R. Svanbäck, E. Wahström, and E. Westman. 2003. Gigantic cannibals driving a whole-lake trophic cascade. *Proceedings of the National Academy of Sciences* 100: 4035–4039.

Polis, G.A., and R.D. Holt. 1992. Intraguild predation: the dynamics of complex trophic interactions. *Trends in Ecology and Evolution* 7: 151–154.

Polis, G.A., and D.R. Strong. 1996. Food web complexity and community dynamics. *American Naturalist* 147: 813–846.

Power, M.E. 1992. Habitat heterogeneity and the functional significance of fish in river food webs. *Ecology* 73: 1675–1688.

Prugh, L.R., C.J. Stoner, C.W. Epps, W.T. Bean, W.J. Ripple, A.S. Laliberte, and J.S. Brashares. 2009. The Rise of the Mesopredator. *Bioscience* 59: 779–791.

Rayner, M.J., M.E. Hauber, M.J. Imber, R.K. Stamp, and M.N. Clout. 2007. Spatial heterogeneity of mesopredator release within an oceanic island system. *Proceedings of the National Academy of Sciences* 104: 20862–20865.

Reinman, M.L., and J.A. Estes. 1990. *The Sea Otter* (Enhydra lutris): *Behavior, Ecology, and Natural History*. Biological Report 90 (14). U.S. Department of the Interior, Fish and Wildlife Service, Washington, DC.

Reisewitz, S.E., J.A. Estes, and C.A. Simenstad. 2005. Indirect food web interactions: Sea otters and kelp forest fishes in the Aleutian archipelago. *Oecologia* 146: 623–631.

Richie, E.G. and C.N. Johnson. 2009. Predator interactions, mesopredator release and biodiversity conservation. *Ecology Letters* 12: 982–998.

Ripple, W.J., and R.L. Beschta. 2003. Wolf reintroduction, predation risk, and cottonwood recovery in Yellowstone National Park. *Forest Ecology and Management* 184: 299–313.

Ripple, W.J., and R.L. Beschta. 2004. Wolves and the ecology of fear: can predation risk structure ecosystems? *Bioscience* 54: 755–766.

Robles, C., and R. Desharnais. 2002. History and current development of a paradigm of predation in rocky intertidal communities. *Ecology* 83: 1521–1536.

Romare, P., and L.A. Hansson. 2003. A behavioral cascade: top-predator induced behavorial shifts in planktivorous fish and zooplankton. *Limnology and Oceanography* 48: 1956–1964.

Rosenheim, J.A. 1998. Higher-order predators and the regulation of insect herbivore populations. *Annual Review of Entomology* 43: 421–447.

Rudolf, V.H.W., 2006. The influence of size-specific indirect interactions in predator–prey systems. *Ecology* 87: 362–371.

Rudolf, V.H.W. 2007. The interaction of cannibalism and omnivory: consequences for community dynamics. *Ecology* 88: 2697–2705.

Rudolf, V.H.W. and J. Armstrong. 2008. Emergent impacts of cannibalism and size refuges in prey on intraguild predation systems. *Oecologia* 157: 675–686.

Sarnelle, O. 1992. Nutrient enrichment and grazer effects on phytoplankton in lakes. *Ecology* 73: 551–560.

Schlosser, I.J. 1987. The role of predation in age- and size-related habitat use by stream fishes. *Ecology* 68: 651–659.

Schmitz, O.J., P.A. Hämback, and A.P. Beckerman. 2000. Trophic cascades in terrestrial systems: a review of the effects of carnivore removals on plants. *American Naturalist* 155: 141–153.

Schmitz, O.J. 2008. Effects of predator hunting mode on grassland ecosystem function. *Science* 319: 952–954.

Schmitz, O.J. and K.B. Suttle. 2001. Effect of top predator species on direct and indirect interactions in a food web. *Ecology* 82: 2072–2081.

Schmitz, O.J., V. Krivan, and O. Ovadia. 2004. Trophic cascades: the primacy of treat-mediated indirect interactions. *Ecological Letters* 7: 153–163.

Sergio, F., T. Caro, D. Brown, B. Clucas, J. Hunter, J. Ketchum, K. McHugh, and F. Hirado. 2008. Top predators as conservation tools: ecological rational, assumptions, and efficacy. *Annual review of Ecology, Evolution, and Systematics* 39: 1–19.

Shapiro, J. 1990. Biomanipulation: the next phase – making it stable. *Hydrobiologia* 200–201: 13–27.

Shurin, J.B., D.S. Gruner, and H. Hillebrand. 2006. All wet or dried up? Real differences between aquatic and terrestrial food webs. *Proceedings of the Royal Society B* 273: 1–9.

Shurin, J.B., E.T. Borer, E.W. Seabloom, K. Anderson, C.A. Blanchette, B. Broitman, S.D. Cooper, and B.S. Halpern. 2002. A cross-ecosystem comparison of the strength of trophic cascades. *Ecology Letters* 5: 785–791.

Sih, A., G. Englund, and D. Wooster. 1998. Emergent impacts of multiple predators on prey. *Trends in Ecology and Evolution* 13: 350–355.

Soulé, M.E. 2010. Conservation relevance of ecological cascades. Pages 337–351 *in* J. Terborgh and J.A. Estes, editors, *Trophic Cascades: Predators, Prey, and the Changing Dynamics of Nature.* Island Press, Washington, DC.

Stachowicz, J.J., J.F. Bruno, and J.E. Duffy. 2007. Understanding the effects of marine biodiversity on communities and ecosystems. *Annual Review of Ecology, Evolution and Systematics* 38: 739–766.

Steffan, S.A., and W.E. Snyder. 2010. Cascading diversity effects transmitted exclusively by behavioral interactions. *Ecology* 91: 2242–2252.

Terborgh, J.A., R.D. Holt, and J.A. Estes. 2010. Trophic cascades: what they are, how they work, and why they matter. Pages 1–18 *in* J. Terborgh and J.A. Estes, editors, *Trophic Cascades: Predators, Prey, and the Changing Dynamics of Nature.* Island Press, Washington, DC.

Thompson, R.M., M. Hemberg, B.M. Starzomski, and J.B. Shurin. 2007. Trophic levels and trophic tangles: the prevalence of omnivory in real food webs. *Ecology* 88: 612–617.

Turchin, P. 2003. *Complex Population Dynamics: A Theoretical / Empirical Synthesis.* Princeton University Press, Princeton, NJ.

Van Buskirk, J. 1992. Competition, cannibalism, and size class dominance in a dragonfly. *Oikos* 65: 455–464.

van de Wolfshaar, K.E., A.M. de Roos, and L. Persson. 2006. Size-dependent interactions inhibit coexistence in intraguild predation systems with life-history omnivory. *American Naturalist* 168: 62–75.

Vance-Charlcraft, H.D., J.A. Rosenheim, J.R. Vonesh, C.W. Osenberg, and A. Sih. 2007. The influence of intraguild predation on prey suppression and prey release: a meta-analysis. *Ecology* 88: 2689–2696.

Veitch, C.R. 2001. The eradication of feral cats (*Felis catus*) from Little Barrier Island, New Zealand. *New Zealand Journal of Zoology* 28: 1–12.

Wade, P.R., V.N. Burkanov, M.E. Dahlheim, N.A. Friday, L.W. Fritz, T.R. Loughlin, S.A. Mizroch, M.M. Muto, D.W. Rice. 2007. Killer whales and marine mammals trends in the north Pacific – a re-examination of evidence for sequential megafauna collapse and the prey-switching hypothesis. *Marine Mammal Science* 23: 766–802.

Williams, T.M., J.A. Estes, D.F. Doak, and A.M. Springer. 2004. Killer appetites: assessing the role of predators in ecological communities. *Ecology* 85: 3373–3384.

Witt, J.W., R.E. Forkner, and R.T. Kraus. 2013 Habitat heterogeneity and intraguild interactions modify distribution and injury rates in two coexisting genera of damselflies. *Freshwater Biology* 58: 2380–2388.

Woodward, G., and A.H. Hildrew. 2002. Body-size determinants of niche overlap and intraguild predation within a complex food web. *Journal of Animal Ecology* 71: 1063–1074.

Zmora, O., A. Findiesen, J. Stubblefield, V. Fenkel, and Y. Zohar. 2005. Large-scale juvenile production of the blue crab, *Callinectes sapidus*. *Aquaculture* 244: 129–139.

Zohar, Y., A.H. Hines, O. Zmora, E.G. Johnson, R.N. Lipcius, R.D. Seitz, D.B. Eggleston, A.R. Place, E.J. Schott, J.D. Stubblefield, and J.S. Chung. 2008. The Chesapeake Bay blue crab (*Callinectes sapidus*): a multidisciplinary approach to responsible stock replenishment. *Reviews in Fisheries Science* 16: 24–34.

Appendix 1 **Problem sets**

1. You survey an annual insect and find 5000 females per acre. One year later, you census the population and find 6000 females per acre. What is the net replacement rate, R, for this population? What size will the population be 3 years from the original census if the population continues to grow at the same rate? 5 years later? 10 years later? In what year would the population reach 100 000?

2. The birth rate for Asia in 1982 was 30/1000, while the death rate was 11/1000. What was the intrinsic rate of increase, assuming a stable age distribution?

   If the population size was 2.67 billion, what was the projected population in 2013?

   The actual population of Asia, according to the Population Reference Bureau, was 4.302 billion in 2013. What explains the discrepancy between the projection and the actual population?

   FYI, the 2013 birth rate is 18/1000 and the death rate is 7/1000.

3. Between1992 and 2011 a human population with a stable age distribution increases from 2.67 to 3.72 billion, what is the r during that time period? Given this r-value, what is the doubling time? What is λ?

4. In 2013 the human population of the world was 7.14 billion. The population of China was 1.36 billion. If its birth rate is 12/1000 and its death rate is 7/1000, what was China's intrinsic rate of increase? In what year would the population of China exceed the 2013 population of the world (7.14 billion) if it continued to grow at the 2013 rate?

5. A yeast population grows as shown below. Time is measured in hours. Find the r_{max} (per hour) and r_a (actual r over the entire time period) for this population.

Time in hours	N (cells per cc)	ln N	ln{(K − N)/N} Use K = 320
0	10		
1	20		
2	40		
3	75		
4	140		
5	210		
6	273		

(continued overleaf)

Introduction to Population Ecology, Second Edition. Larry L. Rockwood.
© 2015 John Wiley & Sons, Ltd. Published 2015 by John Wiley & Sons, Ltd.
Companion Website: www.wiley.com/go/rockwood/populationecology

Time in hours	N (cells per cc)	ln N	ln{(K − N)/N} Use K = 320
7	300		
8	310		
9	315		
10	318		
11	319		
12	317		
13	318		
14	319		
15	315		
16	319		

Hint: Find the natural log of N and complete column 3. Then find the natural log of $(K − N)/N$; then complete column 4. Prepare graphs of columns 2, 3, and 4 versus time. For r_{max} find the slope during the exponential phase of growth (days 0–4, e.g.). For r actual, find the slope of the values computed in column four versus time.

6. **Exercise.** A population has the life table shown below. Find the gross reproductive rate, the net reproductive rate (R_0), p_x, and q_x. Would r be predicted to be positive, negative, or zero? Find r, λ, the stable age distribution, reproductive value by age class, and the life expectancy by age class. Given the population by age class at time zero, project this population one year into the future. Repeat this process, using the Leslie matrix.

Age, x	l_x	m_x	p_x	q_x	n_x at $t = 0$	n_x at $t = 1$
0	1.00	0	0.40	0.60	1000	900
1	0.40	1.0	0.50	0.50	300	400
2	0.20	3.0	0.25	0.75	100	150
3	0.05	2.0	0	1.00	60	25
4	0	0			0	0
Σ		6.0			1460	1475

$$GRR = \sum m_x = 6$$
$$R_0 = \sum l_x m_x = (0.40)\,(1) + (0.20)\,(3) + (0.05)\,(2) = 1.10$$
Since $R_0 > 1$, r should be positive.

To estimate r, we begin by estimating T (generation time).
$$G = \sum x l_x m_x / R_0 = 1.90/1.10 = 1.73$$
Estimate of $r = \ln R_0 / T = \ln(1.1)/1.73 = 0.055$

Next, confirm r with the Euler equation. We use $r = 0.055$ in column 4, below and $r = 0.056$ in column 5. Since 0.056 produces a sum closer to 1.000, we will use 0.056.

Since $\lambda = e^r$, $\lambda = 1.057$

Age, x	$l_x m_x$	$x l_x m_x$	$l_x m_x e^{-rx}$ $r = 0.055$	$l_x m_x e^{-rx}$ $r = 0.056$
0	0	0	0	0
1	0.40	0.40	0.379	0.378
2	0.60	1.20	0.538	0.537
3	0.10	0.30	0.085	0.085
4	0	0	0	0
Σ	1.10	1.90	1.002	1.000

$$\text{Matrix} \qquad t = 0 \qquad t = 1$$

$$\begin{vmatrix} 0.40 & 1.50 & 0.50 & 0 \\ 0.40 & 0 & 0 & 0 \\ 0 & 0.50 & 0 & 0 \\ 0 & 0 & 0.25 & 0 \end{vmatrix} \begin{vmatrix} 1000 \\ 300 \\ 100 \\ 60 \end{vmatrix} = \begin{vmatrix} 900 \\ 400 \\ 150 \\ 25 \end{vmatrix}$$

$$\Sigma = 1460 \quad \Sigma = 1475$$

Stable age distribution with $r = 0.056$

Age, x	$e^{-rx} l_x$	c_x for stable
		age distribution
0	1.00	0.625
1	0.378	0.236
2	0.179	0.112
3	0.042	0.026
Σ	1.599	0.999

Expectation of life

Age, x	l_x	L_x	T_x	e_x
0	1.00	0.700	1.150	1.150
1	0.40	0.300	0.450	1.125
2	0.20	0.125	0.150	0.750
3	0.05	0.025	0.025	0.500
4	0	0	0	0
		$\Sigma = 1.150$		

Reproductive value to calculate reproductive value, we need the following information.

Age, x	l_x	m_x	$e^{-rx}l_x m_x$	$e^{-rx}l_x$
0	1.00	0	0	1.000
1	0.40	1.0	0.378	0.378
2	0.20	3.0	0.537	0.179
3	0.05	2.0	0.085	0.043
4	0	0	0	0
Sum			1.000	

Calculation of reproductive value, V_x.

Age, x	Method 1	Method 2
0	$(0 + 0.378 + 0.537 + 0.085/1.000 = 1.000$	$(0.378 + 0.537 + 0.085)/1.000 = 1.000$
1	$(0.378 + 0.537 + 0.085)/0.378 = 2.64$	$(0.537 + 0.085)/0.378 = 1.64$
2	$0.537 + 0.085/0.179 = 3.47$	$(0.085)/0.179 = 0.47$
3	$(0.085)/0.043 = 1.98$	$(0)/0.132 = 0$

7. Life table problem

Reproductive rate, the intrinsic rate of increase (confirm the value with the Euler equation), and the finite rate of increase. Is this survivorship curve a type I, II or III? Given the age structure at $t = 0$, project the population for two years using the Leslie matrix.

Age, x	l_x	m_x	$l_x m_x$	$x l_x m_x$	p_x	q_x	n_x at $t = 0$	n_x at $t = 1$	n_x at $t = 2$
0	1.000	0			0.025		800		
1		10			0.800		25		
2		20			0.900		16		
3		20					18		
4	0.010	10					10		
5	0	0					0		
Total							869		

Appendix 2 **Matrix algebra: the basics**

For more information on matrices the reader should consult Vandermeer and Golberg (2003), Caswell (1989), Searle (1966) or any basic matrix algebra textbook.

What is a matrix? A matrix is a rectangular array of numbers. Each number in a matrix is known as an **element**. An ordinary single number is known as a **scalar**. The elements of a matrix are usually made up of scalars. A matrix is made up of rows and columns. If there is a single column of numbers, the array is known as a **column vector**. A single row of numbers is known as a **row vector**. In a matrix, if the number of rows equals the number of columns, the result is known as a **square matrix**. Although the rows must be of equal length and the columns must be of equal length in a matrix, the number of rows only equals the number of columns in a square matrix. The size of a matrix is known as its **order** and is known by the number of rows (r) first, followed by the number of columns (c). Thus a matrix is known as "$r \times c$." For example below we have a 4×3 matrix, $|A|$ followed by a column vector (x) and a row vector (y). A scalar, by the way, can be thought of as the element of a 1×1 matrix.

$$|A| = \begin{vmatrix} 1 & 2 & 3 \\ 4 & 5 & 6 \\ 7 & 8 & 9 \\ 0 & 10 & 11 \end{vmatrix} \qquad x = \begin{vmatrix} 1 \\ 4 \\ 7 \\ 0 \end{vmatrix} \qquad y = \begin{vmatrix} 1 & 2 & 3 \end{vmatrix}$$

In a square matrix, if all of the non-diagonal elements are zero, the matrix is described as a **diagonal matrix**. For example, a 4×4 diagonal matrix $|D|$ is shown below.

$$|D| = \begin{vmatrix} 1 & 0 & 0 & 0 \\ 0 & 2 & 0 & 0 \\ 0 & 0 & 3 & 0 \\ 0 & 0 & 0 & -4 \end{vmatrix}$$

Introduction to Population Ecology, Second Edition. Larry L. Rockwood.
© 2015 John Wiley & Sons, Ltd. Published 2015 by John Wiley & Sons, Ltd.
Companion Website: www.wiley.com/go/rockwood/populationecology

A2.1 **Matrix operations**

A2.1.1 *Addition and subtraction*

To add or subtract two matrices, the corresponding elements from the row and column positions are added (or subtracted). For example, matrix $|A|$ and $|B|$ are added as follows.

$$\text{If} \quad |A| = \begin{vmatrix} 1 & 2 & 3 \\ 4 & 0 & 7 \end{vmatrix} \quad \text{and} \quad |B| = \begin{vmatrix} 2 & 5 & 0 \\ 1 & 8 & 2 \end{vmatrix}$$

$$\text{then} \ |A| + |B| = \begin{vmatrix} 1+2 & 2+5 & 3+0 \\ 4+1 & 0+8 & 7+2 \end{vmatrix} = \begin{vmatrix} 3 & 7 & 3 \\ 5 & 8 & 9 \end{vmatrix}$$

Similarly, if matrix B is subtracted from matrix A, we have the following.

$$|A| - |B| = \begin{vmatrix} 1-2 & 2-5 & 3-0 \\ 4-1 & 0-8 & 7-2 \end{vmatrix} = \begin{vmatrix} -1 & -3 & 3 \\ 3 & -8 & 5 \end{vmatrix}$$

It should be obvious that that matrix addition and subtraction can only occur when the two matrices involved are of the same order. Two matrices that are of the same order are said to be "conformable for addition (or subtraction)."

A2.1.2 *The null matrix and the identity matrix*

The null (or zero) matrix, is a matrix equivalent of the scalar number zero. There are many possible null matrices, however, or different orders. Below are a 3×2 null matrix and a square 3×3 null matrix.

$$\begin{vmatrix} 0 & 0 \\ 0 & 0 \\ 0 & 0 \end{vmatrix} \qquad \begin{vmatrix} 0 & 0 & 0 \\ 0 & 0 & 0 \\ 0 & 0 & 0 \end{vmatrix}$$

The equivalent of the number one in algebra is the identify matrix (or the unit matrix). Identified by the letter I sometimes with a subscript for the order. The identify matrix must be a square matrix with all diagonal elements equal to one and all off diagonal elements equal to zero. For example I_3 is shown below.

$$\begin{vmatrix} 1 & 0 & 0 \\ 0 & 1 & 0 \\ 0 & 0 & 1 \end{vmatrix}$$

A2.1.3 *Mutiplication*

1. **By a Scalar**. Multiplication of a matrix by a scalar simply involves multiplying each element, in turn, by the scalar, producing a matrix of the same order as the original matrix. See the example below.

$$|A| = \begin{vmatrix} a_{11} & a_{12} & a_{13} & a_{14} \\ a_{21} & a_{22} & a_{23} & a_{24} \\ a_{31} & a_{32} & a_{33} & a_{34} \\ a_{41} & a_{42} & a_{43} & a_{44} \end{vmatrix}$$

$$3|A| = \begin{vmatrix} 3a_{11} & 3a_{12} & 3a_{13} & 3a_{14} \\ 3a_{21} & 3a_{22} & 3a_{23} & 3a_{24} \\ 3a_{31} & 3a_{32} & 3a_{33} & 3a_{34} \\ 3a_{41} & 3a_{42} & 3a_{43} & 3a_{44} \end{vmatrix}$$

If

$$|A| = \begin{vmatrix} 4 & 5 & 6 & 1 \\ 7 & 8 & 9 & 10 \\ 2 & -1 & -3 & 10 \\ 0 & 1 & -2 & 1 \end{vmatrix}$$

Then

$$3|A| = \begin{vmatrix} 12 & 15 & 18 & 3 \\ 21 & 24 & 27 & 30 \\ 6 & -3 & -9 & 30 \\ 0 & 3 & -6 & 3 \end{vmatrix}$$

2. **Multiplication of a row vector by a column vector.** To multiply a row by a column, the two vectors must be of the same order. The result is a scalar, which is found by using the following formula.

Given row vector $a = \begin{vmatrix} a_1 & a_2 & a_3 \end{vmatrix}$, and column vector $x = \begin{vmatrix} x_1 \\ x_2 \\ x_3 \end{vmatrix}$

Then the product $ax = a_1 x_1 + a_2 x_2 + a_3 x_3$. For example:

If $|a| = \begin{vmatrix} 2 & 4 & 6 \end{vmatrix}$ and $|x| = \begin{vmatrix} -5 \\ 1 \\ 0 \end{vmatrix}$, the result is: $[-10 + 4 + 0] = -6$

3. **Multiplication of a matrix by a column vector.** In order to multiply a matrix by a column vector we repeat the above for each row. That is, each row is multiplied independently by the column vector. The result is a column vector of the same order as the previous column vector. Again, in order to be conformable for multiplication, the number of elements in the row (equal to the number of columns) must equal the number of elements in the column vector. For example:

$$|A| = \begin{vmatrix} 5 & 6 & 8 \\ 2 & 5 & 12 \\ 1 & 3 & 9 \end{vmatrix} \qquad y = \begin{vmatrix} 1 \\ 2 \\ 3 \end{vmatrix} \qquad |A|y = \begin{vmatrix} (5*1)+(6*2)+(8*3) \\ (2*1)+(5*2)+(12*3) \\ (1*1)+(3*2)+(9*3) \end{vmatrix} = \begin{vmatrix} 41 \\ 48 \\ 34 \end{vmatrix}$$

In general, if $|A| = \begin{vmatrix} a_{11} & a_{12} & a_{13} \\ a_{21} & a_{22} & a_{23} \\ a_{31} & a_{32} & a_{33} \end{vmatrix}$ and $y = \begin{vmatrix} x_1 \\ x_2 \\ x_3 \end{vmatrix}$

then:

$$|A|y = \begin{vmatrix} a_{11}x_1 + a_{12}x_2 + a_{13}x_3 \\ a_{21}x_1 + a_{22}x_2 + a_{23}x_3 \\ a_{31}x_1 + a_{32}x_2 + a_{33}x_3 \end{vmatrix}$$

4. **The product of two matrices.** Multiplying two matrices can be thought of as a repetitive exercise in multiplying a matrix by as many column vectors as are present in the second matrix. That is, think of matrix B as a series of column vectors. Again, to be conformable for multiplication the number of columns in matrix A must equal the number of elements in matrix B.

For example, if matrix $|A|$ is as above, and matrix $|B|$ consists of two columns and three rows, the product of $|A||B|$ is shown below.

$$|B| = \begin{vmatrix} x_1 & z_1 \\ x_2 & z_2 \\ x_3 & z_3 \end{vmatrix}$$

$$|A||B| = \begin{vmatrix} a_{11}x_1 + a_{12}x_2 + a_{13}x_3 & a_{11}z_1 + a_{12}z_2 + a_{13}z_3 \\ a_{21}x_1 + a_{22}x_2 + a_{23}x_3 & a_{21}z_1 + a_{22}z_2 + a_{23}z_3 \\ a_{31}x_1 + a_{32}x_2 + a_{33}x_3 & a_{31}z_1 + a_{32}z_2 + a_{33}z_3 \end{vmatrix}$$

For example:

$$|A||B| = \begin{vmatrix} 1 & 0 & 2 \\ 3 & 1 & 7 \\ 4 & 5 & 6 \end{vmatrix} \begin{vmatrix} -1 & 4 & 8 \\ -2 & 0 & 9 \\ -5 & 1 & 5 \end{vmatrix}$$

$$\begin{vmatrix} (1*-1)+(0*-2)+(2*-5) & (1*4)+(0*0)+(2*1) & (1*8)+(0*9)+(2*5) \\ (3*-1)+(1*-2)+(7*-5) & (3*4)+(1*0)+(7*1) & (3*8)+(1*9)+(7*5) \\ (4*-1)+(5*-2)+(6*-5) & (4*4)+(5*0)+(6*1) & (4*8)+(5*9)+(6*5) \end{vmatrix}$$

$$= \begin{vmatrix} -11 & 6 & 18 \\ -40 & 19 & 68 \\ -44 & 22 & 107 \end{vmatrix}$$

A2.2 The laws of algebra

Matrices follow the associative laws of addition and multiplication provided the matrices are conformable. That is, $(A + B) + C = A + B + C = A + (B + C)$. Similarly, $A(BC) = (AB)C = ABC$

The distributive law also holds for matrices provided B and C are conformable for addition (necessarily of the same order) and matrices A and B are conformable for multiplication (hence A and C are also conformable). Thus: $A(B + C) = AB + AC$.

Addition of matrices follows the commutative rules provided they are conformable. If both matrices are of the same order we can write: $A + B = B + A$.

However, multiplication of matrices does not usually follow the commutative rule. In general, $AB \neq BA$. There are particular cases, where $AB = AB$, but this is not true as a general rule.

A2.2.1 The inverse of a matrix

Division in the usual sense does not exist in matrix algebra. The concept of dividing by a matrix is replaced by the idea of multiplication by the inverse of a matrix. In algebra, if we want to divide a number by 2, we have the option of simply multiplying by the inverse of two, that is, 1/2 (0.5). In matrices, instead of dividing y by x, we multiply y times $1/x$. Another property of the inverse is that if we multiply x by its inverse the result is the number one. In matrix algebra if we multiply $|A|$ by $|A|^{-1}$ the result should be the identity matrix, I, which will be of the same order as $|A|$ and $|A|^{-1}$. The above is only true if $|A|$ and $|A|^{-1}$ are square matrices. The inverse of $|A|$ only exists if $|A|$ has a determinant (see next section). It is often difficult to solve for the inverse of a matrix, but computer programs are now available that do the hard work for you.

A2.2.2 *The determinant*

The concept of a determinant also does not exist in traditional algebra. It is also difficult to explain the meaning of a determinant in any intuitive fashion. A determinant is only defined for a square matrix and is the result of a series of calculations resulting in a **scalar**. It is the sum of a series of products of the elements of the matrix and each product is multiplied by +1 or −1 according to certain rules. For example, for a two by two matrix, the determinant is found by taking the difference between the products of the diagonals versus the off diagonals. That is, the determinant of matrix $|A| = (a_{11} * a_{22}) - (a_{12} * a_{21})$

If $|A| = \begin{vmatrix} 7 & 3 \\ 4 & 6 \end{vmatrix}$ the determinant is calculated as: $(7)*(6)-(4)*(3) = 30$.

A determinant for a 3×3 matrix follows similar, but more complicated rules. As the order of the matrices gets larger the calculations become increasingly complex. Again, however, there are computer programs to evaluate the determinant of a matrix.

Appendix 3 List of mathematical symbols used in this book

In all chapters the following are conventionally used.

K a carrying capacity term

N population size

N_t population size at time t

r the intrinsic rate of increase $= b - d =$ growth rate per individual per unit time

r_a the actual intrinsic rate of increase in a given environment

r_m the Malthusian parameter, or r_{max}. The density independent, maximum growth rate of a genotype in a given environment

R^2 the amount of variance explained by a statistical regression model

t time

Symbols used in specific chapters are as follows.

A3.1   Chapter 1

λ the finite rate of increase or the growth rate per unit time $= N_{t+1}/N_t = e^r$ when there is a stable age distribution.

b birth rate per individual per time period (per capita birth rate)

B the number of births per time unit; the litter size or the clutch size

d death rate per individual per time period (per capita death rate)

D the number of deaths per unit time

E the number of emigrants per unit time

I the number of immigrants per unit time

Introduction to Population Ecology, Second Edition. Larry L. Rockwood.
© 2015 John Wiley & Sons, Ltd. Published 2015 by John Wiley & Sons, Ltd.
Companion Website: www.wiley.com/go/rockwood/populationecology

p_i probability of a given λ when computing the geometric mean

$P_{0,t}$ the probability of extinction at time, t

R the net replacement rate or net growth rate per generation in a population with discrete generations

A3.2   Chapter 2

θ (theta)   parameter describing non-linear responses of the logistic equation when $\theta \neq 1$

$\overline{\omega}$ mean weight (mass) per plant

T   tau; measures a lag time effect in the logistic equation.

w_m the maximum potential mass per plant

a' $(R-1)/K$; also a carrying capacity parameter

a a constant of integration

b^* the exponent that relaxes the assumption that population growth follows exact (linear) density dependence (exact compensation). When $b^* = 1$, there is "exact compensation," but when $b^* > 1$, there is "overcompensation," meaning that growth decreases more rapidly than expected with an increase in population density. When $b^* < 1$, there is "undercompensation," meaning that growth decreases less rapidly than expected with an increase in population density.

C the final constant yield

K the carrying capacity of the environment for a given population

MVP   minimum viable population size

R_A the density dependent of "actual" net growth parameter in a population with discrete generations; the net replacement rate or net growth rate per generation

R or R_I the density independent growth parameter in a population with discrete generations

r_a the actual intrinsic rate of increase in a given environment

r_m the Malthusian parameter, or r_{max}. The density independent, maximum growth rate of a genotype in a given environment

A3.3   Chapter 3 D the number of deaths due to density independent factors per unit time

A3.4   Chapter 4

$|A|$ a matrix

A mature adult stages

c_x proportion of the population belonging to an age category, x.

e_x age-specific life expectancy; estimated as: $e_x = T_x/l_x$

G mean generation time

GRR   The Gross Reproductive Rate; the average number of female offspring produced by a female living through all of the reproductive age classes.

I identity matrix

l_x the age specific survivorship; the proportion of the population, measured from age zero, to live to a given age class, x.

L_x the mean survivorship for any particular age interval, and assumes that, on average, an organism dies half way between two age classes. $L_x = (l_x + l_{x+1})/2$

m_x age specific fertility; the mean number of female offspring produced by a female of age, x.

n_x the number of individuals in an age class, x.

p_x age-specific probability of living to the next age class. $p_x = l_{x+1}/l_x$

q_x age-specific probability of death prior to the next age class $= 1 - p_x$

R_0 net reproductive rate in a population with age classes and overlapping generations; the mean number of females produced per female in the population per generation.

S_x the number of survivors to a given age, x, based on a cohort of 1000

T_x is the area under the survivorship curve for an individual of a given age, x, to the age, w, at which the oldest individual dies. T_x is estimated as:

$$T_x = \sum_x^w L_x = \sum_x^w l_x + l_{x+1}/2$$

V_x reproductive value of an individual of the age, x.

x age class

Y young adult stages

A3.5   Chapter 5

α the dispersal ability of each individual organism

β the number of individuals dispersing

ε extinction rate in a metapopulation

A the area of an island or habitat

A_i the area of the habitat patch, i

c the colonization rate in a metapopulation

C the y-intercept in the MacArthur and Wilson species area relationship

C_i the colonization rate or probability per unit time in a metapopulation

d_i the distance from a source population, i

d_{ij} the distance between patches i and j

D a coefficient of diffusion in a random walk movement model

e a parameter related to the probability of extinction per unit time in a patch of a given size

E_i the extinction rate per unit time in a metapopulation

H the total number of available habitat patches

J_i the long term probability of a patch being occupied (the incidence)

M_i the sum of individuals arriving at patch i, from all of the surrounding habitat patches

Pj Pj is 0 for an unoccupied patch and 1 for an occupied patch in the incidence function model

P' the number of habitat patches occupied by a population

P the proportion of available habitat patches occupied by a population

$\hat{P}$ the equilibrium value of P

R_e the total number of possible source species found on the mainland

S the number of species on an island

S_i a measure of connectivity between patches

T the total number of habitat patches available

T_M long-term persistence of a metapopulation

X the rate of change of extinction per unit time with increasing patch size (a measure of environmental stochasticity)

y   efficiency of colonization

z   a constant representing the slope of the line in a log-log plot of area and species number

A3.6   Chapter 6

α the age at first reproduction

ω the age at last reproduction

δ age of maximum reproduction

B   litter size

b   the allometric constant

E   activation energy

G   mean generation time

I   whole-organism metabolic rate as a function of mass (M) and a normalization constant, I_0 or i_0

k   k is the Boltzmann constant

M   mass

S   total reproductive output

T   T is absolute temperature in degrees Kelvin

V_x $l_x m_x$

ϖ mean weight (or mass) per plant

Y   Y is a physiological rate or some other variable dependent on mass; and Y_0 is normalization constant.

A3.7   Chapter 7

$\alpha_{i,j}$ the competition coefficient: the effect of species j on species i

$\hat{p}_i$ the proportion of habitat sites occupied at equilibrium by species i

μ the growth rate on a given substrate of resource, R, in the Michaelis-Menton equation.

μ_{max} the maximum growth rate of the population

b   the efficiency by which each individual converts the resource into new individuals

b_i maximum cell division rate ($=r_{max}$) for species, i

c_i the colonisation rate of species i

C_j the availability of the resource, j.

ε_i the mortality or extinction rate of species i

K_μ or K_i the half-saturation constant for the resource, i; the concentration of the resource that produces half the maximum growth rate.

K_1 the carrying capacity of species one

K_2 the carrying capacity of species two

k_{Ri} k_{Ri} is the supply rate of resource R_i

m   death or mortality rate

N_1 the number of individuals of species one

N_2 the number of individuals of species two

p_i the proportion of available habitat sites or patches occupied by species i

$\hat{p}$ the proportion of habitat sites occupied at equilibrium

q	the rate each individual must consume the resource in order to maintain itself
$R*$	the amount of a resource needed to just sustain the population. That is, $R*$ is the level of the resource needed to balance mortality (growth just offsets mortality)
r_1	the intrinsic rate of increase of species one
r_2	the intrinsic rate of increase of species two
R_i	the quantity or concentration of a resource, i

A3.8 Chapter 8

c_i and c_j	mutualism coefficients that replace the competition coefficients in the Lotka-Volterra competition equations. The term c_i measures the rate at which an individual of N_j benefits the growth rate of population N_i.
K_i*	the carrying capacity for species i after a mutualistic interaction with species j.

A3.9 Chapter 9

α	disease-induced mortality rate
γ	rate at which recovered individuals lose their immunity. That is, the rates at which individuals return to the susceptible class (S) from the recovered (R) class.
ψ	the migration rate between susceptible and infected populations
β	transmission rate of disease from one host to another
δ	probability that an infected disperser infects the resident population in a susceptible patch
A	the average age at which an individual in a population will become infected with a particular disease
b	host birth rate
D	the length of the infectious period
g	infection rate from an "outside source," that is from another species in the patch.
I	infected host density or the proportion of infected patches (both host population and disease present).
m	natural host mortality rate unrelated to disease mortality
N	total host population density
p	the fraction of the population that has been immunized
R	recovered (immune) host density
R_0	the basic reproductive number (BSR) or parameter for a disease. It represents the mean number of new infections caused by a single infective individual, and equals βSD.
S	susceptible host density or proportion of susceptible host patches in the metapopulation model
S_T	the minimum size for an epidemic $= \frac{1}{\beta D}$
v	recovery rate, or the per capita rate of passage from the infected (I) to the recovered (R) classes. This is usually the inverse of the average infectious period
x_I	extinction rates of an infected population in a habitat patch.
x_S	extinction rate of a susceptible population in a habitat patch

A3.10	**Chapter 10**	λ	the finite rate of increase per generation for the prey in the Nicholson-Bailey model
		χ_p	the efficiency by which food (prey) is turned into new predator individuals, or the assimilation efficiency of the predator
		μ_p	the minimum rate of prey consumption necessary for a predator to survive and replace itself.
		ρ_0	the zero term of the Poisson distribution
		a	constant for the parasitoid searching efficiency.
		c	the maximum predation or killing rate when the search and capture components have been minimized; replaces the parameter E.
		d	half saturation parameter equal to half of the maximum predation rate
		E	the searching and capturing efficiency of the predator; a functional response term
		$E_1, E_2,$ and E_3	search and capture components for type 1, 2 and 3 functional responses.
		f or f_n	functional response term; equal to EN in a type I functional response
		$h_1, h_2,$ and h_3	handling and digestion rate components of the type 1, 2 and 3 functional responses.
		K_n	carrying capacity term for the prey population
		K_R	the half saturation constant for a resource
		M	Mass
		M^*	prey population never exposed to predation (prey refuge)
		m_p	instantaneous, density independent mortality rate of the predator population
		n	the number of parasitoids emerging from each host individual
		N	prey population
		N_a	the number of hosts actually parasitized
		N_e	the number of encounters between host or prey species and their parasitoids
		P	the predator population or parasitoid population
		R	the concentration of a resource
		r_n	the maximum growth rate (intrinsic rate of increase) for the prey population

A3.11	**Chapter 11**	θ	a density dependent parameter affecting the herbivore death rate; equal to one at low herbivore densities, but which is greater than one and increases the herbivore death rate at high population densities
		χ_h	the assimilation efficiency of the herbivore population
		χ_p	the assimilation efficiency of the predator population
		κ	the carrying capacity term for the predator population due to territoriality
		b	the half saturation parameter for the herbivore population equal to half of the maximum consumption rate, f.
		c	the maximum killing rate for the predator population when the search and capture components have been minimized
		d	half saturation parameter equal to half of the maximum predation rate, c.
		F	the functional response term for searching efficiency of the herbivore

f the maximum consumption or grazing rate for the herbivore when the search and capture components have been minimized; replaces the parameter F.

h handling and digestion rate components of the functional response of the herbivore

K_v carrying capacity term for the vegetation (plant population)

m_h instantaneous, density independent mortality rate of the herbivore population

n the number of grazers that determines whether plant quality increases or decreases

N the number of herbivores or grazers

P the number of predators

Q average plant quality

q the increase in plant quality in the absence of grazers

r_h herbivore (prey) maximum growth rate

r_v plant maximum growth rate

s_0 predator rate of increase limited by a carrying capacity due to a territory

μ_h the minimum rate of food intake necessary for an herbivore to survive and replace itself

u_0 plant growth rate when V is close to zero and V represents above ground biomass

V plant abundance or biomass

A3.12 **Chapter 12** P Predator

C Consumers (usually equivalent to herbivores)

N Resource (usually equivalent to producers)

E_{pc} Searching efficiency of the predator for the consumer

E_{cn} Searching efficiency of the consumer for the resource

E_{pn} Searching efficiency of the predator for the resource (in omnivory and intraguild predation)

χ_{pc} the efficiency by which consumers are turned into new predator individuals

χ_{cn} the efficiency by which resources are turned into new consumers

m_c morality rate of consumer

m_p mortality rate of predator

b the half saturation parameter for the herbivore population equal to half of the maximum consumption rate, f.

c the maximum killing rate for the predator population when the search and capture components have been minimized

d half saturation parameter equal to half of the maximum predation rate, c.

s s is an intraspecific competition coefficient

K_n resource carrying capacity

K_P predator carrying capacity related to territoriality

r_n maximum growth rates of the resource

r_c maximum growth rates of the consumer

f the maximum consumption or grazing rate for the herbivore when the search and capture components have been minimized

h handling and digestion rate components of the functional response of the consumer

μ_p the minimum rate of prey consumption necessary for a predator to survive and replace itself

μ_c the minimum rate of food intake necessary for a consumer to survive and replace itself

β the benefit the predator receives by causing direct predatory mortality or competitive mortality in the consumer (analogous to χ_{pc}, but in omnivory and intraguild predation)

α is a competition-predation functional response (analogous to E_{pc}, but in omnivory and intraguild predation)

Index

Introduction to Population Ecology, Second Edition. Larry L. Rockwood.
© 2015 John Wiley & Sons, Ltd. Published 2015 by John Wiley & Sons, Ltd.
Companion Website: www.wiley.com/go/rockwood/populationecology

Printed and bound by CPI Group (UK) Ltd, Croydon, CR0 4YY